# Colloid and Interface Science

## VOL. V

Biocolloids, Polymers, Monolayers, Membranes,
and General Papers

Academic Press Rapid Manuscript Reproduction

Proceedings of the International Conference
on Colloids and Surfaces—50th Colloid and Surface Science Symposium,
held in San Juan, Puerto Rico
on June 21–25, 1976

# Colloid and Interface Science

## VOL. V
### Biocolloids, Polymers, Monolayers, Membranes, and General Papers

*EDITED BY*

## MILTON KERKER

Clarkson College of Technology
Potsdam, New York

## Academic Press Inc.

*New York   San Francisco   London   1976*

A Subsidiary of Harcourt Brace Jovanovich, Publishers

ACADEMIC PRESS, INC.
111 Fifth Avenue, New York, New York 10003

*United Kingdom Edition published by*
ACADEMIC PRESS, INC. (LONDON) LTD.
24/28 Oval Road, London NW1

Library of Congress Cataloging in Publication Data

International Conference on Colloids and Surfaces,
  50th, San Juan, P.R., 1976.
  Colloid and interface science.

  CONTENTS:     v. 2.  Aerosols,
emulsions, and surfactants.–v. 3.  Adsorption,
catalysis, solid surfaces, wetting, surface tension,
and water.–v. 4.  Hydrosols and rheology. [etc.]
  1.  Colloids–Congresses.  2.  Surface chemistry
–Congresses.  I.  Kerker, Milton,  II.  Title.
QD549.I6  1976    541'.345     76-47668
ISBN 0–12–404505–7  (v. 5)

# Contents

# CONTENTS

# CONTENTS

CONTENTS

# List of Contributors

**Arthur W. Adamson**, Department of Chemistry, University of Southern California, Los Angeles, California 90007

**Paul Ander**, Seton Hall University, South Orange, New Jersey 07039

**C.D. Armeniades**, Department of Chemical and Electrical Engineering, Rice University, Houston, Texas 77001

**V.A. Arsentiev**, visiting professor from Leningrad Mining Institute, The University of British Columbia, Vancouver 8, British Columbia, Canada

**M. Bender**, Chemistry Department, Fairleigh Dickinson University, Teaneck, New Jersey 07666

**Frederick A. Bettelheim**, Department of Chemistry, Adelphi University, Garden City, Long Island, New York 11530

**B.H. Bijsterbosch**, Laboratory for Physical and Colloid Chemistry, Agricultural University, De Dreijen 6, Wageningen, The Netherlands

**Fred W. Billmeyer, Jr.**, Department of Chemistry, Rensselaer Polytechnic Institute, Troy, New York 12181

**Martin Blank**, Department of Physiology, College of Physicians and Surgeons of Columbia University, 630 West 168th Street, New York, New York 10032

**D.B. Boyer**, Department of Dental Research, Iowa State University, Iowa City, Iowa 52242

**C.K. Butts**, Departments of Public Health and Biochemistry, University of Oregon Health Sciences Center, Portland, Oregon 97201

**J. Chanu**, Universite Paris VII, Laboratoire de Thermodynamique des Milieux Ioniques et Biologiques, 2 Place Jussieu, 75221 Paris–Cedex 05, France

**F.C. Chen**, Chemistry Department, State University of New York, Stony Brook, New York 11794

**T-H. Chiu**, Medical Group, Avco Everett Research Laboratory, Inc., 2385 Revere Beach Parkway, Everett, Massachusetts 02149

**B. Chu**, Chemistry Department, State University of New York, Stony Brook, New York 11794

**Paul W. Chun**, University of Florida, Gainesville, Florida 32610

**Dale H. Cowan**, Department of Macromolecular Science, Case Western Reserve University, Cleveland, Ohio 44106

**E.L. Cussler**, Department of Chemical Engineering, Carnegie-Mellon University, Schenley Park, Pittsburgh, Pennsylvania 15213

**Lars Davidsson**, The Swedish Institute for Surface Chemistry, Drottning Kristinas vag 45, S-114 28 Stockholm, Sweden

**M. Delmotte**, Universite Paris VII, Laboratoire de Thermodynamique des Milieux Ioniques et Biologiques, 2 Place Jussieu, 75221 Paris–Cedex 05, France

**J.N. Desai**, Physical Research Laboratory, Ahmedabad - 380009, India

**Richard J. Ditore**, University of Florida, Gainesville, Florida 32610

**T.E. Dougherty**, Chemistry Department, Fairleigh Dickinson University, Teaneck, New Jersey 07666

**M.E. Duffey**, Department of Chemical Engineering, Carnegie-Mellon University, Schenley Park, Pittsburgh, Pennsylvania 15213

**N. Duzgunes**, Department of Biophysical Sciences, State University of New York at Buffalo, 4234 Ridge Lea Road, Buffalo, New York 14226

**F.R. Eirich**, Polytechnic Institute of New York, 333 Jay Street, Brooklyn, New York 11201

**M.S. El-Aasser**, Emulsions Polymers Institute, Lehigh University, Bethlehem, Pennsylvania 18015

**D. Fennell Evans**, Department of Chemical Engineering, Carnegie-Mellon University, Schenley Park, Pittsburgh, Pennsylvania 15213

**Lidia Roche Farmer**, Gillette Research Institute, 1413 Research Boulevard, Rockville, Maryland 20850

**Janos H. Fendler**, Department of Chemistry, Texas A&M University, College Station, Texas 77843

**Robert J. Fiel**, Department of Biophysics Research, Roswell Park Memorial Institute, 666 Elm Street, Buffalo, New York 14263

**Joseph Finkel**, Department of Chemistry, Adelphi University, Garden City, Long Island, New York 11530

**G.J. Fleer**, Laboratory for Physical and Colloid Chemistry, Agricultural University, De Dieijen 6, Wageningen, The Netherlands

**Mark P. Freeman**, Dorr-Oliver Incorporated, 77 Hayemeyer Lane, Stamford, Connecticut 06820

**H.P.M. Fromageot**, General Electric Research and Development Center, P.O. Box 8, Schenectady, New York 12301

**Yi-Chang Fu**, Miami Valley Laboratories, The Procter & Gamble Co., P.O. Box 39175, Cincinnati, Ohio 45239

**J.G. Gallagher**, Department of Chemical and Electrical Engineering, Rice University, Houston, Texas 77001

**R.B. Gammage**, Health Physics Division, Oak Ridge National Laboratories, P.O. Box X, Oak Ridge, Tennessee 37830

**Anesu Garuba**, Department of Chemistry, The City College of the City University of New York, New York, New York 10031

**A.N. Gent**, Institute of Polymer Science, The University of Akron, Akron, Ohio 44325

**A. Giniger**, 15 Wellesley Avenue, Yonkers, New York 10705

**D.R. Glasson**, School of Environmental Sciences, John Graymore Chemistry Labs, Plymouth Polytechnic, Plymouth PL 48 AA, England

**A.M. Gotto, Jr.**, Department of Chemical and Electrical Engineering, Rice University, Houston, Texas 77001

**Raymond T. Greer**, Department of Engineering Science and Mechanics and Engineering Research Institute, Iowa State University, Ames, Iowa 50010

**J.N. Groves**, General Electric Research and Development Center, P.O. Box 8, Schenectady, New York 12301

**D.G. Hall**, Unilever Research Laboratory, Port Sunlight, England

**G.R. Hamed**, Institute of Polymer Science, The University of Akron, Akron, Ohio 44325

**J. Fred Hazel**, Department of Chemistry, University of Pennsylvania, Philadelphia, Pennsylvania 19174

**P.J. Herley**, Department of Materials Science, State University of New York, Stony Brook, New York 11790

**H. Herman**, Department of Materials Science, State University of New York, Stony Brook, New York 11790

**Willie L. Hinze**, Chemistry Department, Wake Forest University, Winston-Salem, North Carolina 27109

**M. Hittmeier**, Empire State Paper Research Institute, State University of New York, College of Environmental Science and Forestry, Syracuse, New York 13210

**A.W. Horton**, Departments of Public Health and Biochemistry, University of Oregon Health Sciences Center, Portland, Oregon 97201

**Horst W. Hoyer**, Department of Chemistry, Hunter College, The City University of New York, New York, New York 10021

**S. Igbal**, Emulsions Polymers Institute, Lehigh University, Bethlehem, Pennsylvania 18015

**Y.H. Kao**, Department of Materials Science, State University of New York, Stony Brook, New York 11790

**Robert L. Kay**, Department of Chemistry, Carnegie-Mellon University, Pittsburgh, Pennsylvania 15213

**Milton Kerker**, Clarkson College, Potsdam, New York 13676

**B. Khazai**, Chemistry Department, Fairleigh Dickinson University, Teaneck, New Jersey 07666

**Dae M. Kim**, Department of Chemical and Electrical Engineering, Rice University, Houston, Texas 77001

**Alexander Kowblansky**, Seton Hall University, South Orange, New Jersey 07039

**Peter Laggner**, Institute für Rontgenfeinstruk-turforschung der Oster-reichischen Akademi der Wissenschaften und des Forschungszentrums Graz, Steyrergasse 17, A-8010 Graz, Austria

**E.P. Lancaster**, Empire State Paper Research Institute, College of Environmental Science and Forestry, State University of New York, Syracuse, New York 13210

**D.M. Lederman**, Medical Group, Avco Everett Research Laboratory, Inc., 2385 Revere Beach Parkway, Everett, Massachusetts 02149

**Beatrice B. Lee**, Department of Physiology, College of Physicians and Surgeons of Columbia University, 630 West 168th Street, New York, New York 10032

**James C. Lee**, Graduate Department of Biochemistry, Brandeis University, Waltham, Massachusetts 02154

**K.H. Lee**, Department of Chemical Engineering, Carnegie-Mellon University, Pittsburgh, Pennsylvania 15213

**J. Leja**, Department of Mineral Engineering, The University of British Columbia, Vancouver 8, British Columbia, Canada

**Harold I. Levine**, Department of Biophysics Research, Roswell Park Memorial Institute, 666 Elm Street, Buffalo, New York 14263

**C.H. Lin**, Department of Materials Science, State University of New York, Stony Brook, New York 11790

**Tom Lindström**, Swedish Forest Products Research Laboratory, Box 5604, S-114 86 Stockholm, Sweden

**Nan-I Liu**, Department of Chemistry, Carnegie-Mellon University, Pittsburgh, Pennsylvania 15213

**P. Luner**, Empire State Paper Research Institute, State University of New York, College of Environmental Science and Forestry, Syracuse, New York 13210

**J.D. Morrisett**, Division of Atherosclerosis and Lipoprotein Research, Baylor College of Medicine and Methodist Hospital, Houston, Texas 77025

**Susan K. Nevin**, Department of Chemistry, Hunter College, The City University of New York, New York, New York 10021

**E. Nyilas**, Medical Group, Avco Everett Research Laboratory, Inc., 2385 Revere Beach Parkway, Everett, Massachusetts 02149

**C.B. Ohki**, Department of Biophysical Sciences, State University of New York at Buffalo, 4234 Ridge Lea Road, Buffalo, New York 14226

**S. Ohki**, Department of Biophysical Sciences, State University of New York at Buffalo, 4234 Ridge Lea Road, Buffalo, New York 14226

**Gary R. Olhoeft**, Petrophysics Laboratory, United States Department of the Interior, Geological Survey -Stop 964, Box 25046 - Denver Federal Center, Denver, Colorado 80225

**James J. O'Malley**, Webster Research Center, Xerox Corporation, Webster, New York 14580

**B.A. Pethica**, Clarkson College of Technology, Potsdam, New York 13676

**Eugene P. Pittz**, Graduate Department of Biochemistry, Brandeis University, Waltham, Massachusetts 02154

**John A. Quinn**, Department of Chemical and Biochemical Engineering, University of Pennsylvania, Philadelphia, Pennsylvania 19174

**Owen M. Rennert**, University of Florida, Gainesville, Florida 32610

**O.K. Rice**, Department of Chemistry, The University of North Carolina, Chapel Hill, North Carolina 27514

**Kelvin Roberts**, The Swedish Institute for Surface Chemistry, Drottning Kristinas vag 45, S-114 28 Stockholm, Sweden

**Alejandro Romero**, Department of Chemistry, Texas A&M University, College Station, Texas 77843

**Henri L. Rosano**, Department of Chemistry, The City College of the City University of New York, New York, New York 10031

**M. Rosoff**, 15 Wellesley Avenue, Yonkers, New York 10705

**Eugene E. Saffen**, University of Florida, Gainesville, Florida 32610

**L.S. Sandell**, Empire State Paper Research Institute, College of Environmental Science and Forestry, State University of New York, Syracuse, New York 13210

**Robert A. Sasso**, Seton Hall University, South Orange, New Jersey 07039

**B.J.R. Scholtens**, Laboratory for Physical and Colloid Chemistry, Agricultural University, De Dreijen 6, Wageningen, The Netherlands

**A.R. Schuff**, Departments of Public Health and Biochemistry, University of Oregon Health Sciences Center, Portland, Oregon 97201

**Dinesh O. Shah**, Department of Chemical Engineering, University of Florida, Gainesville, Florida 32610

**H.S. Shah**, Regional College of Engineering and Technology, Surat, India

**Howard S. Sherry**, Mobile Research and Development Corporation, Research Department, Paulsboro, New Jersey 08066

**Jayesh B. Shukla**, University of Florida, Gainesville, Florida 32610

**Vida Slawson**, Laboratory of Nuclear Medicine, University of California at Los Angeles, Los Angeles, California 90024

**L.E. Smith**, Institute for Materials Research, National Bureau of Standards, Washington, D.C. 20234

**Christer Söremark**, Swedish Forest Products Research Institute, Box 5604, S114-86 Stockholm, Sweden

**D. Dewey Solomon**, Department of Macromolecular Science, Case Western Reserve University, Cleveland, Ohio 44106

**Hans Stabinger**, Institute für Rontgenfeinstruk-turforschung der Oster-reichischen Akademi der Wissenschaften und des Forschungszentrums Graz, Steyrergasse 17, A-8010 Graz, Austria

**Junzo Sunamoto**, Department of Chemistry, Texas A&M University, College Station, Texas 77843

**Bun-Ichi Tamamushi**, Nezu Chemical Institute, Musashi University, 1-26 Toyotama-kami, Nerima-ku, Tokyo 176, Japan

**W. Jape Taylor**, University of Florida, Gainesville, Florida 32610

**Serge N. Timasheff**, Graduate Department of Biochemistry, Brandeis University, Waltham, Massachusetts 02154

**S.I. Tu**, Department of Chemistry, State University of New York, Stony Brook, New York 11794

**D.B. Vaidya**, Gujarat College, Ahmedabad–380006, India

**J.W. Vanderhoff**, Emulsion Polymers Institute, Lehigh University, Bethlehem, Pennsylvania 18015

**John Van Dusen**, Webster Research Center, Xerox Corporation, Webster, New York 14580

**A.G. Walton**, Department of Macromolecular Science, Case Western Reserve University, Cleveland, Ohio 44106

**F.F.Y. Wang**, Department of Materials Science, State University of New York, Stony Brook, New York 11790

**Paul F. Waters**, Department of Chemistry, The American University, Washington, D.C. 20016

**James H. Whittam**, The Gillette Company, Personal Care Division, Gillette Park 1-T-1, South Boston, Massachusetts 02106

**Jaime H. Wong**, Department of Chemical and Biochemical Engineering, University of Pennsylvania, Philadelphia, Pennsylvania 19174

**Joel L. Zatz**, College of Pharmacy, Busch Campus, Rutgers, New Brunswick, New Jersey 08903

# Preface

This is the fifth volume of papers presented at the International Conference on Colloids and Surfaces, which was held in San Juan, Puerto Rico, June 21–25, 1976.

The morning sessions consisted of ten plenary lectures and thirty-four invited lectures on the following topics: rheology of disperse systems, surface thermodynamics, catalysis, aerosols, water at interfaces, stability and instability, solid surfaces, membranes, liquid crystals, and forces of interfaces. These papers appear in the first volume of the proceedings along with a general overview by A. M. Schwartz.

The afternoon sessions were devoted to 221 contributed papers. This volume includes contributed papers on the subjects of biocolloids, polymers, monolayers, membranes, and general papers. Three additional volumes include contributed papers on aerosols, emulsions, surfactants, adsorption, catalysis, solid surfaces, wetting, surface tension, water, hydrosols, and rheology.

The Conference was sponsored jointly by the Division of Colloid and Surface Chemistry of the American Chemical Society and the International Union of Pure and Applied Chemistry in celebration of the 50th Anniversary of the Division and the 50th Colloid and Surface Science Symposium.

The National Colloid Symposium originated at the University of Wisconsin in 1923 on the occasion of the presence there of The Svedberg as a Visiting Professor (see the interesting remarks of J. H. Mathews at the Opening of the 40th National Colloid Symposium and also those of Lloyd H. Ryerson in the Journal of Colloid and Interface Science 22, 409, 412, (1966)). It was during his stay at Wisconsin that Svedberg developed the ultracentrifuge, and he also made progress on moving boundary electrophoresis which his student Tiselius brought to fruition.

The National Colloid Symposium is the oldest such divisional symposium within the American Chemical Society. There were no meetings in 1933 and during the war years 1943–1945, and this lapse accounts for the 50th National Colloid Symposium occurring on the 53rd anniversary.

However, these circumstances brought the numerical rank of the Symposium into phase with the age of the Division of Colloid and Surface Chemistry. The Division was established in 1926, partly as an outcome of the Symposium. Professor Mathews gives an amusing account of this in the article cited above.

The 50th anniversary meeting is also the first one bearing the new name

Colloid and Surface Science Symposium to reflect the breadth of interest and participation.

There were 476 participants including many from abroad.

This program could not have been organized without the assistance of a large number of persons and I do hope that they will not be offended if all of their names are not acknowledged. Still, the Organizing Committee should be mentioned: Milton Kerker, Chairman, Paul Becher, Tomlinson Fort, Jr., Howard Klevens, Henry Leidheiser, Jr., Egon Matijevic, Robert A. Pierotti, Robert L. Rowell, Anthony M. Schwartz, Gabor A. Somorjai, William A. Steele, Hendrick Van Olphen, and Albert C. Zettlemoyer.

Special appreciation is due to Robert L. Rowell and Albert C. Zettlemoyer. They served with me as an executive committee which made many of the difficult decisions. In addition Dr. Rowell handled publicity and announcements while Dr. Zettlemoyer worked zealously to raise funds among corporate donors to provide travel grants for some of the participants.

Teresa Adelmann worked hard and most effectively both prior to the meeting and at the meeting as secretary, executive directress, editress, and general overseer. She made the meeting and these Proceedings possible. We are indebted to her.

Milton Kerker

# Colloid and Interface Science

---

## *VOL. V*

### Biocolloids, Polymers, Monolayers, Membranes, and General Papers

# INTERACTION OF BLOOD PLATELETS
# WITH SYNTHETIC COPOLYPEPTIDE FILMS

A. G. Walton, D. Dewey Solomon and Dale H. Cowan*
Department of Macromolecular Science and *Cleveland
Metropolitan General Hospital, Case Western Reserve University
Cleveland, Ohio   44106

ABSTRACT


The process of adhesion and release of radioactive serotonin by platelets adsorbed on various copolypeptide films
has been studied.  The films are coated on glass beads and a
kinetic method of measurement has been employed.  Results from
this study show that adhesion is essentially independent of
the composition of the substrate (which consists of two components chosen from L-glutamic acid, γ-benzyl-L-glutamate, L-
leucine, L-lysine and L-phenylalanine), but is dependent upon
the duration of interaction and the adsorption of material
produced as the platelets undergo release.

Studies of the surface, show fibrous texture at the ultrastructural level but the texture does not seem to affect the
adsorbtion, platelet release, or contact angle measurements to
any great extent.

It is concluded that adsorption is effected primarily by
entropic effects and that the release process is initiated by
modification of platelet membranes and activation of the contractile mechanism due to mismatch of interfacial energetics
between platelet and substrate.

INTRODUCTION

Interaction of polymeric materials with blood involves a complex interplay between the adsorption and activation of various proteins and cellular components in blood. One of the important entities involved in the initiation of blood clotting is the blood platelet. In general blood platelets exist in a milieu of proteins which are in delicate balance with the behavioral properties of the platelet. In contact with plastic materials, the blood clotting process involves adhesion of the platelet (to a proteinaceous layer) and subsequent deformation and release of clot promoting factors and entities (1-8). The mechanism of interaction of platelets with proteins and plastics is therefore of central concern in the design of prosthetic cardiovascular materials and understanding of the surface induced blood clotting process.

In this context it is logical to examine the behavior of platelets in the presence of the most simple protein analogs, e.g polyamino acids and polypeptides. These materials provide a method of studying the role of specific amino acid residues in promoting platelet adhesion, aggregation and release.

Furthermore, they provide a means of studying the involvement of ionic and hydrophobic interactions at an interface. The materials explored in this paper are random copolypeptides of two components i.e $(A_x B_y)_n$ where A is $\gamma$-benzyl-L-glutamate, L-glutamic acid or L-lysine and B is L-leucine. This combination is such that films of materials may be made whose surface charge at neutral pH is positive (lys), negative (glu) or netural (benzyl glutamate).

In the general context of cell adhesion to plastics there seems to be some indication that with nucleated cells (e.g fibroblasts), the initial contact adhesion may be dictated by interfacial energetics including the type of interactions mentioned above (9). Indeed it has been claimed that there is a correlation between the critical surface tension of plastics, platelet adhesion and blood clotting (4, 10), although the mediation of blood proteins is now realized as representing a distinct complicating factor in correlating surface energy on one hand and whole blood clotting on the other.

To our knowledge platelet interactions with insoluble polypeptide films have not been reported previously, although soluble polyamino acids have been studied in conjunction with their influence on blood clotting (11) and platelet aggregation and release (12-18).

MATERIALS AND METHODS

A. Washed Platelet Preparation

Whole blood was obtained from healthy donors. No volunteer ingested any drug known to affect platelet function during the preceeding two weeks. Blood was collected using a two-syringe technique and anticoagulated with 1.1 ml of EDTA solution per 10 ml of blood, 0.054 M EDTA in 0.056 M sodium chloride. The blood was immediately centrifuged at 4°C for 3 minutes at 1500 x g to obtain platelet rich plasma (PRP). The PRP was incubated 45 minutes with [$^3$H]-serotonin (5-hydroxytryptamine binoxalate [$1,2-^3$H(N)], New England Nuclear), $1.75 \times 10^{-4}$ millimoles/μCi. The final concentration was 1.4 μCi. The PRP was then centrifuged at 2000 x g for 15 minutes and the platelets resuspended in 25 ml of buffer containing NaCl (140 mM), EDTA (3.3 mM, Tris (30 mM)), and glucose (5 mM) at pH 7.4. This resuspension was centrifuged again at 2000 x g for 15 minutes. The platelets were suspended in the final resuspension buffer to approximately 250,000 platelets/μl. The final resuspension buffer was identical to the initial resuspension buffer except for a change

3

in the final EDTA concentration (0.3 mM). Experiments were completed within four hours from the time the blood was drawn.

Determination of possible platelet lysis was made using a cytoplasmic specific label $^{51}$Cr (as sodium chromate), 100 lambda with specific activity of 50-500 Ci/mg (New England Nuclear). The same basic procedure which was used for the serotonin labelled washed platelets was also used for the chromium labelling. Platelet lysis was found to be less than 10%.

B. Copolypeptide Substrates

The copolypeptides used in this work were prepared by standard N. Carboxyanhydride polymerization and were mainly provided by Prof. T. Hayashi of Kyoto University, Japan. Other materials were prepared and provided by Prof. J. Anderson and co-workers at Case Western Reserve University and by Mr. M. VanDress. All have viscosity molecular weights in excess of 100,000.

C. Preparation of Polypeptide Coated Glass Beads

All blood platelet experiments were carried out in jacketed glass columns (0.5 x 30 cm) with teflon mesh column supports and stopcocks (Glenco Scientific Inc., Houston, Texas). Each column was packed with 3.1 g of ∿0.5 mm glass beads (Thomas Scientific Co., Philadelphia, Pa.). The beads were placed in the column and the polymer was dissolved in an appropriate solvent to the extent of 1-2% by weight, and poured onto the column and quickly drawn onto the beads by means of suction at the base of the column. This method of introducing the polypeptide solution onto the glass beads helped to prevent air bubbles and channeling. After 20 minutes the solution was drained and the column was slowly air dried. A very slight vacuum was applied to the base of the column to prevent contamination. The drying procedure was continued for 30 minutes.

After drying, the columns were filled with 37°C buffer solution (final resuspension buffer) in the same manner as the polypeptide solution. The buffer was allowed to flow for several minutes before the platelets were introduced.

## D. Affinity Chromatography

The columns were kept at 37°C and the flow rate was adjusted to about 1.0 ml/min. A continuous head of buffer was maintained throughout the experiment. Two ml of washed platelets were introduced slowly near the top of the glass beads. The columns were allowed to flow for 15 minutes. The effluent was then collected, depending on the experiment, either in 1.00 ml aliquots or in a single aliquot. The aliquots were mixed gently and sampled for radioactive counting.

## E. Measurement of Release of Platelet Serotonin

The release of intracellular serotonin determined from radioactive counts of platelet suspensions and of supernatants prepared from platelet suspensions before and after passage through the columns. The initial total counts available ($T_0$, intra + extracellular counts) were measured in two 0.2 ml aliquots of the final platelet resuspension applied to the columns. The initial extracellular counts ($E_0$) were determined from the radioactivity in the supernatant obtained from centrifuging the final platelet suspension for 10 minutes at 12,000 x g. The total ($T_f$) and extracellular ($E_f$) counts in the effluent from each column were similarly determined. Scientiverse (Fisher Scientific, Cleveland, Ohio) was used as the scintillation fluid (10 ml scintillation fluid + 0.5 ml water/0.2 ml sample.

The amount of increase in extracellular labelled serotonin was calculated by the following:

$$\% \text{ Release} = \frac{E_f - E_o}{I_o} \times 100$$

where % Release is assumed to equal the percent increase in the level of labelled serotonin in the extracellular fluid, $E_f$ = final effluent (extracellular) counts corrected for volume effects, $E_o$ = extracellular initial counts, and $I_o$ = Total initial counts − $E_o$.

Adhesion of platelets was calculated by:

$$\% \text{ Adhesion} = \left[ 1 - \frac{I_f}{I_o} \right] \times 100$$

where $I_f$ = Total final counts−extracellular final counts.

The same equations for the chromium label were used to check lysis.

F.  Contact Angle Determination

Contact angle measurements were made using a conventional goniometer shadow angle method.  The device was originally constructed under the guidance of Prof. A. Mann.  Small water droplets were allowed to contact the material for at least 10 min. prior to a measurement and the average of several serarate measurements of the advancing angle on the same and separate samples is reported.

Table 1

Contact Angles of Water on Copolypeptide Films

| Material | Casting Solvent | Conformation | Contact Angle(Aver) |
|----------|-----------------|--------------|---------------------|
| GL 1:1 | $CHCl_3$ | $\alpha$-helix | 80° |
| GA 4:1 | TFA | $\alpha$ + (random) | 71 |
| GA 1:1 | $CHCl_3$ + Tr TFA | $\alpha$-helix | 58 |
| GP 4:1 | $CHCl_3$ | $\alpha$-helix | 63.5 |
| ZL 4:1 | $CHCl_3$ | $\alpha$-helix | 53.5 |
| ZL 1:1 | $CHCl_3$ | $\alpha$-helix | 60 |
| ZL 1:4 | $CHCl_3$ + Tr TFA | $\alpha$-helix | 63 |
| G(OH)V 4:1 | TFA | $\alpha$ + (random) | 37 |
| G(OH)V 1:1 | TFA | $\alpha$ + (random) | 38 |
| G(OH)V 1:4 | TFA | $\beta$-sheet | 47 |
| G(OH)L 1:1 | TFA | $\alpha$-helix | 40 |

Key    G = $\gamma$-benzyl-L-glutamate    Z = $\epsilon$carbobenzoxy-L-lysine

       L = L-luecine                  V = L-valine

       A = L-alanine          G(OH) = L-glutamic acid

*Note that conformation of the G(OH) polypeptides may change
 slightly on hydration.  Anhydrous film conformations are
listed.

## G.  Chain Conformation in the Polypeptide Films

        The conformation of surface adsorbed films was establish-
ed primarily by film circular dichroism spectropolarimetry and
transmission infrared spectroscopy.  These methods are not
sensitive to small admixtures of  $\alpha$-helix and random chain
forms and the precominant conformation only is recorded.

RESULTS

A. <u>Characterization of the Surfaces</u>

Examination of the beads by scanning electron microscopy
(fig 1) shows a uniform film which is interrupted only in
regions where the adjacent glass beads are in contact. At
high magnification in the transmission microscope, however,
replicated surfaces show the surface layer to consist of
fibrous material interwoven into a dense sheet (fig 2). The
surface is, therefore, far from uniform at the molecular level.
(Calculations of platelets adsorbed/cm$^2$ which are reported in
the text are based on the assumption that the available surface area is the same as that of the uncoated beads).

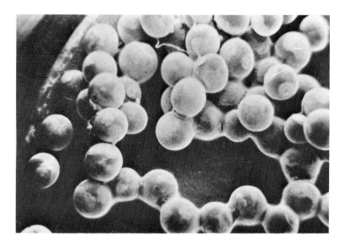

Fig 1  Scanning electron micrograph of glass beads coated with
a film of copoly γ-benzyl-L-glutamate/L-leucine ratio
1:1. Actual size of beads is 0.5 mm. Non coated
regions can be seen where beads were formerly in
contact.

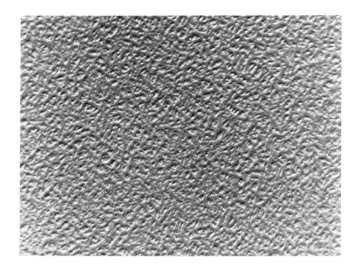

Fig 2  Transmission electron micrograph of copoly γ-benzyl-L-glutamate/leucine,ratio 1:1 (mag x 10,000) (kindly provided by P. H. Geil).

Contact angle measurements have been carried out on many of the copolypeptide films.  However, we were unable to prepare films suitable for contact angle work from the lysyl copolymers even though platelet adhesion to coated beads was acceptably reproduceable.  On the other hand we were able to study water contact angles on a number of other copolypeptide films in which the γ-benzyl-L-glutamate of L-glutamic acid were paired with L-alanine, L-valine or L-phenylalanine and the L-leucine moiety was combined with carbobenzoxy-L-Lysine. These results are presented in Table 1 and appear to compare favorably with the data of Baier et al. (9).

It is evident that the negatively charged (glutamic acid containing) polypeptides interact more strongly with water droplets, reducing their contact angle.

Scanning electron microscopy on platelets adhering to the copolypeptide coated spheres appears to show (fig 3), that the platelets are fairly well dispersed on the surface (at low surface density) and do not form large clusters.  This result

9

was quite unexpected since platelet aggregation is usually
conceived as an important step in the release process. These
data suggest that either the platelets, once adsorbed, are not
free to diffuse across the surface or the platelet/surface
interaction is favored over the platelet/platelet interaction.

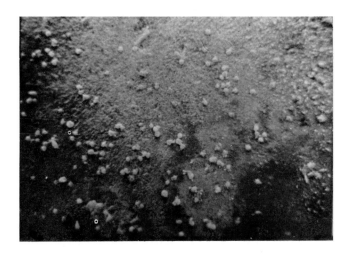

Fig 3   Scanning electron micrograph of (glutaraldehyde fixed)
platelets on the surface of a polypeptide (same as
fig 2) film.  Platelet diameter is approximately 2μ.
Very little surface aggregation is evident.

B.   Adhesion of Platelets

The rate of release is shown in fig 4.  Most of the
serotonin release occurred (on glass beads and the γ-benzyl-
L-glutamate, L-leucine (1:1) copolymer) in the first four
minutes after application of the platelets to the columns.
By fifteen minutes the release reaction was essentially (99%)
complete.  All subsequent data were obtained after fifteen
minutes delay from injection of the platelets.  Neither
siliconizing treatment of the column jacket nor addition of up
to 2.0 mM $Ca^{2+}$ to the column buffer affected the rate or extent of serotonin release.

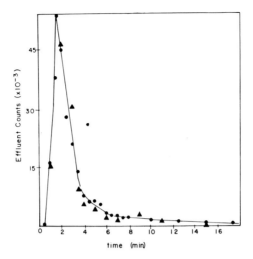

Fig 4   Emergence of radioactive serotonin from a glass bead/
polypeptide column.  Evidently non adhering platelets
pass through the column in approximately two minutes.
Essentially all radioactive material has emerged from
the column in fifteen minutes, the time used as a standard in all experiments.

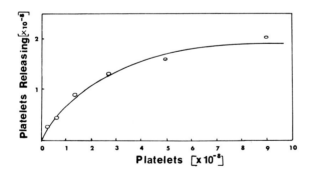

Fig 5   Release of serotonin measured as a function of platelet
dosage.  The experiment is run in successive "doses".
The curve approaches a level corresponding to "monolayer" coverage calculated on the basis of bead surface
area and a platelet area of $4\mu^2$.  Substrate surface was
(Glu(Bz):Leu), 1:1.

It is of some importance to know whether platelets adhere to a surface at random or whether surface adhesion is "nucleated" by preadsorbed cells. To examine this, successive "doses" of platelets were applied to the columns and the number adhering calculated. Fig 5 shows that the surface coverage appears to approach a limiting value, probably close to a monolayer. A further and rather surprising observation is that the higher the surface density of adsorbed platelets, the lower (on a percentage basis) the release of serotonin. This effect is demonstrated in fig 6 where at low surface coverage platelets undergo essentially 100% release, whereas at close to monolayer coverage, the percentage release is quite low. Since many of the reports in the literature on adhesion or release of platelets in contact with commercial polymers do not indicate specified levels of platelet dosage, it may be that a similar effect has been overlooked.

The extent of adherence of platelets and of serotonin release from platelets applied to columns composed of different polypeptide substrates are presented in Table 2. Platelet adherence appeared to be essentially independent of the polypeptide composition however, there did appear to be a relationship between platelet release and surface composition. The negatively charged glutamic acid surface was associated with the highest values for release. The basic, positively charged free lysyl surfaces did not promote a significant amount of release over that observed with control columns of uncoated glass beads. This relationship was less evident in the presence of preadsorbed albumin. The relatively low amount of release observed with uncoated glass has been noted previously (8). (This is <u>not</u> to say that glass is desirable in terms of its compatibility with whole blood).

Table 2. Effect of Polypeptide Film Composition on the Adhesion and Serotonin Release of Washed Blood Platelets

| Material | Casting Solvent | Measurements | %Platelets Adhering* | $^{3}$H-Serotonin Release(%) |
|---|---|---|---|---|
| Glass (Reference) | - | 12 | 30.4 ± 4.8 | 12.6 ± 2.2 |
| Glu (Bz) | CHCl$_3$ | 1 | 19.2 | 18.4 |
| Glu (Bz):Leu 4:1 | CHCl$_3$ | 2 | ≠ 27.2 | 18.1 |
| Glu (Bz):Leu 1:1 | CHCl$_3$ | 3 | 42.8 ± 13.9 | 20.8 ± 1.1 |
| Glu (Bz):Leu 1:4 | CHCl$_3$ | 3 | 33.0 ± 1.7 | 18.1 ± 1.6 |
| Glu (OH):Leu 1:4 | TFA | 3 | 41.4 ± 7.6 | 25.8 ± 1.4 |
| Lys:Leu 1:99 | TFA | 1 | 41.6 | 18.3 |
| Lys:Leu 1:9 | TFA | 1 | 32.1 | 16.6 |
| Lys:Phe 1:4 | TFA | 1 | ≠ 38.4 | 13.2 |

* Total dosage 5 x 10$^8$ platelets.

TFA = trifluoroacetic acid

≠ Average values

*Variances listed are $\Sigma|$ Measurement-Average$|$/No. of measurements

It seems likely that the numbers of platelets adhering and the mode of adhesion of platelets are affected by the contact time and number of contacts made between platelet and substrate in the experimental flow procedure. Consequently a number of experiments have been performed using different flow rates of platelet suspensions through the column. Although only a small range of flow rates could be achieved it may be seen (fig 7) that as the flow rate increases, a slight decrease in number of platelets adhering occurs but this decrease is accompanied by a slight increase in the amount of serotonin released (on a percentage basis).

DISCUSSION

Although the polypeptide films appear to be fibrous at the ultrastructural level, the reproducibility in contact angle and adsorption/release behavior of platelets suggest that the effect of the microstructure of the films is at best subtle when compared with macroscopic observation. The co-polypeptides always contain at least one hydrophobic residue component to render them insoluble and it is thus not surprising that the films are not wetted by water. We have not at this stage studied critical wetting phenomena or the effect of conformation on interfacial properties since we seek only preliminary information on hydrophilic character. In fact the choice of different casting solvents, in addition to changing conformation, also changes morphology and microstructure, often in a drastic manner. These effects are often difficult to separate in terms of their effect on wetting and only the solvents effecting the most coherent films have been used and the films explored.

In general, the contact angles are of the order expected and decrease with the increase in free surface ion

14

concentration. For the hydrophobic materials it is possible
in principle to calculate dispersive contributions by the
Fowkes method (19) for the components. This approach will be
explored elsewhere.

The generally conceived view of the platelet release
reaction is that platelets undergo aggregation and morphologic
changes prior to releasing intracellular constituents. The
adsorption "isotherm" experiments appear to show that the
number of platelets adhering to the surface is a function of
the free surface area of copolypeptide film. Thus, adsorbed
platelets are not acting as sites for further adhesion. It
is, of course, possible that platelets adhering to the substrate diffuse together to form small clusters which then
undergo release. A strong argument against this model is
provided by the data in fig 6 which show that the less dense
the surface layer of platelets, the more likely is the release
reaction. In fact, approaching monolayer coverage (calculated)
the probability of release is down to 5%. We conclude that
isolated platelets adhere to the polymer-coated beads, undergo morphologic constituents such as serotonin. They probably
do not diffuse across the surface (at least within the time
framework studied here). When crowded together, pseudopod
formation may be impaired and platelet release is inhibited.
Alternatively, products may be released that inhibit further
adsorption of platelets. In view of the available surface
area, the latter conclusion seems more likely.

The adsorption/release type isotherm of fig 5 also suggests that platelets do not desorb after release since no limit
to platelet release would be encountered with increasing dosage under those circumstances.

A puzzling feature is the observation that only a small
proportion of platelets from any "dose" adhere. Possible
reasons for this may be  a) not all platelets are viable

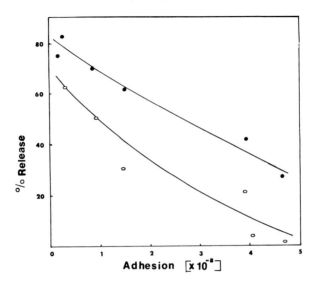

Fig 6   Percentage of total possible serotonin released as a
function of the number of platelets retained by the
column (normally assumed to be adhering but some may
be physically trapped) for two separate platelet
preparations.   (Same substrate composition as for
previous figs).

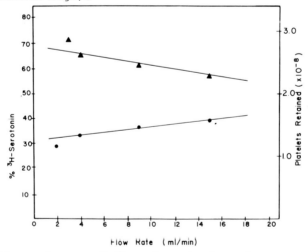

Fig 7   Effect of rate of flow through the column on platelet
retention and release.   With increased flow rates the
number of platelets retained (adhering) decreases but
the relative number of adsorbed platelets releasing
serotonin, increases.   Polypeptide film composition was
the same as for previous figs.

b) platelets adsorb reversibly but after release, remain ir-
reversibly attached   c) platelet release produces products
(proteins, glycoproteins, polysaccharides) which adhere
competitively with further platelets   d) platelets adhere as a
function of contact time.   Of these possibilities it can be
shown that   a) is not correct since preadsorption of proteins,
particularly albumin and fibrinogen produce very large changes
in platelet adsorption and release (2, 20).   All of the
preceding data seem to be consistent with the following
mechanism.

1. Most platelets probably contact the surface at some
   stage during the kinetic flow procedure.

2. Some mechanical damage to the platelets may occur
   during passage through the column but with the slow
   flow rates used in this study the effect is small
   compared with surface interaction effects.

3. Time of contact with the surface is probably important.   Short term contact leads to reversible adsorption, long term contact leads, in general, to platelet
   release reactions.

4. Platelet interactions with the surface determine the
   contact time.   Strong interactions lead, in general
   to increased adsorbance and release.

5. Released protein probably interferes with further
   uptake of platelets.

Within the preceding framework it seems relevant to
attempt to assess the surface interaction forces which determine platelet surface residence time and release processes.

Platelet Adhesion and Release

The data show that adhesion for a given platelet dose
and flow rate is remarkably independent of the amino acid
residues in the substrate and does not seem to be governed by
surface charge of hydrophobic content (within a limited range).
From the present data then it would appear that adhesion is
driven entropically i.e the free energy of adsorption arises

mainly from the disruption of hydration layers around the platelet and substrate. One implication of such an assessment is that minimum adhesion is likely to occur on substrates whose interfacial structure is similar to that of the platelet i.e strongly hydrated with "ordered" water. Such a concept is supported by the observation that adsorption on highly hydrated proteins or hydrogels is minimal.

We are led then to the concept that, foreign surfaces, of the type examined here, readily affect the behavior of platelets by causing adhesion and often serotonin release. We propose a tentative model for platelet adhesion as follows.

a) The platelet possesses an outer cell membrane which is readily modified in its fluidity under the influence of perturbing forces. Such modifications can originate in the modification of surface tension as the platelet adheres to a foreign surface.

b) A change in membrane fluidity, initiated by surface adsorbtion, can be the result of compressive or expansive forces, and results in the mobilization of an entity or entities which become free to activate platelet contractility.

c) The extent of contractile activation in part determines whether serotonin and other materials are released. The preceding concepts are expressed in visual form in fig 8. It can be seen that release will occur if the interfacial tension of the platelet changes significantly on adsorbtion. Release, is thus a reflection of the platelet's ability to adjust its membrane to conform to the energetic requirements of the substrate.

The release reaction occurs, we feel, through two main mechanisms. The first involves chemical stimulation of the contractile protein in or underlying the platelet membrane, which then causes the morphologic changes preceding the release process. The second, generally more relevant to

interaction with foreign surfaces, is the mechanical activation mentioned above. Thus any surface which causes the
membranes to expand or contract in order to attain a minimum
interfacial free energy, releases embedded enzymes and acti-
vates the contractile process. The result of this concept is
that minimum release will occur when the interfacial energy
of the adsorbed platelet is identical to that in free (plasma)
solution. Again such a concept requires a "gentle" substrate,
such as hydrated protein, lipid etc. We shall report elsewhere
that artificial "proteinaceous" surfaces having such properties are feasible. In fact such surfaces resemble those of
the arteries and veins to which platelets are exposed "in
vivo".

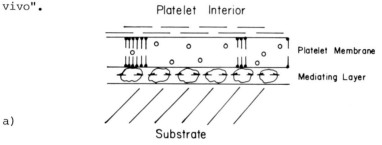

a)

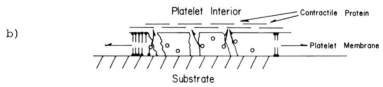

b)

Fig 8 a)  Model for mechanical distortion of platelet membrane
          by interfacial forces, allowing molecular entities
          to activate contractile protein.
     b)   Mediation of a mobile phase between platelet surface
          and solid substrate, minimizing disruptive forces on
          the membrane. The intermediate phase may arise from
          released or coated protein or may be structured
          solvent.

SUMMARY

A study of the adhesion of washed blood platelets to
random copolypeptide films has shown that adhesion is

i)    Essentially independent of the amino acid compositions
      examined.

ii)   Dependent upon, and inhibited by, preadsorbtion of
      platelets.

iii)  Slightly dependent on flow rate and exposure time.
The reaction in which radioactive serotonin is released is

i)    Slightly dependent on substrate composition, basic,
      positively charged substrates causing somewhat less
      release.

ii)   Decreased for increasing adsorbed platelet density.
The above information is interpreted as indicating that
platelet adhesion is a function of contact time and the
(free) energy of adhesion, which may be mianly entropic in
origin.   The release reaction apparently involves proteins
and other materials which compete with subsequent platelet
adhesion and which render the surface somewhat more passive.

The model presented provides some insight into design of
more platelet compatible materials.

ACKNOWLEDGEMENTS


We are pleased to acknowledge the help of Mr. G. Sullivan
and Dr. A. Mann with contact angle measurements, the experimental aid of Mr. W. Holaday and Mr. M. VanDress, Drs. J. M.
Anderson, P. H. Geil, T. Hayashi, C. R. McMillin and P. Vasko,
and discussions with Dr. D. F. Gibbons.   The main financial
support for this work was provided by the National Institutes
of Health under grants NHLI 15195 and NIGMS 07225 (Research
Service Award to D. Dewey Solomon) and NIAAA 00272.   We also

appreciate the financial support of the American Heart Assoc., Northeast Ohio Affiliate.

REFERENCES

1. M. F. Glynn, H. Z. Movat, E. A. Murphy and J. F. Mustard, J. Lab. Clin. Med., 65, 179 (1963).
2. M. F. Glynn, M. A. Packham, J. Hirsh and J. F. Mustard, J. Clin. Invest. 45, 103 (1966).
3. J. F. Mustard, M. F. Glynn, E. E. Nishizawa, and M. A. Packham, Fed. Proc. 26, 106 (1967).
4. D. J. Lyman, J. L. Brash, S. W. Chaikin, K. G. Klein, and M. Carini, Trans. Amer. Soc. Artif. Int. Organs, 14, 250 (1968).
5. E. F. Luscher and P. Massini, Thromb. Diath. Haem., 27, 121 (1972).
6. D. J. Lyman and S. W. Kim, Fed. Proc., 30, 1658 (1971).
7. S. W. Kim, R. G. Lee, H. Oster, D. Lentz, L. Coleman, J. D. Andrade and D. Olsen, Trans. Amer. Soc. Artif. Int. Organs, 20, 449 (1974).
8. S. Berger, E. W. Salzman, E. W. Merrill, and P. S. L. Wong in "Platelets" Ed. M. G. Baldini and S. Ebbe, Grune and Stratton Pub. Co., New York (1974).
9. R. E. Baier, "Adhesion in Biological Systems", Ed. R. S. Mauly, Academic Press, New York (1970), p. 15.
10. J. L. Brash and D. J. Lyman, J. Biomed. Mater. Res., 3, 175 (1969).
11. A. deVries, A. Schwager, and E. Katchalski, Biochem. J., 49, 10 (1951).
12. C. A. P. Jenkins, M. A. Packham, R. Kinlough-Rathbone, and J. F. Mustard, Blood, 37, 395 (1971).
13. T. H. Spaet, J. Cintron, N. Spivack, Proc. Soc. Exp. Biol., 111, 292 (1962).
14. W. Schneider, W. Kubler, R. Gross, Thrombos. Diathes. Haemorrh., 19, 207 (1968).
15. R. C. Keller, R. Mueller-Eckhardt, J. H. Kayser, and H. U. Keller, Int. Arch. Allergy, 33, 239 (1968).
16. K. A. Grottman, Thrombos. Diathes. Haemorrh., 21, 450 1969).
17. Th. Pfleiderer, R. Brossmer, Ibid., 18, 674 (1967).
18. P. Massini, L. C. Metcalf, and E. F. Luscher, Haemostasis, 3, 8 (1974).
19. F. M. Fowkes, Advan. Chem. Series, 43, 99 (1964).
20. D. D. Solomon, D. H. Cowan, and A. G. Walton (in preparation).

# INTERACTION OF POLYAMINES WITH THE SICKLING ERYTHROCYTE[1]

Paul W. Chun[2], Owen M. Rennert, Eugene E. Saffen,
Richard J. DiTore, Jayesh B. Shukla, W. Jape Taylor,
and Dinesh O. Shah
*University of Florida*

*The electrokinetic properties of normal and sickling red
blood cells examined in 1.5% glycine buffer as a function of
pH show distinct differences in the zeta potentials of the two
erythrocytes. At the physiological pH (7.4), the zeta potential of a solution of normal red blood cells is about -52 mv,
whereas that of sickling erythrocytes in solution is -45 mv.*

*The electrophoretic mobilities of normal red blood cells
in solution at pH 7.4 were found to be -1.2 μm sec$^{-1}$v$^{-1}$ cm in
5% sucrose, -2.0 μm sec$^{-1}$v$^{-1}$ cm in 5% sorbitol, -3.6 μm sec$^{-1}$
v$^{-1}$ cm in 1.5% glycine, and -1.3 μm sec$^{-1}$v$^{-1}$ cm in M/15 phosphate buffer, with a standard deviation of ±0.61. Glycine
buffer was found to be best suited for measuring the electrophoretic properties of the red blood cell.*

*The surface charge density variation of sickling erythro-
cytes in 1.5% glycine buffer, pH 7.4, based on the Debye-
Hückel exponential model for an electrical double layer, was
determined to be -70.4 x 10$^3$ stat coulombs/cm$^2$ at a normal
distance coordinate of 8 Å toward the cell surface from the
shearing plane. Based on the Gouy-Chapman model, we calculate
the surface charge density variation for normal erythrocytes
to be -1421 x 10$^3$ stat coulombs/cm$^2$ at a normal distance coordinate of 7 Å, as compared to -1765 x 10$^3$ stat coulombs/cm$^2$
for sickling erythrocytes at 8 Å. From our results, the
electrical double layer field would be 3.2 x 10$^6$ v/cm, with*

[1]This work was supported by National Science Foundation Grants
BMS 71–00850–A03 and PCM 76–04367, National Cystic Fibrosis
Research Foundation, NIH TGAM 05680–04, and, in part, by
General Research Support, College of Medicine, University of
Florida. We are grateful for the assistance provided by the
University of Florida computing center.
[2]Address all correspondence to Dr. P. W. Chun, Department of
Biochemistry and Molecular Biology, College of Medicine, Box
J245 JHMHC, University of Florida, Gainesville, Florida 32610.

$\Delta\psi$ = -225 mv (difference in surface potentials) and the thickness of the double layer being 7 Å for normal erythrocytes. For sickling erythrocytes, this measurement would be 3.0 x 10⁶ v/cm, where $\Delta\psi$ = -243 mv, and the electrical double layer would be 8 Å thick.

Measurement of the zeta potential as a function of pH reveals a 10-15% reduction in the case of the sickling red blood cells. Measurement of the interfacial surface viscosity of the sickling erythrocytes shows it to be ten times greater than that of normal red blood cells.

In sickling red blood cells, the polyamine content determined by amino acid analysis is found to be five or six times greater than that of normal erythrocytes. The polyamine content of whole blood taken from 24 patients having sickle cell anemia has been found to be approximately ten times greater than that of whole blood taken from normal donors.

The difference in the electrokinetic properties of the normal and sickling red blood cells in glycine buffer may be attributed in part to a variation in the polyamine content of the two types of erythrocytes.


I.  INTRODUCTION

A great deal of data on the surface alteration and deformation of the red blood cell has been obtained by examining the electrokinetic behavior of its electrical double layer (1-8), rheological flow behavior (9-13), recognition and binding isotherms of a variety of external chemical substances to the surface receptors of the erythrocytes (14-16), fluorescence labeling to reduce the negative surface charge (17,18), and electron micrographic examination of sickled red blood cells under deoxy conditions (19).

Polyamine analysis in bone marrow samples of leukemic and nonleukemic conditions has revealed increased concentrations of putrescine, spermidine, and spermine in proliferating cells (20-24), and elevation and alteration of the polyamine content of blood in the inherited disorder, cystic fibrosis (25-27).

Variations in their distribution patterns indicate that spermidine and spermine may fulfill different functions in the human body (28,29). It has been determined, however, that ornithine decarboxylase is the controlling enzyme in the pathway of polyamine biosynthesis (20,21). Induction of ornithine decarboxylase and the concentration of polyamines are closely linked to cellular proliferation, with a potential relationship existing between polyamines as a "third messenger" and cyclic nucleotide metabolism (30-32).

We have previously reported a notable elevation in the polyamine content of sickling whole blood when compared to whole blood samples from normal donors. We have found differences in the electrokinetic properties of the two types of erythrocytes in 1.5% glycine buffer which may be attributed, in part, to a variation in their polyamine content (33-35).

To date, no findings have been reported on the polyamine content of sickling red blood cells. Little is known about the possible role of polyamines in the deformation and rigidity of the red blood cell in vitro or in vivo.

In this communication, we report on our determination of the magnitude and possible mechanism of the interaction between polyamines and normal or sickling erythrocytes, considering the electrokinetic properties of sickling red blood cells as a function of polyamine concentration, measurement of the apparent interfacial surface viscosity, and examination of the pulse nuclear magnetic resonance spectra of sickling erythrocytes.

## II. MATERIALS AND METHODS

### A. Isolation of Red Blood Cells

Ten to fifteen ml each of whole blood from donors with normal (hematocrit values were 40-45 as compared with 18-30 for erythrocytes from sickling donors) and homozygous sickling hemoglobin were collected in Heparin tubes and centrifuged at 900 r.p.m. in an HN-S centrifuge (International, Damon) for 20 minutes prior to decanting the serum and leukocytes. The cells were washed eight times in 10 ml of isotonic 0.9% NaCl solution and centrifuged for 20 minutes at 1000 r.p.m. after each washing. A greater degree of adhesion was observed for the sickle cells in solution during this washing procedure. All subsequent procedural operations were completed during a two-day period. Prolonged standing for greater lengths of time at 4°C resulted in a leaching of hemoglobin from the cells.

### B. Electrophoretic Mobility of Erythrocytes

The electrophoretic mobility of the red blood cell was measured in 1.5% glycine buffer, as a function of pH as well as polyamine concentration, with a Riddick Zeta Meter (Zeta Meter, Inc., New York). The zeta potential was computed from the dielectric constant of the medium, the current intensity, dimensions of the cell, and the electrophoretic mobility of

the cells, using the Helmholtz-Smoluchowski equation (36).
The concentration of red blood cells in suspension was diluted
to an equivalent of $1 \times 10^{-4}$ M/heme for each sample, using the
ratio of molar extinction at 576 to 541 nm of 1.066.

    A known concentration of polyamines (Sigma Chemical Co.,
St. Louis) was incubated with washed red blood cells at 37°C
for two hours prior to measurement of electrophoretic mobility.
After measurement of the electrophoretic mobility, the red
blood cells were rewashed and the polyamine content determined.

C.    Measurement of the Apparent Interfacial Surface Viscosity
      of the Sickled Erythrocytes

    After red blood cells were washed eight times with 0.9%
NaCl (approximately $1 \times 10^9$ cells/ml), ten ml of red blood
cells were diluted to 25 ml with 0.9% NaCl solution. Using
the interfacial viscometer constructed in our laboratory, it
was possible to determine the ratio between shear stress and
shear rate in the plane of the interface, primarily to examine
the rheological flow behavior from the angular displacement of
the bob corresponding to the angular velocity, shear rate.
The apparent interfacial viscosity is then found by calculating the ratio of the coordinate, the mean shear stress (37,38).

D.    Examination of the Pulse Nuclear Magnetic Resonance
      Spectra of Sickling Erythrocytes

    The washed cells in either 0.9% NaCl in $H_2O$ or $D_2O$ were
packed into quartz G-S tubes specially designed for pulse
nuclear magnetic resonance studies, and measuring 7.5 cm by
1 cm in diameter. The cells were tightly packed and centrifuged at 800 r.p.m. until no further sedimentation occurred.
Centrifugation was continued for approximately 30 minutes until all extraneous water that was not tightly bound to the red
blood cells had been removed. At this point, the maximum
hematocrit value of 74~75% is reached. All subsequent Praxis
pulse NMR studies of the resulting solution were computed
within one day.

1.    *Measurement of $T_1$*
    $T_1$ was measured using a PR103, 10 MHz pulse NMR spectro-
meter with a 90°-90° pulse program. Provided $T_1 \gg T_2$, the
free induction following the first 90° pulse decays to zero
more rapidly than the magnetization along the z axis reaches
its equilibrium value, $M_0$. Hence, a second 90° pulse permits
a sampling of $M_z$ at a variable delay time, $\tau$. The magnetization rapidly decreases to a steady-state value dependent on $\tau$

and $T_1$. When $M_z = 0$ at $t = 0$, the Bloch plot (39) becomes $\ln(A_\infty - A_\tau)$ versus $\tau$, giving a straight line from which $T_1$ can be determined from the slope. $A_\tau$ is the initial amplitude of the free induction decay following the 90° pulse at time $\tau$, and $A_\infty$ is the limiting value of the amplitude at $\tau$ for a very long interval between the 90° and 90° pulse.

## 2.   Measurement of $T_2$

The spin-spin relaxation time, $T_2$, was measured using a 90°-180° pulse program similar to Hahn's procedure (40), providing for the application of a 90°, $\tau$, 180° sequence and the observation at a time $2\tau$ of a free induction "echo". The net magnetization along the z axis, $M_z$, is the vector sum of individual macroscopic magnetization, $M_i$, arising from nuclei in different parts of the sample and, hence, experiencing slightly different values of the applied field. $M_0$ loses coherency along $M_x$, and each $M_i$ decreases in magnitude during transverse relaxation in the time $T_2$. Thus, echo amplitude depends on $T_2$ and this quantity may in principle be determined from a plot of peak echo amplitude as a function of $\tau$. It is necessary to carry out a separate pulse sequence for each value of $\tau$ and to wait between sequences an adequate time for restoration of equilibrium.

A 60 MHz nuclear magnetic resonance spectrometer was also used to determine the chemical shift of water in normal and sickling red blood cells. Results were compared with those from a Praxis pulse NMR spectrometer.

## E.   Tryptic Digestion

Whole white sickling ghost erythrocyte membranes isolated from red blood cells (41) were digested with trypsin for 20 hours and two-dimensional peptide mapping was performed, as described elsewhere (42,43).

## F.   SDS Gel Electrophoresis

One ml samples of white erythrocyte ghost membrane material were dissolved in 33% acetic acid, 6 M urea, by a method similar to that of Fairbanks et al. (44). Gel composition, the staining procedure, and sample preparation for the gel are according to the method of Fairbanks et al.

## G.   Separation of Erythrocytes on a Ficoll Density Gradient

Twenty and thirty percent by weight of Ficoll type 70

($M_w \simeq 70,000$), obtained from Sigma Chemical Company, was dissolved in 291 mosm BSG buffer, pH 7.4 (1.22 gm $Na_2HPO_4$, 0.129 gm $NaH_2PO_4$, $2H_2O$, 8.1 g NaCl, 2 gm of glucose, and 7% BSA [Bovine serum albumin, Sigma Chemical Co.]) to 1 liter with $dH_2O$. The refractive indexes of 20% and 30% Ficoll solutions were measured by refractometer and found to be $n^{20} = 1.3840$, $\rho^{20} = 1.0820$ and $n^{30} = 1.3822$, $\rho^{30} = 1.1276$, respectively. For separation of erythrocytes, 20 ml of each solution were layered on each other in 8.7 x 3.1 cm cellulose nitrate tubes. Two ml of washed red blood cells were layered on top of the 20% gradient. The tube was filled with light mineral oil and centrifuged in the Spinco L-65 Beckman centrifuge with SW 25.1 rotor for 60 minutes at 15,000 r.p.m. at 10°C.

## H.   Analysis of Polyamines

One ml of washed red blood cells (after counting the number of red blood cells) was extracted with an equal volume of 10% sulfosalicylic acid. Extraction was repeated twice. The supernatant was lyophilized and resuspended in 0.5 ml of 0.2 N sodium citrate buffer, pH 2.2, and then filtered through a millipore filtration apparatus.

Polyamine analysis was performed using a Durrum D-500 high-pressure chromatograph coupled with a digital data PDP8/M coupler. The separatory system utilized a Durrum DCX12-8 cation exchange resin, a sulfonated polystyrene polymer with a bead diameter of 8±1 μm, and 12% cross linkage, packed to a height of 11.5 cm in a stainless steel 1.75 mm column. A four-buffer elution system of sodium citrate is used: 0.067 M, pH 6.17; 0.3 M, pH 4.68; 0.35 M, pH 4.68; 0.38 M, pH 4.68.

Polyamine detection involved fluorometric quantitation utilizing Durrum Fluoropa (0-phtholaldehyde). A 500 pmole polyamine standard of putrescine and spermidine gave an optical density deflection of 0.6, and spermine gave a deflection of 0.4 units.

## III.  POLYAMINE CONTENT OF SICKLING ERYTHROCYTES

Based on the described isolation procedure for white ghost erythrocyte membranes, complete removal of hemoglobin from the HbS erythrocyte proved to be more difficult than for HbA cells. A greater degree of adhesion of hemoglobin to the sickling red blood cell membrane may account for this phenomenon. In the case of HbA, the total yield from a 10 ml sample of erythrocytes is 936 ± 50 mg, as compared to 639 ± 30 mg for HbS.

5.6% SDS gel electrophoretic patterns of normal and

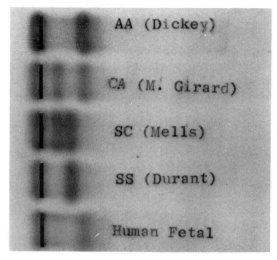

Fig. 1. Starch gel electrophoretic patterns of various forms of hemoglobin at pH 8.6 (vernal buffer, μ = 0.1) (45).

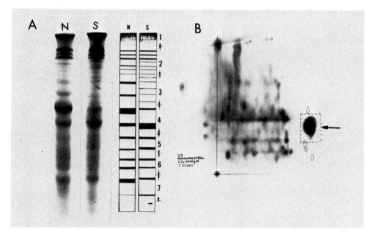

Fig. 2. A. 5.6% SDS polyacrylamide gel electrophoretic patterns of normal and sickling white ghost membrane dissolved in 33% acetic acid, 6 M urea.

B. Two-dimensional high-voltage electrophoretic patterns of the soluble tryptic digest of white ghost membrane of sickling erythrocytes at pH 4.8 in buffer (1.25% pyridine - 1.25% acetic acid) for one hour at 2000 volts.

N and S represent normal and sickling erythrocyte ghosts, respectively. Numbers indicate the group of bands representing a certain molecular weight range. Group 7 - $M_W$ of 12,000-20,000; Groups 5 and 6 - $M_W$ of 20,000-45,000; Group 4 - $M_W$ of 50,000-70,000; Group 3 - $M_W$ of 80,000-100,000.

sickling white ghost erythrocyte membranes are shown in Figure 2A. We observed at least 18 different protein bands in both normal and sickling erythrocytes, with major variations appearing in a protein class band at a molecular weight of approximately 60,000~100,000 daltons.

The lyophilized product of the soluble portion of the white ghost membrane for the two types of erythrocytes, following 20 hours of trypsin digestion, was subjected to peptide mapping, as shown in Figure 2B. No differences could be observed by this technique alone.

However, when the polyamine contents of the large spots (indicated by arrows) were analyzed, the sickling erythrocytes were found to contain both putrescine and conjugated spermidine (formyl spermidine)(Figure 3B), while the normal erythrocytes were found to contain no polyamines (Figure 3D). The presence of these polyamines was even more pronounced in the insoluble product of the sickling erythrocytes (Figure 3A), while none appeared in the normal erythrocyte (Figure 3C).

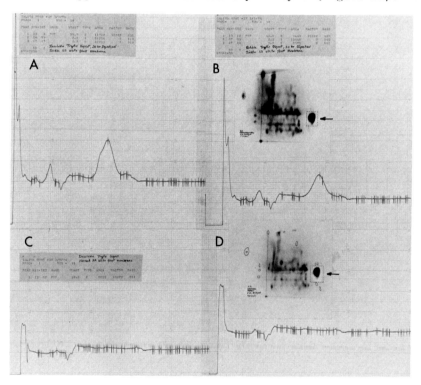

Fig. 3. *Polyamine analysis of the soluble and insoluble products of normal and sickling red blood cells, following 20 hours of tryptic digestion.*

Fig. 3. A. Polyamine content (formyl conjugated) of the insoluble tryptic digest of sickled white ghost membrane.
B. Polyamine content of the spot, indicated by arrow, from the peptide map of sickled white ghost membrane.
C. Polyamine content of the insoluble tryptic digest of normal white ghost membrane.
D. Polyamine content of the spot, indicated by arrow, from the peptide map of normal white ghost membrane.

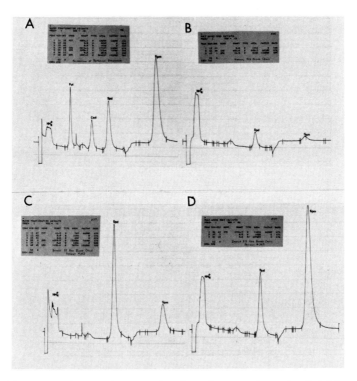

Fig. 4. Polyamine analysis of normal and sickling erythrocytes.
A. Polyamine standard (500 pmole/ml) of putrescine (Put), cadaverine (Cad), spermidine (Spd), and spermine (Spm).
B. Normal red blood cells.
C. Homozygous SS patient, age 26.
D. Homozygous SS patient, age 2.

The polyamine content of 500 pmoles of polyamine standard (putrescine, cadaverine, spermidine, and spermine)(Figure 4A) is compared with that of normal red blood cells in Figure 4B, while the polyamine contents of red blood cells from two homozygous sickling patients are shown in Figures 4C and 4D.

31

The quantity of spermidine and spermine present in both sickling samples is five to six times that of the normal red blood cells (Figure 4B and Table 1). The polyamine content of these abnormal erythrocytes was higher than that of the normal red blood cells. In some instances, we also noted a considerable elevation of the putrescine content. No differences were observed, however, in the polyamine content of normal red blood cells or the erythrocytes of cystic fibrosis patients (Table 1).

TABLE 1

Polyamine Content of Erythrocytes (nmoles/$10^9$ cell)

| | Putrescine | Spermidine | Spermine |
|---|---|---|---|
| Normal AA Rbc (N = 9) | 0.007~ | 1.39 ± 0.46 | 0.9 ± 0.27 |
| Normal AA Rbc (Prosthetic heart valve with hemolysis) (N = 1) | 0.009~ | 1.64 ± ~ | 1.20 ± ~ |
| Sickle SS Rbc (N = 2) | 0.01 | 6.81 ± 0.40 | 6.95 ± 0.30 |
| (N = 1)[1] | 15.1±--- | 40.8 ± ---- | 4.29 ± ---- |
| HbAc Rbc[2] (N = 1) | 0.005~ | 3.91 ± ---- | 0.6 ± ---- |
| HbSA Rbc[2] (N = 2) | tr | 1.94 ± 0.42 | 1.64 ± 0.31 |
| HbSC Rbc[3] (N = 1) | 6.73±-- | 5.11 ± ---- | 0.47 ± ---- |
| Cystic Fibrosis Rbc[2] (N = 4) | tr | 1.55 ± 0.25 | 0.99 ± 0.27 |

[1]Leukocyte count of this donor was twice as high as that of other homozygous sickle cell patients examined. Hematocrit value of 18%.

[2]Sample had a hematocrit value of 45%.

[3]Sample had a hematocrit value of 34%.

We have previously reported that the spermidine content of whole blood from 24 patients with sickle cell anemia was approximately ten times greater than that of whole blood from normal donors: 35.97 ± 17.9 nmoles/ml, as compared to 3.87 ± 1.29 nmoles/ml from normal donors. For spermine, the difference is three-fold: 13.52 ± 5.41 nmoles/ml, as compared to 4.01 ± 1.37 nmoles/ml from normal donors (33–35).

In the leukocytes, the spermine content is ten times (322.5 ± 12.1 nmoles/$10^9$ cell) greater in blood from sickle cell anemia patients than that from normal donors (35.95 ± 0.91 nmoles/$10^9$ cell). For spermidine, the difference is 16-fold (238 ± 11.9 nmoles/$10^9$ cell for sickle cell, as compared with 15.33 ± 0.5 nmoles/$10^9$ cell for normal)(33,35).

Since the erythrocytes do not have a mechanism for the synthesis of polyamines, we speculate that there may be several contributing factors involved. First, chronic reticulocytosis characteristic of low-grade anemia (as in sickle cell anemia) may result in the intercalation of polyamines into the membrane of immature red blood cells. At the same time, the bone marrow is stimulated to produce an increased number of leukocytes and platelets. We have previously suggested that some equilibration may take place between the polyamines of the leukocytes and the immature red blood cells. In view of the hemolysis characteristic of sickling red blood cells, a third possibility may be that some form of polyamine equilibration occurs in these abnormally shaped erythrocytes.

## IV. POLYAMINES – RED BLOOD CELL INTERACTION

At the physiological pH 7.4, the zeta potential of the normal red blood cell in 1.5% glycine buffer was found to be $-52$ mv ($-3.6$ $\mu$m sec$^{-1}$v$^{-1}$cm), whereas that of the sickling erythrocyte is $-45$ mv ($-3.2$ $\mu$m sec$^{-1}$v$^{-1}$cm)(46). In 5% sucrose solution at pH 7.4, the zeta potential of the normal erythrocyte is $-19$ mv ($-1.2$ $\mu$m sec$^{-1}$v$^{-1}$cm), as compared with $-26$ mv ($-1.9$ $\mu$m sec$^{-1}$v$^{-1}$cm) in 5% sorbitol (Figure 5).

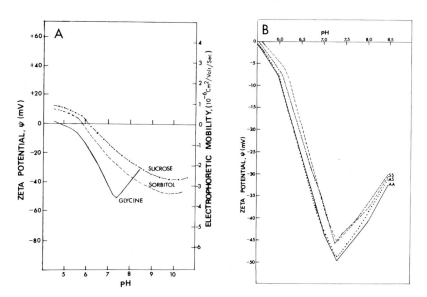

*Fig. 5. A. The electrophoretic mobility and zeta potential of the erythrocyte as a function of pH in 1.5% glycine, 5% sucrose, and 5% sorbitol, with a standard deviation of 5%. Results represent the mean average of data for five normal blood samples at 22°C.*

*B. Comparison of the zeta potential as a function of pH of red blood cells from normal donors (AA) and persons with various hemoglobin disorders (AS and SS Rbc) in 1.5% glycine buffer at 22°C.*

The surface charge density variation of sickling erythrocytes in 1.5% glycine buffer, pH 7.4, based on the Debye-Hückel exponential model for an electrical double layer, was determined to be $-70.4 \times 10^3$ stat coulombs/cm$^2$ at a normal distance coordinate of 8 Å toward the cell surface from the shearing plane (Figure 6). Based on the Gouy-Chapman equation (4,47), we calculate the surface charge density variation for normal erythrocytes to be $-1421 \times 10^3$ stat coulombs/cm$^2$ at a normal distance coordinate of 7 Å, as compared to $-1765 \times 10^5$ stat coulombs/cm$^2$ for sickling erythrocytes at 8 Å. From our results, the electrical double layer field would be $3.2 \times 10^6$ v/cm, with $\Delta\psi = -225$ mv, and the thickness of the double layer being 7 Å for normal erythrocytes; whereas, $\Delta\psi = -243$ mv, and the electrical double layer would be 8 Å thick, as seen in Figures 7A and 7B.

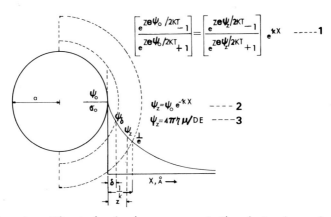

Fig. 6. *Electrical phenomena at the interface for evaluation of surface potential of the electrical double layer. Equation 3 is the Helmholtz-Smoluchowsky equation, where D is the dielectric constant (89.5 in our case), μ is the electrophoretic mobility of the red blood cell, and η is the viscosity of the medium in poses.* $\psi_\delta$ = *Stern potential.* δ, z, *and 1/k are Stern, zeta, and Debye-Hückel distances, respectively. The zeta potential lies between 1/k, which is 37.2% of exponential, and* δ. X, *in* Å, *is varied toward the cell surface from the shearing plane where* $\psi_z$ *is measured (48).*

The binding of polyamines as a function of concentration to normal and sickling erythrocytes was analyzed by Langmuir-type binding isotherms based on the surface potential variation for an electrical double layer. The surface charge density ($\sigma_o$)(Figures 7A and 7B) of the univalent electrolytes of the flat electrical double layer is treated according to Verwey and Overbeek (46,47), where the zeta potential ($\psi_z$) within

34

the double layer is a function of only the normal distance coordinate, X, in Å. The flat double layer model for evaluation of $\psi_0$ takes the form of Equation 1, shown in Figure 6, where $y = Ze\psi_z/KT$ and $y_0 = Ze\psi_0/KT$. Z is the valence of the ions, e is the charge of electrons, K denotes the Boltzman constant, and 1/k is the Debye-Hückel length. In those cases where $y_0 \ll 1$, $\psi_0 \ll 25.7$ mv and Equation 1 is reduced to Equation 2 of this figure.

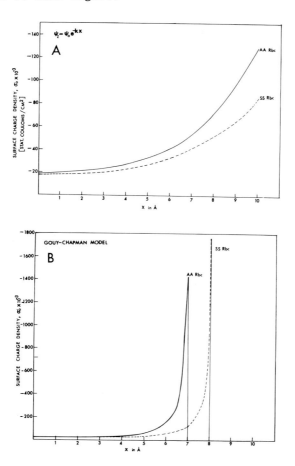

Fig. 7. A. Plot of surface charge density variation of the Gouy-Chapman model, based on the Debye-Hückel exponential function, $\psi_0$, as a function of the normal distance coordinate, X, in Å, for normal (AA) and sickling (SS) red blood cells.
B. Plot of surface charge density variation of the Gouy-Chapman model as a function of the normal distance coordinate. Graph has been truncated close to the surface (7 Å for AA Rbc; 8 Å for SS Rbc, dotted line).

1.    The first method of examining polyamine binding to the red blood cell (Rbc + polyamines $\rightleftharpoons$ Rbc:polyamines) is based on the Langmuir-type of linear binding isotherm shown below:

$$1/\psi_o = 1/\psi_m K_a C + 1/\psi_m$$

where $\psi_o = \psi_m e^{kx}$ is the electrical potential at the surface of the red blood cell, $\psi_m$ is the zeta potential measured after polyamine binding, C is the concentration of polyamines, and $K_{app}^{ex/es}$ is the apparent binding constant of polyamine to the red blood cell as a function of the normal distance coordinate X, plotting $1/\psi_o$ versus $1/C$.

2.    The second method of analysis is based on a Langmuir-type expression, except that the fraction of red blood cells liganded with polyamine is measured, plotting $1/(\psi_m^o - \psi_z^o)$ versus $1/C$.

$$1/(\psi_m^o - \psi_z^o) = -1/K_a \psi_z^o C - 1/\psi_z^o$$

where $\psi_z = -52$ mv and $-45$ mv for normal and sickling erythrocytes, respectively.   $\psi_z^o$ is the zeta potential at zero concentration of polyamine, which is equivalent to the total zeta potential of the red blood cells.   $\psi_m^o$ is the zeta potential of the unbound red blood cells after the addition of polyamines.   Thus, the fraction of the zeta potential influenced by polyamine binding to the red blood cell is $(\psi_z^o - \psi_m^o)/\psi_z^o$.

At the physiological pH 7.4, the zeta potential of normal red blood cells in 1.5% glycine buffer was found to vary with the addition of polyamine, as seen in Figure 8.  Addition of spermidine or spermine at concentrations up to 5.0 x $10^{-3}$ M to the normal red blood cells reduces the zeta potential by approximately 20 mv.  The reduction for sickling erythrocytes is only 5 mv.  The effect of spermidine, as may be seen in this figure, is far greater than that of $CaCl_2$ and procaine, although both would appear to undergo biphasic or sigmoidal binding with the red blood cell.  It may be seen from these results that little or no further polyamine binding occurs after the addition of spermidine to sickling erythrocytes.

Figure 8a-A shows a computer plot of $1/(\psi_o e^{-kx})$ versus $1/C$, the polyamine concentration, for normal and sickling erythrocytes.  The nonlinearity of the plot indicates that multiple independent binding sites are probably present in the normal and sickling red blood cells.  The limited binding of polyamine to the sickling red blood cells, at normal distance coordinates of 3 to 8 Å, which is apparent in the hyperbolic curves, is consistent with the change in zeta potential of only 5 mv.  The sigmoidal curves exhibited for

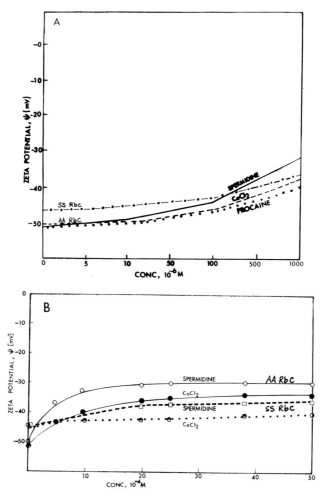

*Fig. 8. A & B. The zeta potentials of normal (● or 0) and sickling (···· or ----) erythrocytes as a function of ligand concentration in the presence of spermidine and CaCl₂. Cells were incubated with ligand at 37°C for two hours. Zeta potential is measured in 1.5% glycine buffer, pH 7.4.*

the normal red blood cell, at various distance coordinates, strongly indicate that polyamine binding may be biphasic, that is, both esotropic and exotropic in nature.

In plotting Gouy-Chapman's surface potential, $1/(\psi_m^0 - \psi_z^0)$, versus the concentration of spermidine, $1/C$, for normal and sickling red blood cells, the biphasic nature of polyamine binding becomes apparent in both cases, as may be seen in Figures 8a-B and 8a-C.

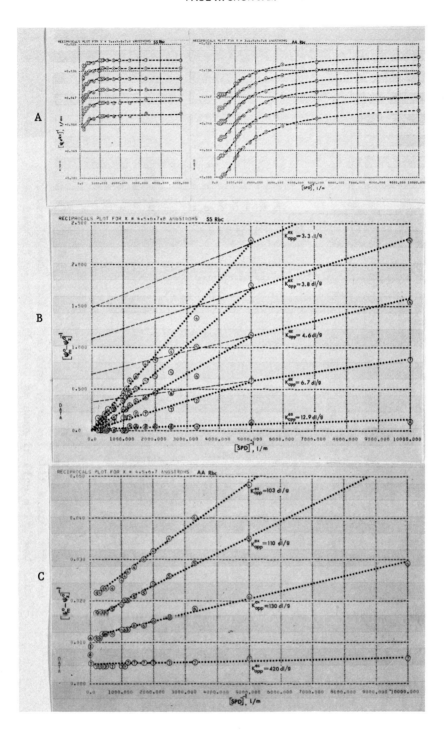

*Fig. 8a. A. Reciprocal plot of $1/(\psi_0 e^{-kx})$ versus $1/(Spd)$ as a function of spermidine concentration for sickling (SS) and normal (AA) erythrocytes. Zeta potential values are an average of ten measurements for each of several red blood cell samples. Circle with number represents the normal distance coordinate, X, in Å.*
*B. Reciprocal plot of $1/(\psi_m^0 - \psi_z^0)$ versus $1/(Spd)$ for sickling erythrocytes. $\psi_m^0$ values are computed from the flat double layer model, based on Equation 1 in Figure 6, varying the distance coordinate from 4 Å to 8 Å toward the cell surface from the shearing plane.*
*C. Reciprocal plot of $1/(\psi_m^0 - \psi_z^0)$ versus $1/(Spd)$ for normal erythrocytes, varying the normal distance coordinate from 4 Å to 7 Å.*

The apparent exotropic binding constants of polyamine with the sickling erythrocytes are 3.3, 3.8, 4.6, and 6.7 dl/g at X distances of 4 to 7 Å, respectively. The esotropic binding constant is 12.9 dl/g at 8 Å (see Figure 8a-B). The surface charge density variations of the sickling red blood cells, prior to the addition of polyamine, were found to be $-26.9$, $-36.2$, $-54.3$, $-106$, and $-1765 \times 10^3$ stat coulombs/cm$^2$ at normal distance coordinates of 4 to 8 Å.

As seen in Figure 8a-C, the apparent exotropic binding constants of polyamine with normal red blood cells are 103, 110, and 130 dl/g at normal distance coordinates of 4 to 6 Å, while the esotropic binding constant was found to be 420 dl/g at 7 Å. The surface charge density variations of the normal erythrocytes were determined to be $-34.9$, $-58.4$, $-122.8$, and $-1421.9 \times 10^3$ stat coulombs/cm$^2$ at normal distance coordinates of 4 to 7 Å. We may surmise from these results that the electrical double layer of the sickling erythrocyte is one angstrom thicker than that of the normal red blood cell.

If polyamines are adsorbed on the red blood cell by electrostatic attraction alone, then the surface charge density, $\sigma_0$, should be equivalent to the Gouy–Chapman calculations of the same value. Under the experimental conditions studied, $\psi_\delta$, the Stern layer, should change very little since the adsorption surface density of the polyamines should be directly proportional to the square root of the concentration of added polyamines.

Our studies indicate, however, that in sickling red blood cells the thickness of the electrical double layer increases by one angstrom, measuring 8 Å, as compared with 7 Å for the normal red blood cell. This suggests that polyamines in vivo will strongly interact in the Stern plane as a result of electrostatic or hydrophobic bonding or intercalation into the red blood cell membrane, causing a relocalization of the

surface charge density distribution or perturbation of the
Stern layer. This phenomenon we have designated as "esotropic" interaction to differentiate it from "exotropic" interaction in which there is little alteration of the Stern
layer.

The variation of $K_{app}^{ex}$ as a function of the thickness of
the electrical double layer which we have observed indicates
that the binding affinity of polyamines to the normal red
blood cell is approximately 30 times greater than to the sickling erythrocyte. This is consistent with our earlier reported findings that the polyamine content of the sickling red
blood cell is five to six times that of the normal erythrocyte, effectively limiting any further binding with polyamines.

## V.   FICOLL GRADIENT CENTRIFUGATION OF NORMAL AND SICKLING ERYTHROCYTES

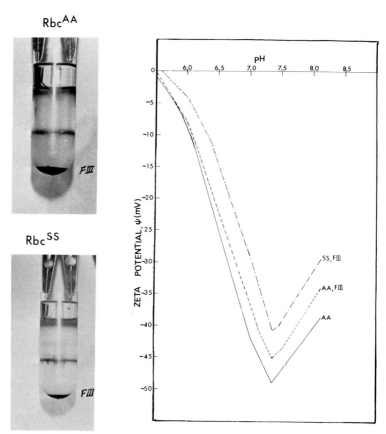

*Fig. 9. Fractionation of normal and sickling red blood cells by Ficoll density gradient. Top, normal erythrocyte; bottom, sickling erythrocyte. Fraction III in each case was used to determine the zeta potential as a function of pH in 1.5% glycine buffer. AA represents normal erythrocytes and SS represents the sickling erythrocytes.*

Step Ficoll gradient centrifugation of normal erythrocytes yields three distinct red blood cell fractions, the top fraction having a density of 1.0820; the second (middle band), 1.1276; and the bottom fraction, 1.1810. Determination of the zeta potential of the major Fraction III, shown in Figure 9, reveals a reduction of 5 mv from that of the normal washed red blood cells. We observed a further 7 mv reduction in the zeta potential of Fraction III isolated from sickling erythrocytes. This reduction may be attributed to the influence of Ficoll on the medium. The magnitude of the change, however, is consistent with that which we observed in unfractionated samples of normal and sickling erythrocytes.

Experiments to characterize this Fraction III, with emphasis on the age of the cells which make up the population and their polyamine content, are currently underway in our laboratory.

## VI. THE APPARENT INTERFACIAL VISCOSITY OF THE SICKLING ERYTHROCYTE

An interfacial viscometer was constructed in our laboratory based on a modification of the design of Brown et al. (37) and Karam (38), as shown in the accompanying schematic diagram (Figure 10). The interfacial viscosity curves for normal and sickling erythrocytes are shown in Figures 11A and 11B. Figure 11A shows the time-dependent deflection of the bob at various rates of rotation of the cup for samples of normal and sickling red blood cells. At 30 minutes, differences of 10-15 degrees are observed between normal and sickling erythrocyte samples.

The nonNewtonian flow behavior exhibited by both normal and sickling erythrocytes may be clearly seen in Figure 11B, plotting deflection of the cup after 20 minutes have elapsed.

For nonNewtonian fluids, the apparent interfacial viscosity is a function of the shear rate; hence, a rheogram of mean shear rate versus mean shear stress conveys more information about the red blood cell system.

Karam's expression for apparent interfacial viscosity as a function of the mean shear rate is $\bar{D} = (2R_bR_c\omega)/(R_c^2 - R_b^2)$. The apparent interfacial viscosity is then found by

41

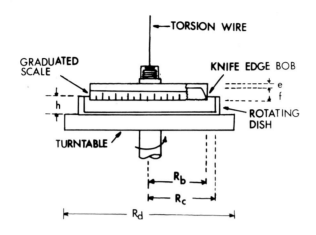

Fig. 10. Schematic diagram of the interfacial viscometer, a modification of the design of Brown *et al.* (37) and
Karam (38), showing the turntable, rotating dish, and knife-
edged bob suspended by a torsion wire. The instrument itself
is shown in the accompanying photograph (dotted line defines
area shown in diagram above). $R_b$ is the radius of the knife
edge of the bob, 2.0 cm; $R_c$ is the radius of the cup, 2.5 cm;
$R_d$ is the diameter of the turntable, 20.7 cm; while e = 3 mm,
f = 5 mm, and h = 10 mm. The torsion constant of the wire, K,
was determined by measuring the period of oscillation time, t,
with an object of known moment of inertia, I, suspended by the
wire, $K = 4\pi^2 \Delta I / t^2 - t_0^2$, where $t_0$ is the first measure of the
chuck and damping ring.

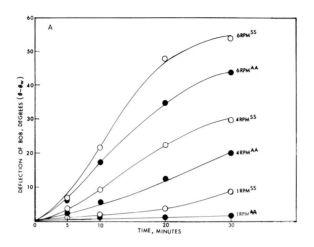

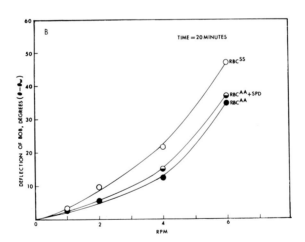

Fig. 11. A. Plot of deflection of the bob in degrees $(\theta - \theta_w)$ versus elapsed time at a given velocity of the turntable for samples of normal and sickling red blood cells.

B. Plot of deflection of the bob in degrees $(\theta - \theta_w)$ versus velocity of the turntable (r.p.m.) at an elapsed time of 20 minutes for samples of normal and sickling erythrocytes and normal cells in the presence of spermidine. All samples exhibit nonNewtonian flow behavior.

calculating the ratio of the coordinates. The mean shear stress, $\bar{\tau}$, is given by $\bar{\tau} = K(\theta - \theta_\omega)/(2\pi R_b R_c)$, where $\omega$ = angular velocity in degrees/sec and $(\theta - \theta_\omega)$ = angle of deflection of the bob, in degrees.

Evaluation of the apparent interfacial viscosity (in surface poise, dyne sec/cm) for an ideal Newtonian flow from the angular displacement of the bob corresponding to the angular velocity (shear rate) was formulated by Reiner (37).

In examining the polyamine content of blood samples from 24 donors, we found a ten-fold difference in the polyamine content of normal and sickling erythrocytes (33-35). Addition of 300 nmoles of spermidine per ml of normal red blood cells $(1.89 \times 10^9$ cells/ml) caused a variation of 4~5 degrees in the angle of deflection of the bob during interfacial viscosity measurements, as seen in Figure 11B.

The increase in the interfacial viscosity with time under constant stress apparent from these results is rheotropic, as opposed to another frequently observed example of nonNewtonian flow behavior under unsteady-state conditions, the theotropic decrease in viscosity with time. Our measurements, then, would seem to support earlier reports that the normal red blood cell exhibits nonNewtonian flow behavior. Our results show significant differences in the apparent interfacial viscosity of normal and sickling erythrocytes, as shown here in tabular form* (Table 2).

TABLE 2

The Apparent Interfacial Viscosity of the Red Blood Cells

| RPM | \multicolumn{5}{c}{Normal Erythrocyte} | | | | | \multicolumn{5}{c}{Sickling Erythrocyte} | | | | |
|---|---|---|---|---|---|---|---|---|---|---|
| | $(\theta-\theta_\omega)$ | $\bar{D}$ | $\bar{\tau}$ | Log $\bar{D}$ | Log $\bar{\tau}$ | $(\theta-\theta_\omega)$ | $\bar{D}$ | $\bar{\tau}$ | Log $\bar{D}$ | Log $\bar{\tau}$ |
| 1 | 0.20 | 26.6 | 0.13 | 1.43 | -0.89 | 3.5 | 26.6 | 2.38 | 1.43 | 0.38 |
| 2 | 6.30 | 53.3 | 4.11 | 1.73 | 0.61 | 8.7 | 53.3 | 5.92 | 1.73 | 0.77 |
| 4 | 12.1 | 106.6 | 7.89 | 2.03 | 0.90 | 22.7 | 106.6 | 15.44 | 2.03 | 1.19 |
| 6 | 34.8 | 159.8 | 22.7 | 2.20 | 1.36 | 47.6 | 159.8 | 32.37 | 2.20 | 1.51 |

*Results represent the mean average of data for five blood samples, with an experimental error of ±3.2%. Least square analysis for Log $\bar{D}$ versus Log $\bar{\tau}$ gives

$$\eta_s^{AA} = 9.94 \times 10^{-3}(\bar{D})^{0.5} \text{ and } \eta_s^{SS} = 20.3 \times 10^{-3}(\bar{D})^{0.5}$$

$$\Delta\eta_s^{(SS-AA)} = 10.4 \times 10^{-3}(\bar{D})^{0.5}$$

K in these experiments was 20.48 dynes-cm for normal and 21.36 dynes-cm for sickling cells.

## VII. PULSE NUCLEAR MAGNETIC RESONANCE SPECTRA OF RED BLOOD CELLS

This pulse method has proven to be the most versatile for measuring $T_1$, the spin-lattice relaxation time, or net relaxation of the proton to the preferred orientation (i.e., ground level) after the radio frequency (rf) is applied (i.e., spin flipping or longitudinal relaxation time). At equilibrium, nuclei are distributed among the energy levels according to the Boltzman distribution. After the application of rf energy, the nuclear spin system returns to equilibrium with its lattice by a first-order relaxation process (39,40,49-51).

To account for the process that causes nuclear spins of the water to come to equilibrium with each other, the spin-spin relaxation time, $T_2$, can be measured by application of low radio frequency energy. $T_2$ is generally related to a Lorentzian line shape, with a line of full width at half maximum intensity (39,40).

The chemical shift of water in both normal and sickling red blood cells measured by 60 MHz nuclear magnetic resonance spectrometer was approximately 5.3 $\delta$ (ppm), as shown in Figure 12A and 12B. The same chemical shifts were obtained for samples washed with $D_2O$, as seen in Figure 12C and 12D. The nuclear magnetic resonance spectra were integrated and the spin-spin relaxation time, $T_2$, was determined from the full line width at half maximum intensity. The results were found to be 45 msec ± 8% in both cases.

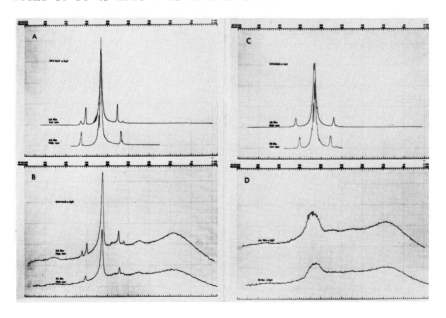

*Fig. 12. Chemical shift of $H_2O$ in red blood cells (washed with 0.9% NaCl in $H_2O$ or in $D_2O$) measured by 60 MHz nuclear magnetic resonance spectrometer, as compared with Tetramethyl silane, $Si(CH_3)_4$, standard 0 δ.*
   *A. Normal erythrocytes, chemical shift of $H_2O$ at 5.3 δ at high and low r.p.m.*
   *B. Sickling erythrocytes, chemical shift of $H_2O$ at 5.3 δ at high and low r.p.m.*
   *C. Normal erythrocytes (washed with $D_2O$), chemical shift of $H_2O$ at 5.3 δ; with SS erythrocytes, chemical shift of $H_2O$ at 5.3 δ at high r.p.m.*
   *D. Normal and sickling erythrocytes (washed with $D_2O$) both showing chemical shift of $H_2O$ at 5.3 δ, without spinning the sample.*

In $D_2O$, additional peaks which may be attributed to the membrane were observed at 1.9 δ and 3.4 δ. Integration of these peaks showed no difference in the normal and sickling red blood cell samples.

Figure 13A shows a plot of the pulse echo reading versus a delay time which varied from 8 to 30 msec for measurement of the spin-spin relaxation time, $T_2$, in normal and sickling red blood cell samples in $H_2O$. No differences were observed within the range of experimental error.

However, measurements of the spin-lattice relaxation time, $T_1$, shown in Figure 13B, revealed that $T_1$ for the sickling red blood cell sample in $H_2O$ was approximately 100 msec slower than $T_1$ for normal erythrocytes. A similar experiment was attempted in $D_2O$. However, due to the limitations of the instrument, no detectable signal was obtained. When cells were not packed to a maximum hematocrit value of 74%, however, this variation in $T_1$ for the two types of erythrocyte was not observed.

Since the nuclear spin relaxation time is dependent upon the molecular orientation time and, consequently, upon the details of molecular diffusion, both the rotational and the translational modes of the $H_2O$ molecules can affect the magnitude and direction of the internuclear vector for nuclei in different molecules in a surrounding liquid and, thus, cause fluctuations in the interaction energy between the various magnetic fields (or dipoles). These fluctuations lead to a dipole-dipole spin-lattice relaxation, spin coupling, and other interactions. The intensity and magnitude of these interaction energies, which couples the nuclear procession frequency to the molecular motion, determines the value of $T_1$ and $T_2$.

From our examination of the spin-spin relaxation time, $T_2$, it appears that the proton-proton interaction of water in

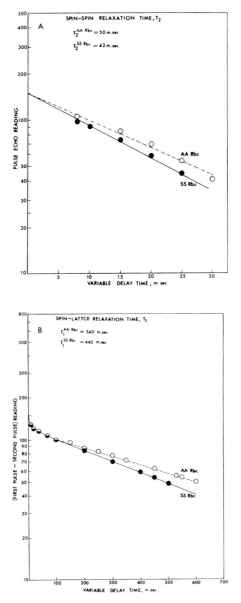

Fig. 13. A. The spin-spin relaxation time, $T_2$, of normal and sickling erythrocytes in 0.9% NaCl in $H_2O$.
B. The spin-lattice relaxation time, $T_1$, of normal and sickling erythrocytes in 0.9% NaCl in $H_2O$.

a solution of packed normal and sickling red blood cells is similar.

The difference of approximately 100 msec, which we observed in the spin-lattice relaxation times of the two types of erythrocytes, may be due to some differences at the surface of the red blood cell membrane.

We have presented, in the previous section (VI), evidence for the presence of polyamine in or on the sickling red blood cell membrane, resulting in a reduction in the surface charge potential ($\psi_o$) of 13% over that of the normal erythrocyte (34,35). The polyamine results in a greater adhesion at the membrane surface and increased hydrophobicity in the sickling red blood cell, as measured by the apparent interfacial viscosity.

The polyamine's nitrogens, each with a nuclear spin I=1, have a quadrapole moment and, therefore, set up an electrical field. Such alterations in the electrical field at the surface of sickling erythrocytes would account for lower values for $T_1$ in the sickling red blood cell and are compatible with the $H_2O$ molecules at the membrane surface having a greater mobility than in the normal erythrocyte. Both the presence of the polyamine and the shorter $T_1$, which we observed in these pulse nuclear magnetic resonance experiments, would be compatible with an increased mobility of $H_2O$ at the surface of the sickling red blood cells.


VIII.   CONCLUSION

To date, despite wide discussion in the literature, there has been no direct experimental evidence that the sickling of the human erythrocyte is a result of some alteration in the hemoglobin molecule itself. Based on our findings, we theorize that the aggregation of sickle cell hemoglobin may in fact be membrane-facilitated, with the erythrocyte membrane itself playing a role in the sickling process.

Our results indicate that there are distinct differences in the electrokinetic properties of normal and sickling red blood cells, which may be attributed in part to a variation in the polyamine content of the two types of erythrocytes.

Although the mechanism of polyamine binding to the red blood cell remains to be determined, it is apparent that the interaction of polyamines results in a specific alteration of the Stern electrical layer, an alteration which varies in normal and sickling erythrocytes.

This suggests that polyamines in vivo will strongly interact in the Stern plane as a result of electrostatic or hydrophobic bonding into the red blood cell membrane,causing a relocalization of the surface charge density distribution or perturbation of the Stern layer. These alterations may be attributed to several electrokinetic factors, among them adsorption density (moles/$cm^2$) of the red blood cell, coulombic interaction in the electrical double layer, and the contribution of the cohesive potential (or energy) of ligand to the surface potential of the red blood cell.

IX.  REFERENCES

1.  Mehaishi, J. N. and Seaman, G.V.F., Biochim. Biophys. Acta 112, 154 (1966).

2.  Cook, G.M.W., Heard, D. H., and Seaman, G.V.F., Nature (London) 191, 44 (1961).

3.  Haydon, D. A., Biochim. Biophys. Acta 50, 450 (1964).

4.  Overbeek, J.Th.G. and Wiersema, P. H., in "Electrophoresis" (M. Bier, Ed.), Vol. II, pp. 1-52.  Academic Press, New York, 1967.  See Debye, P. and Hückel, E., Phys. Z. 25, 97 (1924).

5.  Gingell, D., J. Theor. Biol. 30, 121 (1971).

6.  Heard, D. H. and Seaman, G.V.F., J. Gen. Physiol. 43, 635 (1960).

7.  Brooks, D. E., Millar, J. S., Seaman, G.V.F., and Vassar, P. S., J. Cell. Physiol. 69, 155 (1967).

8.  Seaman, G.V.F., in "The Red Blood Cell" (D.MacN. Surgenor, Ed.), Vol. 2, Chapter 27, pp. 1135-1229, 1975. See Chapman, D. L., Phil. Mag. 25, 475 (1913).

9.  Roscoe, R., Brit. J. Appl. Phys. 9, 280 (1952).

10.  Brinkman, H. C., J. Chem. Phys. 20, 571 (1952).

11.  Dintenfass, L., in "Blood Microrheology; Viscosity Factors in Blood Flow, Ischaemia and Thrombosis".  Appleton, New York, 1971.

12.  Chien, S., Usami, S., Dillenback, R. J., and Gregersen, M. I., Am. J. Physiol. 219, 136 (1970).

13.  Chien, S., in "The Red Blood Cell" (D.MacN. Surgenor, Ed.), 2nd Edition, p. 1031.  Academic Press, New York, 1975.

14.  Tillack, T. W., Scott, R. E., and Marchesi, V. T., J. Exp. Med. 135, 1209 (1972).

15.  Marchesi, V. T., Tillack, T. W., Jackson, R. L., Segrest, J., and Scott, R. E., Proc. Natl. Acad. Sci. 69, 1445 (1972).

16.  Jackson, R. L., Segrest, J. P., Kahane, I., and Marchesi, V. T., Biochemistry 12, 3131 (1973).

17.  Larsen, B., Nature 258, 345 (1975).

18.  Gordon, J. A. and Marquardt, M. D., Nature 258, 346 (1975).

19.  Bessis, M., "Living Blood Cells and Their Ultrastructure", Translated by R. I. Weed, Springer-Verlag, 1973.

20. Tabor, C. W. and Tabor, H., <u>Ann. Rev. Biochem.</u> 45, In Press (1976).

21. Tabor, H. and Tabor, C. W., <u>Pharm. Rev.</u> 16, 245 (1964).

22. Russell, D. H., "Polyamines in Normal and Neoplastic Growth", Raven Press, New York, 1973.

23. Cohen, S. S., "Introduction to the Polyamines", Prentice Hall, Inc., New Jersey, 1971.

24. Bachrach, U., "Function of Naturally Occurring Polyamines", Academic Press, New York, 1973.

25. Lundgreen, D., Farrel, P., and DiSant'agnese, P., <u>Clinica. Chemica. Acta</u> 62, 357 (1975).

26. Rennert, O., Frias, J., and LaPointe, D., in "Fundamental Problems of Cystic Fibrosis and Related Diseases" (J. Mangos and R. Talamo, Eds.), pp. 44-52, IMB, 1973.

27. Rennert, O., Miale, J., Shukla, J., Lawson, D., and Frias, J., <u>Blood</u>, In Press (1975).

28. Seiler, N., in "Polyamines in Normal and Neoplastic Growth" (D. H. Russell, Ed.), Raven Press, New York, 1973.

29. Seiler, N., "George and Elizabeth Frankel GAP Conference on Polyamines", Bethesda, Maryland, 1976.

30. Russell, D. H. and Stambrook, P. J., <u>Proc. Natl. Acad. Sci.</u> (U.S.A.) 72, 1482 (1975).

31. Byus, C. V., Costa, M., Sipes, I. G., Brodie, B. B., and Russell, D. H., <u>Proc. Natl. Acad. Sci.</u> (U.S.A.), In Press (1976).

32. Wright, R., Buehler, B., Schott, S., and Rennert, O., <u>Biochem. Biophys. Res. Comm.</u>, In Press (1976).

33. Chun, P. W., Rennert, O. M., Saffen, E. E., and Taylor, W. J., <u>Biochem. Biophys. Res. Comm.</u> 69, 1095 (1976).

34. Chun, P. W., Rennert, O. M., Saffen, E. E., and Taylor, W. J., <u>Fed. Proc.</u>, In Press (1976).

35. Chun, P. W., Rennert, O. M., Saffen, E. E., Shukla, J. B., Taylor, W. J., and Shah, D. O., ICCS Colloid and Surface Science Symposium Abstract, 1976.

36. Weiser, H. B., "Colloid Chemistry", John Wiley & Sons, New York, 1949.

37. Brown, A. G., Thuman, W. C., and McBain, J. W., <u>J. Colloid and Int. Sci.</u> 8, 491 (1953).

38. Karam, H. J., <u>J. Appl. Pol. Sci.</u> 18, 1693 (1974).

39. Block, F., Hansen, W. W., and Packard, M., <u>Phys. Rev.</u> 69, 127 (1946).

40. Hahn, E. L., <u>Phys. Rev.</u> 76, 1059 (1948).

41. Dodge, J. T., Mitchell, C., and Hanahan, D. J., <u>Arch. Biochem. Biophys.</u> 100, 119 (1963).

42. Taylor, W. J., Easley, C. W., and Kitchen, H., <u>J. Biol. Chem.</u> 247, 7320 (1972).

43. Fried, M. and Chun, P. W., Comp. Biochem. Physiol. 39B, 523 (1971).

44. Fairbanks, G., Steck, T. L., and Wallach, D.F.H., Biochemistry 10, 2606 (1971).

45. Chernoff, A. I. and Pettit, N. M., Blood 24, 750 (1964).

46. Derjaguin, B. V. and Landau, L. D., Acta. Phys-Chim. U.S.S.R. 14, 633 (1941).

47. Verwey, E.J.W. and Overbeek, J.Th.G., "Theory of the Stability of Lyophobic Colloids", Amsterdam, Elsevier, 1948.

48. Osipow, L. I., "Surface Chemistry, Theory and Industrial Applications", R. E. Krieger Publishing Company, New York, 1972.

49. Torrey, H. C., Phys. Rev. 76, 1059 (1948).

50. Emsley, J. W., Feeney, J., and Sutcliffe, L. H., "High Resolution NMR Spectroscopy", Chapter 2, Pergammon Press, Oxford, 1965.

51. Farrar, T. C. and Beeker, E., "Pulse and Fourier Transform NMR, Introduction to Theory and Methods", Academic Press, 1976.

# DIFFERENTIAL SCANNING CALORIMETRY AND THE KINETICS OF THE THERMAL DENATURATION OF DNA

Horst W. Hoyer and Susan K. Nevin
*The City University of New York, Hunter College,*
*Department of Chemistry*

## I.ABSTRACT

*Differential scanning calorimetric studies on DNA gels in the concentration range of 0.6 to 6% in phosphate buffer at pH 7.0 and in the temperature range of 99 to 113°C demonstrate that the thermal denaturation of calf thymus DNA follows a first order rate equation at the lower concentrations and lower temperatures. Over this temperature range the specific rate constant increases from 0.018 to 0.13 min.$^{-1}$, corresponding to an activation energy of 49 kilocalories per mole.*

## II. INTRODUCTION

Several investigators have recently published calorimetric studies on deoxyribonucleic acid (DNA)[1-4]. We have recently reported differential scanning calorimetric studies on DNA[5] in which both thermal and kinetic information on the thermal denaturation of calf thymus DNA gels was obtained over a pH range of 5.4 to 8.2. One unusual aspect of the kinetic data was the need to assign an apparent second order to the rate equation for this thermal denaturation. The present research was undertaken in order to reexamine the kinetics of the process by an alternate and more direct approach, one which would bypass the assumptions[6] in the previous kinetic study. We examined gels of calf thymus DNA in phospate buffer at a pH of 7.0 over a concentration range of 0.6 to 6.1% and established the order of the reaction by fitting the data directly to the rate law equation

1.        $rate = kC^x$

where k is the specific rate constant, C the concentration and x the order of the reaction.

III.  EXPERIMENTAL

All of the studies herein reported were made with
the differential scanning calorimeter cell of the
duPont 900 differential thermal analyser.  This DSC
cell has been characterized by Baxter[7] and will not
be further described here.  Checks were made periodically to monitor the rate of temperature rise and to
assure calibration of the instrument for enthalphy
changes and for transition temperatures.  Most of the
thermograms in this study were recorded at a heating
rate of 22°C per minute and at maximum sensitivity of
0.01°C per inch.

The DNA used was the sodium salt of calf thymus DNA,
Sigma batch number 101C9520 with a reported average
molecular weight of 1,350,000.  Samples were weighed
directly into phosphate buffer containing 0.75 moles
of phosphate per liter, giving concentrations of approximately 0.1 to 1.0 milligrams of DNA per 15 micro-
liters.  These solutions were prepared in small aluminum cups supplied by duPont Co. and were hermitically
sealed with a special die press. In most cases the
seal held under the pressures generated by heating the
solutions to 120°C, corresponding to an internal pressure of approximately 2 atmospheres.  Similar aluminum
cups filled with an equal volume of buffer solution
served as references.  Leakage or rupture was readily
detected by a sudden and substantial shift of the base-
line.

IV.  PROCEDURE AND RESULTS

In the DSC experiment a sample and an inert reference of approximately the same heat capacity are
heated at a constant rate.  Heat energy is supplied
to the system in order to maintain zero temperature
difference between the reacting sample and the inert
reference and is recorded as a thermogram showing heat
input as a function of temperature (time).  Thus the
total area A of the thermogram is proportional to $\Delta H$,
the heat transferred to the sample.  If it is assumed
that this heat is directly proportional to the number
of moles of reactant, n, it follows that the rate of

the reaction in terms of moles reacting per unit time, is given by

2. $$Rate = \frac{-dn}{dt} = \frac{n_o}{A}\frac{dH}{dt}$$

where $n_o$ is the initial number of moles of reactant and dH/dt is the instantaneous height of the thermogram above the base line. Rewriting equation 2 in terms of concentrations and combining it with the rate equation for a reaction involving one component gives

3. $$\frac{C_o}{A}\frac{dH}{dt} = kC^x$$

All quantities in equation 3 except k and x are available from the thermogram, Figure 1. The concentration of DNA at

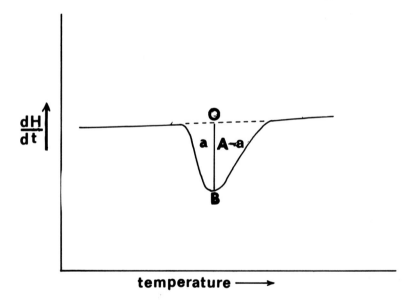

Fig. 1. Typical DNA Thermogram

the start of the reaction, $C_o$, is proportional to the total area A while that of the undenatured DNA at any particular time (and temperature) is proportional to A-a, the total area minus that portion already swept out by the reaction. The length of the line OB is proportional to dH/dt. Proportionality constants are determined by calibration against appropriate samples.

Each thermogram thus provides information concerning the concentration of unreacted DNA existing at a particular temperature as well as the rate of the reaction at that concentration and temperature. If a series of such thermograms is run with different initial concentrations, then each would contain information concerning the rates of the reaction over a range of temperature and concentrations. Isothermal rates and the corresponding concentrations can then be extracted from this series of thermograms and the order of the reaction and the rate constant determined with the aid of equation 3. In practice, plots of log $(C_0/A)(dH/dt)$ versus log $C$ at a series of constant temperatures were used to establish the order of the reaction. Estimates of the specific rate constant were obtained by extrapolation to a concentration of unity. Fourteen thermograms were determined with values of $C_0$ ranging from 0.63% DNA to 6.1%. The resulting values of $C_0/A$ are plotted in Figure 2 against initial concentration and are an indication of the reproducibility of the system, which is seen to be in order of $\pm$ 14%. The progressive decreases in the value of $C_0/A$ implies an increase in the $\Delta$ H of the helix-coil transition of DNA with increasing concentration.

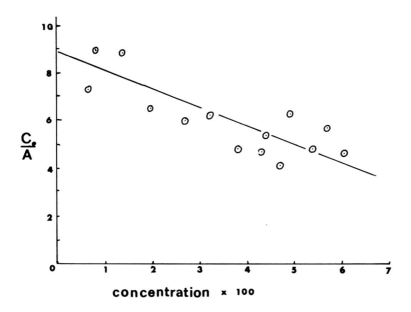

Fig. 2. *Initial concentration versus ratio of initial concentration divided by thermogram area. Concentration units are in grams of DNA per milliliter of solution.*

Values of the reaction rate, $(C_o/A)(dH/dt)$, and of helix concentration C were then determined at intervals of $2^o$ for each of the fourteen thermograms from 99°C to 115°C. These values are plotted in Figures 3, 4 and 5. Because of the inaccuracies in measuring small areas, rates involving reactions of less than 5% conversion or more than 95% were not included in these summaries.

Inspection of the figures shows that the slope, and therefore the order of the reaction, is unity at 99°C over the entire range of concentrations. Except at the higher temperatures of 111°C and 113°C all of the other rate plots start out with an initial slope of unity at the lower concentrations and then at higher concentrations show departure from this initial slope. For the three highest temperatures (Figure 5) the slope is essentially constant at 0.7 over the entire range, a value approximately that of the other curves in the high concentration region.

Specific rate constants were estimated from Figures 3, 4

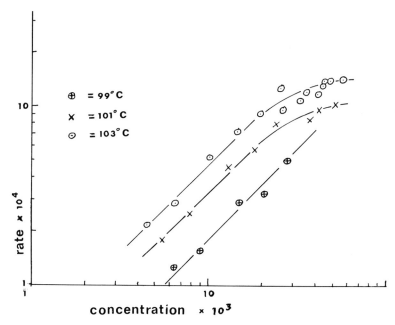

*Fig. 3. Plots of reaction rate versus concentration at 99°C, 101°C and 103°C. Concentration units are in grams of unreacted DNA (helix form) per milliliter of solution.*

and 5 by extrapolation of the linear first order slope to unity concentration. These values are tabulated in Table 1

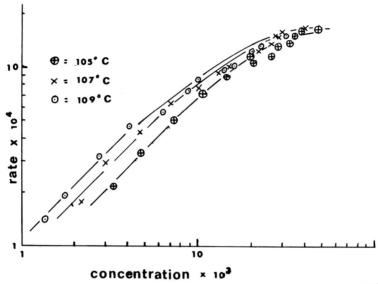

Fig. 4. Plots of reaction rate versus concentration at 105°C, 107°C and 109°C. Concentration units are in grams of unre-acted DNA(helix form) per milliliter of solution.

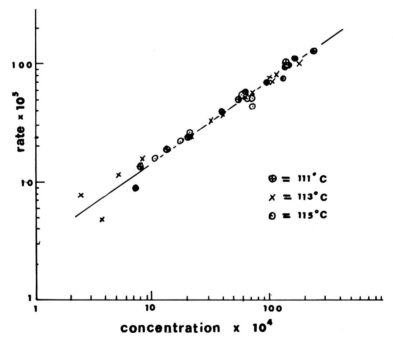

Fig. 5. Plots of reaction rate versus concentration at 111°C, 113°C and 115°C. Concentration units are in grams of unre-acted DNA (helix form) per milliliter of solution.

and correspond to an Arrhenius plot activation energy of 49 kilocalories per mole of kinetic unit.

TABLE I
*Specific Rate Constants of Thermal Denaturation of DNA*

| Temperature ($^{O}K$) | $k,min^{-1}$ |
|:---:|:---:|
| 372 | $1.8 \times 10^{-2}$ |
| 374 | $3.3 \times 10^{-2}$ |
| 376 | $4.5 \times 10^{-2}$ |
| 378 | $6.8 \times 10^{-2}$ |
| 380 | $8.8 \times 10^{-2}$ |
| 382 | $11.0 \times 10^{-2}$ |
| 384 | $13.0 \times 10^{-2}$ |

## V. DISCUSSION

The reactions generally believed to be involved in the thermal denaturation of DNA are (1) breakage of hydrogen bonds, (2) disruption of stacking reactions, (3) unwinding of the helix and (4) strand separation[8]. The unwinding process is believed to follow almost immediately after the energy absorbing processes which is then followed by the somewhat slower process of strand separation. The DSC experiment responds to the energy absorbing processes (1) and (2). Our data demonstrate that these start out as first order kinetic processes at the lower temperatures and concentrations. At higher temperatures and corresponding higher conversions to the denatured form the plot is no longer linear, an observation which can be explained as due to either a decrease in the apparent order of the reaction or to a decrease in the rate constant. The latter explanation seems more reasonable and is consistent with the observation of Record and Zimm[9] that the rate constant decreases as the extent of denaturation increases.

Comparison of our results with kinetic studies reported by others using different techniques is complicated by the much higher concentrations and temperatures necessitated by our studies. Because of the small heat effects of the transition we were forced to use DNA concentrations from ten to one hundred times those normally used in previous kinetic studies

using absorption or viscosity techniques.  These higher concentrations resulted in increased ionic strengths of the solutions, a factor known to shift the denaturation to higher temperatures[10].  For this reason our transition temperatures,
taken as the midpoint of the transition, were in the range of
104 to 106°C, twenty to twenty five degrees higher than those
of workers using the more sensitive techniques of ultraviolet
absorbance or viscosity.

Peacocke and Walker[11] studied the thermal denaturation of
DNA by observing the changes in optical density.  They observed first order kinetics with a specific rate constant for
calf thymus DNA of 1.0 min$^{-1}$ at 82.5°C and at a concentration
of 2.0 mg/ml.  While they did not study the kinetics of calf
thymus DNA over a range of temperatures they did so for herr-
ing sperm DNA.  For the temperatures of 72.5, 75.0, 77.5 and
83.0°C the values for the specific rate constants for the denaturation of this DNA were 0.03, 0.095, 0.12 and 0.42 min$^{-1}$.
Thus one can expect that at the higher temperatures prevail-
ing in our studies of calf thymus DNA the specific rate constant should be greater than 1.0 min$^{-1}$.  However our more concentrated solutions are also much more viscous and this would
be expected to decrease the rate constant.  Davison[11] demonstrated that the time for strand separation in bacteriophage
DNA in aqueous glycerol solution was directly proportional to
the viscosity.  Comparison is further complicated by the observation of Record and Zimm[9] that decreasing the ionic strength
of the solution decreases the denaturation rate constant.  Because of these differences in experimental conditions, any
quantitative comparison between our differential thermal studies and denaturation studies by spectral or viscosity methods
is impossible.  Unfortunately it is also impossible for us to
perform the differential scanning calorimeter experiments at
the concentrations used by other investigators since such concentrations would necessitate equipment approximately one hundred times as sensitive as our Model 900 Thermoanalyzer.

VI.  REFERENCES

1.  Bunville, L.G., Geiduschek, E.P., Rawitscher, M.A. and
    Sturtevant, J.M., Biopolymers, 3, 213 (1965).
2.  Privalov, P.L., Ptitsyn, O.B., Birshtein, T.N.,
    Biopolymers, 8, 559 (1969).
3.  Barber, R., Biochemica et Biophysica Acta, 238, 60, (1971).
4.  Shaio, D.D.F. and Sturtevant, J.M., Biopolymers, 12,1829
    (1973).
5.  Hoyer, H.W. and Nevin, Susan in "Analytical Calorimetry"
    (Roger S. Porter and Julian F. Johnson, Eds.) Vol.3,p.465.
6.  Borchardt, H.J. and Daniels, F., J. Am. Chem. Soc.,
    79, 41, (1957).

7.  Baxter, R.A., in "Thermal Analysis" (R.F. Schwenker and P.D. Garn, Eds.), Academic Press (1969).
8.  Elson, Elliot L. and Record, M. Thomas, Jr., Biopolymers 13, 797 (1974).
9.  Record, M. Thomas and Zimm, Bruno H., Biopolymers, 11,1435 (1972).
10. Marmur, J. and Doty, P., J. Mol. Biol., 5, 109 (1962).
11. Peacocke, A.R. and Walker, I.O., J. Mol. Biol., 5, 560 (1962).
12. Davison, P.F., Biopolymers, 5, 715 (1967).

CHARACTERIZATION OF PHOSPHATIDYLCHOLINE VESICLES
BY QUASIELASTIC LASER LIGHT SCATTERING*

F. C. Chen, S. I. Tu and B. Chu
Chemistry Department
State University of New York at Stony Brook
Stony Brook, New York, 11794

## Abstract

Experiments are reported on the angular dependence of decay times of concentration fluctuations of phosphatidylcholine vesicles. While unfractionated vesicle suspensions formed by ultrasonic irradiation under nitrogen in
aqueous solutions of different ionic strengths and pH
contain small amounts of relatively polydispersed large
fragments which cannot be removed easily by Millipore
filtration, vesicle suspensions fractionated by a Sepharose 4B column (1.5x40 cm) are quite uniform in size.
The fractionated vesicle suspensions should be a more
appropriate model membrane system which can provide a
large specific surface area for binding, kinetic, and
spectroscopic studies. Results on the stability of
vesicle suspensions and the effects due to different
intra- and intervesicular media as well as antibiotics,
such as valinomycin, and proton transport mediators, such
as carbonyl cyanide m-chlorophenyl hydrazone and dinitrophenol are presented.

*Work supported by the National Institutes of Health, the
National Science Foundation, and the Research Foundation
of the State University of New York.

I.  INTRODUCTION

In a recent quasielastic laser light scattering (QLLS) study
of phosphatidylcholine vesicle suspensions,[1] we have observed
a trace amount of about one part per thousand of relatively
polydispersed large fragments contaminating the smaller vesicles. The trace contaminants had an estimated average size of
about $1.1 \times 10^3$ Å while the smaller vesicle spheres prepared
under those conditions had an average diameter of about
$3.1 \times 10^2$ Å. As the phosphatidylcholine vesicles were formed
by ultrasonic irradiation and subsequent filtration using a
VCWP 01300 Millipore filter of 1000 Å nominal pore size, the
presence of those large fragments suggests that the vesicle
membrane must be flexible. Furthermore, it will be difficult

63

to obtain vesicle suspensions of essentially uniform size without a trace of the large fragments by means of Millipore filtration since the use of filters with pore-sizes comparable to vesicle diameters will invariably block the filtration process. Although QLLS is an excellent method for studying the trace amount of those large particles which cannot be detected easily by other more standard techniques, such as ultracentrifugation and gel permeation chromatography, the same large fragments so distort the measured time correlation function that data analysis of multiexponential decay curves becomes ambiguous. Consequently, we need to prepare vesicle suspensions of uniform size before QLLS can be utilized fruitfully as an analytical tool.

In this report, we want to show that fractionation of vesicle suspensions by means of a Sepharose 4B column can indeed yield vesicles of relatively uniform size, that vesicles remain stable for periods of 2-3 days and then aggregation occurs, and that perturbation of vesicle membranes, by antibiotics, such as valinomycin and proton-transport mediators, such as carbonyl cyanide m-chlorophenyl hydrazone (CCCP) and dinitrophenol do not change the size of vesicles.

## II. EXPERIMENTAL DETAILS

### A. Materials
Egg yolk L-$\alpha$-phosphatidlycholine (Type V-E) in chloroform-methanol (9:1) solution and carbonyl cyanide m-chlorophenyl hydrazone (CCCP) were purchased from Sigma Chemical Company. Valinomycin (A-grade) was obtained from Calbiochem. 2,4-dinitrophenol (DNP) was a Fischer product which was recrystallized twice from benzene solution. All other reagents used were of highest purity available.

### B. Vesicle preparation
A 0.5-ml chloroform-methanol solution containing 50mg of egg-yolk lecithin was evaporated to dryness under nitrogen at room temperatures in a 15-ml Corex centrifuge tube. The residual solvents were removed by a mild vacuum. 2.5ml of an aqueous solution were then added over the dry film of lecitin. The contents were mixed by vigorous vortexing for about 2 minutes. The resultant milky suspension was sonicated for 1.5-2.5 hours at $10^{\circ}$C using a Branson W185 sonifier operating at 40 watts. The vesicle suspension was centrifuged at 48,000 g for 30 minutes to remove undispersed particles and then fractionated by means of a Sepharose 4B column (1.5x40cm) which was equilibrated with a suitable solution corresponding to the intervesicular medium. The effluent was fractionally collected at 0.5-ml aliquots with an ISCO

fraction collector.

Different types of vesicle suspensions were prepared by the above procedure. Table 1 lists the intra- and inter-vesicular media used.

Table 1
Intra- and Inter-Vesicular Media

| Designation[a,b] | Intravesicular Medium | Intervesicular Medium |
|---|---|---|
| A | 150 mM KCl | 300 mM glucose |
| C | " | " |
| D | 75 mM KCl + 150 mM glucose | 150 mM KCl |
| E | Both media contain 50 mM sodium phosphate and 175 mM glucose | |
| F | Same as E | |
| (i) | pH = 7.0 | pH = 6.0 |
| (ii) | pH = 7.0 | pH = 7.0 |
| (iii) | pH = 7.0 | pH = 8.5 |

(a) The first alphabet denotes the lot number. Thus, F- and D- denote the 6th and the 4th set of preparations, respectively.
(b) Lot No. - (pH) - aliquot location-additive.

Valinomycin, dinitrophenol, and carbonyl cyanide m-chlorophenyl hydrazone were dissolved in ethanol. A $10\mu l$ of each was added to a vesicle suspension when indicated. The final concentrations of valinomycin, dinitrophenol, and carbonyl cyanide m-chlorophenyl hydrazone were $2.25 \times 10^{-6}$, $1.4 \times 10^{-5}$, and $1.6 \times 10^{-6}$ M, respectively, in the vesicular solutions for light scattering measurements.

C. Light scattering measurements

The light scattering photometer has the same optical design as the previous report for vesicle studies.[1] However, decay rates of the time-dependent concentration fluctuations of vesicle suspensions were measured by means of a 96-channel single-clipped photon correlation spectrometer.[2-5] All measurements were performed at $25^{\circ}$C.

D. Data analysis

The time-dependent correlation function from a single-clipped correlator has the form

$$G^{(2)}(\tau) = A(1 + \beta|g^{(1)}(\tau)|^2) \tag{1}$$

where $A = \langle n \rangle \langle n_k \rangle N^*$ with $\langle n \rangle$, $\langle n_k \rangle$, and $N^*$ being the mean un-clipped and clipped photocounts per sample time and the total number of samples, respectively, $\tau$ is the delay time and $\beta$ is treated as an unknown parameter.

For a monodisperse system, the first-order electric field correlation function $g^{(1)}(\tau)$ has the form

$$|g^{(1)}(\tau)| = \exp(-\Gamma\tau) \tag{2}$$

where $\Gamma = DK^2$ with D and K being the translational diffusion coefficient and the magnitude of the momentum transfer vector, respectively. For a polydisperse system, we used the truncated quadratic form of the method of cumulants[2,3]:

$$|g^{(1)}(\tau)| = \exp(-\bar{\Gamma}\tau)(1 + \frac{\mu_2\tau^2}{2}) \tag{3}$$

where $\bar{\Gamma} = \int_o^\infty \Gamma G(\Gamma)d\Gamma$ and $\mu_2 = \int_o^\infty (\Gamma-\bar{\Gamma})^2 G(\Gamma)d\Gamma$ with $G(\Gamma)$ being the normalized distribution function of decate rates. Eq. (3) remains valid even for a polydisperse system with a very broad distribution so long as $\bar{\Gamma}\tau_{max}$ remains sufficiently small. $\tau_{max}$ is the maximum delay-time range used in the fitting procedure.

When the vesicle suspensions exhibit minimal angular dissymetry, $\bar{\Gamma}/K^2$ can be identified with the z-average diffusion coefficient $D_z$ even at finite K since the particle scattering factor $P(K) \sim 1$. We have also used the Stokes-Einstein relation $D = k_B T/3\pi\eta d$ to compute the diameter (d) of the vesicle spheres where $k_B$, T and $\eta$ are the Boltzmann constant, the absolute temperature and the viscosity, respectively.

## III. RESULTS AND DISCUSSION
### A. Polydispersity effects - fractionated vs. unfractionated
Elution patterns of phosphatidylcholine dispersions reveal a bimodal distribution.[6] Fig. 1 shows a typical plot of $A\beta|g^{(1)}(\tau)|^2$ versus $\tau$ for an unfractionated sample (A). The dots denote the measured values: The solid line (1) represents force-fitting the time-correlation function of the signal using Eq. (2).

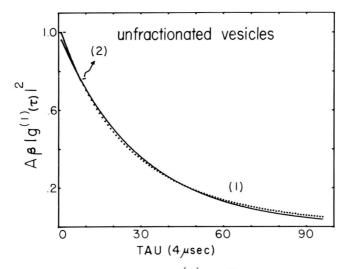

Fig. 1  A typical plot of $A\beta |g^{(1)}(\tau)|^2$ versus $\tau$ for an un-
fractionated sample A . T = 25°C; $\theta$ = 135°. Dots represent the
measured values.  The solid lines are

(1) $g^{(1)}(\tau) \propto \exp\ (-4.23\mathrm{x}10^3\tau)$

(2) $g^{(1)}(\tau) \propto \exp\ (-4.98\mathrm{x}10^3\tau)\ [1 + 3.77 \times 10^6\tau^2]$
    with $\tau$ expressed in sec.

Clearly, the unfractionated vesicle suspension is not of uniform size.  A least-squares fitting of $g^{(1)}(\tau)$ using the first
eight delay times by means of Eq. (3) shows that $\bar\Gamma = 4.98\mathrm{x}10^3$
$\sec^{-1}$ which differs from $\Gamma = 4.23\mathrm{x}10^3\ \sec^{-1}$ by about 15%.  On
the other hand, the samples which have been fractionated using
a Sepharose 4B column show remarkable uniformity in the vesicle size distribution.

Fig. 2 shows a typical plot of the time-correlation function
of the signal for a fractionated sample F-(i)-15 where the
last two digits denote the 15th aliquot in the fractionation
procedure.  The dots again denote the measured values.  Now
Eq. (2) represents the measured data quite well.  The difference in the magnitudes of linewidth between Figs. 1 and 2
does not only mean that we have taken out the larger fragments
in the fractionated sample F but also depends upon which
function we used as well as a slight variation in the ultrasonic irradiation time.  Fig. 3 shows plots of $\bar\Gamma/K^2$ versus

$\sin^2(\theta/2)$ for the fractionated F-(i)-15 and the unfractionated A samples.

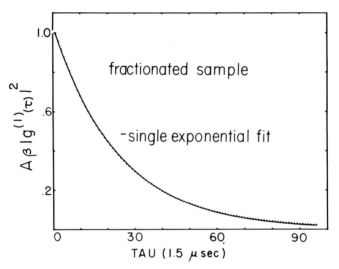

Fig. 2 A typical plot of $A\beta|g^{(1)}(\tau)|^2$ versus $\tau$ for a frac-
tionated sample F-(i)-15. T = 25°C; $\theta$ = 135°. Dots represent
the measured values. The solid line is $g^{(1)}(\tau) \propto \exp$
$(-1.38 \times 10^4 \tau)$ with $\tau$ expressed in sec.

The fractionated sample (F-15) obeys the relation $D_z = \bar{\Gamma}/K^2$
and reveals a small value of $\mu_2/\bar{\Gamma}^2$. By fitting the entire 96
data points to Eq. (3) at two different delay times, we have
compared the measured data with the truncated Eq. (3) at two
different delay-time ranges. Thus, variation of $\bar{\Gamma}$ and $\mu_2/\bar{\Gamma}^2$
at different delay times means that the system has a broad
size distribution. For the unfractionated sample, we note
that $\mu_2/\bar{\Gamma}^2 \sim 0.4$-$0.6$ which confirms that the sample is indeed
very polydisperse. It should also be noted that the values
of $\mu_2/\bar{\Gamma}^2$ increases with decreasing scattering angle for the
unfractionated sample indicating the presence of large fragments whose scattering effects increase with decreasing scattering angle. Furthermore, for the unfractionated sample,
$P(K) \neq 1$. Consequently, we cannot obtain $D_z$ at finite scattering angles under those circumstances. Then, the diameter

from the Stokes-Einstein relation becomes a rough estimate which we shall refer to as an apparent diameter $d_a$.

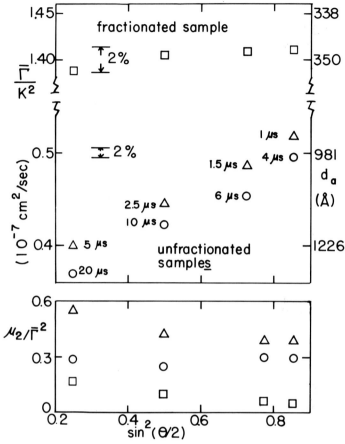

Fig. 3 Plots of $\bar{\Gamma}/K^2$, $d_a$, and $\mu_2/\bar{\Gamma}^2$ versus $\sin^2(\theta/2)$ for a fractionated F-(i)-15 and an unfractionated (A) sample.

Fig. 4 shows a plot of $D_z$ and $\mu_2/\bar{\Gamma}^2$ as a function of a few of the aliquots taken during fractionation collection. The agreement of $D_z$ and $\mu_2/\bar{\Gamma}^2$ measured at two different delay times again confirms the validity of Eq. (3). The variation in d shows that our fractionation procedure is separating vesicle suspensions of different sizes. The small values of $\mu_2/\bar{\Gamma}^2$ suggest that the fractionated samples are quite uniform in size almost approaching the uniformity of "monodisperse" latex spheres. The values of $\mu_2/\bar{\Gamma}^2$ also vary with aliquot number, and under the best of conditions, we have obtained $\mu_2/\bar{\Gamma}^2$ values of about 0.05.

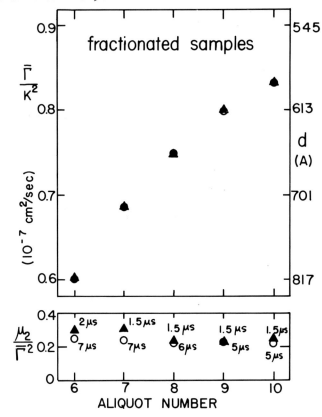

Fig. 4 A plot of $D_z$, d, and $\mu_2/\bar{\Gamma}^2$ versus aliquot number. T = 25°C; θ = 90°.

## B. Stability of vesicles

Vesicle suspensions are not stable. Fig. 5 shows increases of vesicle sizes for samples C-24, C-24-V and C-24-D as a function time. The last alphabets -V and -D represent samples with trace amounts of valinomycin and 2,4-dinitrophenal(DNP). The addition of DNP shows virtually no effect while valinomycin seems to inhibit the change appreciably. Figs. 6 and 7 show detailed time-correlation functions for C-24, C-24-D and C-24-V at various time intervals after sample preparation.

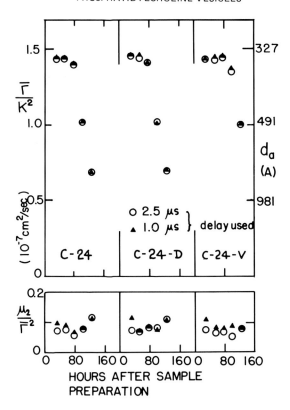

Fig. 5  Plots of $D_z$ and $\mu_2/\bar{\Gamma}^2$ as a function of time interval after sample preparation.

In Fig. 6, there are actually 10 time-correlation functions plotted. During the first five days (2,3,4,5), C-24 and C-24-D are indistinguishable. We can observe only inconsequential variations during the 6th day. In Fig. 7, we have plotted a total of 6 time-correlation functions. The three time-correlation functions of C-24-V are difficult to see and are indicated by 4V, 5V, and 6V while those of C-24 are represented by 4,5, and 6, respectively. Thus, vesicle suspensions seem to remain noticeably more stable in the presence of valinomycin. Fig. 5 also reveals a remarkable stability in the values of $\mu_2/\bar{\Gamma}^2$ for all the samples. It means that as the vesicles become larger, the variance has remained the same. We have also noted an increase in the scattered intensity as the vesicle size increases. In the first approximation, there are three possible processes to this size increase. These could be due to (1) aggregation, (2) formation of larger but single bilayer vesicles, and (3) formation of multilamellar vesicles.

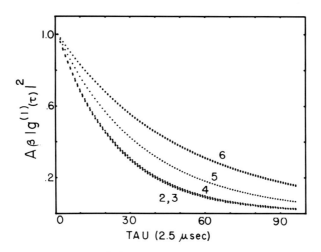

Fig. 6  Plots of $A\beta \left| g^{(1)}(\tau) \right|^2$ versus $\tau$.  $T = 25^{\circ}C$, $\theta = 90^{\circ}$. Samples of C-24 and C-24-D measured at 2,3,4,5,6 days after preparation.

The formation of multilamellar vesicles is not likely because it yields a larger than expected $\mu_2/\bar{\Gamma}^2$ value.  The formation of larger but single bilayer vesicles will produce larger intensity increases than that of aggregation.  Our results seem to indicate a gradual uniform aggregation of vesicles as the main decay process even though more quantitative intensity measurements are necessary to resolve this interesting question unambiguously.

C.  Effects of antibiotics and of pH
Valinomycin has been used to increase the $K^+$ permability and CCCP and DNP are known as proton carriers.  We have added trace amounts of valinomycin, CCCP and DNP to vesicle suspensions.  Figs. 6-9 show plots of detailed time-correlation functions of vesicle suspensions with and without valinomycin and DNP.

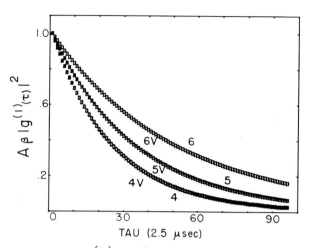

Fig. 7   Plots of $A\beta |g^{(1)}(\tau)|^2$ versus $\tau$.   T = 25°C, θ = 90°.
Samples of C-24 and C-24-V measured at 4,5,6 days after prep-
aration
□  C-24          ●   C-24-V.

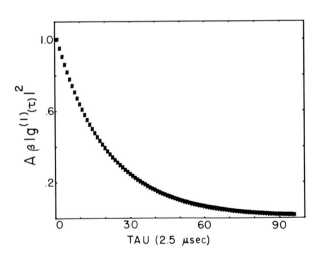

Fig. 8   Plots of $A\beta |g^{(1)}(\tau)|^2$ versus $\tau$.   T = 25°C, θ = 90°.
□ D-11              X D-11-D

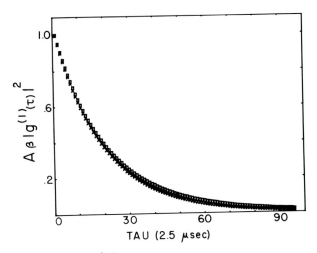

Fig. 9 Plots of $A\beta\left|g^{(1)}(\tau)\right|^2$ versus $\tau$.  $T = 25^\circ C$, $\theta = 90^\circ$.

▢ D-11          ✗ D-11-V

Table 2
Effect of dinitrophenol and valinomycin on vesicle size

| Sample | Delay time used (μsec) | Time after preparation (hour) | d (Å) | $\mu_2/\bar{\Gamma}^2$ |
|---|---|---|---|---|
| D-11 | 2.5 | 24 | 285 | 0.11 |
|  | 1.0 | 24 | 288 | 0.10 |
|  | 2.5 | 48 | 286 | 0.13 |
|  | 1.0 | 48 | 289 | 0.10 |
| Ave. |  |  | 288 | 0.11 |
| D-11-D | 2.5 | 24 | 275 | 0.08 |
|  | 1.0 | 24 | 278 | 0.07 |
|  | 2.5 | 48 | 276 | 0.07 |
|  | 1.0 | 48 | 276 | 0.07 |
| Ave. |  |  | 277 | 0.07 |
| D-11-V | 2.5 | 24 | 283 | 0.10 |
|  | 1.0 | 24 | 286 | 0.07 |
|  | 2.5 | 48 | 285 | 0.12 |
|  | 1.0 | 48 | 288 | 0.07 |
| Ave. |  |  | 286 | 0.09 |

Table 3
Effect of pH on vesicle size

| Sample[a] | d (Å) | $\mu_2/\bar{\Gamma}^2$ |
|---|---|---|
| E-(ii)-11 | 482 | 0.12 |
| E-(i)-11 | 482 | 0.14 |
| E-(ii)-11-C | 484 | 0.12 |
| E-(i)-11-C | 480 | 0.12 |
| F-(ii)-15 | 346 | 0.08 |
| F-(iii)-15 | 349 | 0.11 |
| F-(ii)-15-C | 349 | 0.09 |
| F-(iii)-15-C | 357 | 0.11 |
| F-(ii)-11 | 423 | 0.10 |
| F-(iii)-11 | 428 | 0.13 |
| F-(ii)-11-C | 423 | 0.10 |
| F-(iii)-11-C | 424 | 0.09 |

a)  C denotes carbonyl cyanide m-chlorophenyl hydrazone (CCCP)

The results are summarized in Tables 1 and 2.

In conclusion, we have been able to prepare phosphatidylcholine vesicles which are uniform in size.  The vesicle
suspensions are stable at room temperatures for a couple of
days and then uniform aggregation occurs.  The membrane is
flexible but not elastic.  Effects of antibiotics and of pH
are negligible with respect to vesicular size.  However, the
presence of valinomycin seems to inhibit vesicle aggregation.
Finally, QLLS should be an extremely useful tool for characterizing vesicle sizes as the linewidth measurements take
about less than ten minutes to perform using a single-clipped
photon correlation spectrometer.

IV.  REFERENCES

*Work supported by the National Institutes of Health, the National Science Foundation, and the Research Foundation of the State University of New York.

1.  F. C. Chen, A. Chrzeszczyk, and B. Chu, J. Chem. Phys., 64, 3403 (1976).
2.  B. Chu, "Laser Light Scattering" (Academic, New York, 1974).
3.  "Photon Correlation and Light Beating Spectroscopy," edited by H. Z. Cummins and E. R. Pike (Plenum, New York, 1974).
4.  E. R. Pike, Riv. Nuovo Cimento (Ser. 1) 1, 277 (1969).
5.  R. Foord, E. Jakeman, C J. Oliver, E. R. Pike, R. J. Blagrove, E. Wood, and A. R. Peacock, Nature 277, 242 (1970).
6.  C. H. Huang, Biochemistry 8, 344 (1969).

# THERMODYNAMICS OF NATIVE PROTEIN/FOREIGN SURFACE INTERACTIONS. II. CALORIMETRIC AND ELECTROPHORETIC MOBILITY STUDIES ON THE HUMAN γ-GLOBULIN/GLASS SYSTEM[1]

E. Nyilas, T-H. Chiu and D.M. Lederman
*Avco Everett Research Laboratory, Inc.*

*Thromboembolic phenomena induced by surfaces foreign to blood represent severe limitations to the application of implantable or paracorporeal artificial organs. The initial event in blood/foreign surface interactions is the adsorption of native plasma proteins involving, in most cases, surface-induced conformational changes.*

*In a study undertaken to elucidate the mechanism and energetics of native plasma protein adsorption, one of the strongest known procoagulants, viz., glass has been utilized as the first model adsorbent. The $25^O$ and $37^OC$ isotherms of human-γ-globulin, adsorbed on a glass powder from a buffer partially simulating normal plasma, indicated irreversible multilayer sorption with greater adsorptivities at the higher temperature. Using a custom-built thermistorized isothermal-jacketed microcalorimeter routinely capable of resolving $\pm 1 \times 10^{-5}$ $^OC$ in 100 ml of liquid volume, the net heats of γ-globulin sorption were determined as a function of bulk protein concentrations under identical conditions. The exothermic net heats obtained per mole of sorbed protein displayed maxima at the completion of the first monolayer. The electrophoretic mobility of glass particles, coated with known amounts of γ-globulin adsorbed under identical conditions, was measured as a function of increasing protein coverages. The results obtained in all 3 sets of independent experiments are consistent with each other and indicate that, on glass, γ-globulin undergoes conformational changes which are relatively the most extensive for protein molecules directly attached to the surface.*

---

[1]The research and development upon which this publication is based was performed with support from the Biomaterials Program, Devices and Technology Branch, Division of Heart and Vascular Diseases, National Heart and Lung Institute, under Contract NIH-N01-HB-3-2917.

## I. INTRODUCTION

Any surface other than the intact natural endothelium is recognized by native blood as a "foreign" surface. As a result of this, a series of processes generally referred to as contact activation is initiated at the molecular level. Depending upon a number of factors, including hemodynamic conditions and the effective molecular structure of the contact surface, the initial interaction can lead to thromboembolic phenomena involving the cellular components of blood. This sequence of untoward re-actions severely limits the potential capabilities of artificial organs, such as circulatory assist devices, blood oxygenators, etc., needed to sustain the life of patients in many cases.

Based on diffusion kinetics, it can be shown[1] that the probability of plasma proteins colliding with a surface freshly exposed to blood is much higher than that of any of the cellu-lar components. Since both the bulk number density and diffu-sion coefficient of most plasma proteins are greater than those of any of the cellular components, it can be expected that the attachment of plasma proteins to the surface should overwhelm-ingly precede that of the blood cells. A number of experiment-al observations[2-4] have indicated that surfaces exposed under controlled conditions become "primed" by a layer of sorbed plasma proteins and platelets settle on the top of that layer. From the heterogeneous flux of the large number of different proteins colliding with the surface under competitive conditions, some are deflected and others become spontaneously adsorbed with or without concurrent conformational changes. Thus, the composi-tion and the state of the sorbed protein layer are primary de-terminants of the initiation of thromboembolic phenomena at the cellular level. In this sense, the conditioning layer of plasma proteins mitigates the effects of the original surface on the rest of the blood.

The incidence of adsorption of a particular native plasma protein depends upon the overall energy balance of its adsorp-tion process. The overall process of the spontaneous adsorption of a native protein is a complex sequence of changes involving the transformation of the solvated native molecule into a structure which is in an energetically stable state with respect to its neighbors, the surface and the interfacial layer. In this transi-tion, the protein or some of its segments first perturb and/or displace a number of solvent molecules in the interfacial layer that can be more, equally or less ordered than the bulk liquid. This is followed by the establishment of a number of direct at-tachment points between the protein and the surface. As the working hypothesis of this study,[1] it has been assumed that the interaction energy evolved in these processes can be utilized in part or, in entirety, to trigger conformational changes in the sorbed protein leading to the formation of additional

protein-to-surface bonds with the concurrent displacement of
additional solvent molecules until a steady state is attained.
The triggering of adsorption-induced conformational changes appears quite possible since the energies of most of the noncovalent
intramolecular structural bonds in native proteins are equal or,
comparable to the energies of bonds which can be established in
the direct attachment of an initially native protein to a contact surface. Based on this hypothesis, the interaction energy
arising in the overall sorption process of a native protein can
be considered as a characteristic quantity determining the extent, to which a particular material will induce conformational
alterations in the sorbed protein.

In the present work, the enthalpy changes resulting from
the overall adsorption process of native human plasma proteins
on various microparticulate adsorbents serving as model surfaces
have been determined by direct microcalorimetric measurements.
Microelectrophoretic mobility determinations have been made to
study the properties and structure of the sorbed protein layers.
As model adsorbates, 4 human plasma proteins were chosen whose
known concentrations in plasma and established diffusion coefficients[5] indicate that these proteins exhibit the relatively greatest flux to a contact surface and are likely to become
the major constituents of the conditioning layer described
earlier. One of the selected proteins is $\gamma$-(7S)-globulin whose
adsorptive properties on one of the strongest procoagulants
known, viz., glass have been studied under conditions partially
simulating plasma environment and are described below.


II. EXPERIMENTAL MATERIALS AND METHODS

Human $\gamma$-(7S)-Globulin ($\gamma$-GLB). The highest purity commercially available protein[2] (Cohn Fraction II), estimated to be 95% pure
by electrophoresis, was used without any further purification.
Although native $\gamma$-GLB molecules contain a total of about 1,555
amino acid units forming 2 so-called heavy and 2 light chains
linked together via disulfied bridges,[6] each chain has a micro-
heterogeneous region toward its amino-group terminal due to variable primary structures. Thus, the isoelectric point of $\gamma$-GLB
is not a single value but falls into the range of 5.8 - 7.0.[5]
To alleviate potential variances, a homogenized stock of the protein obtained by pooling batches of $\gamma$-GLB having different lot
numbers was used in all experiments. $\gamma$-GLB solutions prepared
for all experiments were in the same standardized sodium acetate/
HCl buffer having a pH = 7.2 and an ionic strength of 0.05.

---

[2]Nutritional Biochemicals Div., ICN Pharmaceuticals, Inc.,
Cleveland, Ohio.

Glass Powder Adsorbent.   This material was prepared by the
continuous ball milling of commercially available glass microbeads.[3]   Following the separation of particles $\leq$ 1 $\mu$ by fractionation, the powder obtained was repeatedly extracted at room temperature with 6N. HCl and exhaustively washed with dist. $H_2O$
until the pH of effluents was 6.0 - 7.0.   Figure 1 displays a

GLASS POWDER AT 20,000 X

*Fig. 1.   Scanning Electron Micrograph of Glass Powder Ad-
sorbent having a Specific Surface Area of 9.85 $m^2/g$.*

SEM of these particles, which were found to be nonporous, but
whose surfaces are jagged with irregularities in the order of
0.05 $\mu$ or less.   The composition of this adsorbent is specified
elsewhere.[7]   B.E.T. multipoint $N_2$ adsorption at -195$^O$C gave
the specific surface area of the powder, $\Sigma$=9.85 $m^2/g$.   As determined by direct microcalorimetric measurements at 25$^O$ and
37$^O$C, the heats of immersion of this adsorbent into water,
$h_I$(SLW), are 203 $\pm$ 5 and 286 $\pm$ 9 erg/$cm^2$, respectively.   The
heats of immersion into the standard buffer, $h_I$(SLB), are 252 $\pm$ 6
and 347 $\pm$ 10 erg/$cm^2$ at the respective temperatures.   As measured in the standard buffer, the microelectrophoretic mobility
of the particles, $\mu_e$ (25$^O$C) = - 3.20 $\pm$ 0.21 ($\mu$-$sec^{-1}$-$v^{-1}$-cm).

$\gamma$-GLB Adsorption Isotherms and Data Reduction.   The adsorptivities ($\sigma$) of $\gamma$-GLB from the standard buffer were determined at both 25$^O$ and 37$^O$C, in the initial bulk concentration

---

[3]Type A, Class V, "Close Sized Unispheres," Microbeads Div.,
Cataphote Corp., Jackson, Miss.

$(C_O)$ range of 0-2.0 mg/ml, in duplicate runs.  After equilibration for 2 hr at the selected temperatures, the adsorbent was
removed by high speed centrifugation and the σ's were determined
by monitoring the changes in the optical density of γ-GLB solutions at 278 nm.  The ratio of the total surface area available
for adsorption to the total protein solution volume, $\Sigma'/V$, was
found to affect the adsorptivity.  Thus, in all of the experiments, γ-GLB was adsorbed at a fixed $\Sigma'/V = 98.5$ cm$^2$/ml.

As an approximate method for the characterization of the
shape of sorbed γ-GLB molecules, their mean nominal cross-sectional
area at the completion of the first monolayer, $\bar{a}_1$, has been derived from the adsorbance, $\bar{\sigma}_1$ (g/cm$^2$) corresponding to that coverage.  Assuming that the partial specific volume of the hydrated
protein, $\bar{V}$, is not appreciably changed in the adsorbed state, it
can be shown that $\bar{a}_1 = \pi M/2\sqrt{3}\ \bar{\sigma}_1 N_O = 0.907\ M/\bar{\sigma}_1 N_O$ where M is the
protein molecular weight taken as $1.61 \times 10^5$ for γ-GLB, $N_O$ rep-
resents Avogadro's number and $\pi/2\sqrt{3}$ is the appropriate packing
factor.  Similarly, the mean nominal thickness of the first
sorbed monolayer has been computed as $\bar{\delta}_1 = 3\sqrt{3}\ \bar{\sigma}_1/\pi d = 1.654\ \bar{\sigma}_1/d$
where d, the density of the protein is taken as $(\bar{V})^{-1}$, and
$3\sqrt{3}/\pi$ is the cubic packing factor.

Heat of Immersion and Heat of Protein Sorption Measurements.
All of these experiments were performed with a custom built
isothermal-jacketed microcalorimeter system utilizing some of the
design principles described in the literature.[8,9]  The instrument can be operated at more than one temperature and it employs
2 highly sensitive and matched thermistors positioned in the
opposite arms of a conventional bridge.  By optimizing circuit
parameters and with a chopper amplifier, resolutions of $1 \times 10^{-5}$°C
in 100 ml of aqueous sample can be routinely attained.  A detailed description of the design and operation of this instrument has been presented elsewhere.[10]  In a typical experiment,
0.10 g of the adsorbent is placed in a thin-walled spherical bulb
(1.6 cm dia.), outgassed and sealed under vacuum.  The cell containing 100.0 ml of a sample is equilibrated, with the bulb in
position, in the heat sink.  After 2 successive determinations
of the nominal heat capacity of the cell and its contents, the
bulb is imploded by slowly lowering it against the spiked stainless steel axle of the magnetic stirrer, which flushes the adsorbent into the vortex of the stirred liquid.  As determined in
separate control experiments, the rupturing of an evacuated bulb
having the size specified above releases $0.0597 \pm 0.0030$ and
$0.0847 \pm 0.0076$ cal at 25° and 37°C, respectively.  All calorimetric heats of immersion or protein sorption reported here have
been corrected for the energy of bulb breaking.

Microelectrophoretic Mobility Measurements.  These experiments were performed in a precision bore glass capillary cell
having a ground optical polished flat as described in the literature.[11-14]  The apparatus was calibrated with fresh washed

human erythrocytes according to published methods.[14] Mobility measurements with glass particles, coated with amounts of sorbed γ-GLB, were carried out in the standard buffer, at temperatures identical to those at which the protein was sorbed. For each $C_0$ value where γ-GLB was adsorbed, the mobilities of an average of 12-14 separate protein-coated particles were determined in both directions by reversing polarities. All mobility values obtained at 37°C have been corrected to 25°C according to standard methods.

In general, the experimental conditions given above represent only items considered to be the most salient. Accounts of other experimental details found necessary to ascertain reproducibility have been described in published reports.[10,15]

III. RESULTS AND DISCUSSION

The studies reported here have been designed to apply experimental conditions approximating those involved in blood/foreign surface interactions. Thus, in all experiments, γ-GLB was adsorbed at pH = 7.2 from a buffer to partially simulate the plasma environment. The ionic strength of this buffer, corresponding to about one-third of that of plasma, was selected: (a) to allow for sufficient protein solubilities, at both 25° and 37°C, with the capacity of maintaining a constant pH, and (b) to maintain the effects of electrolytes at a level that minimizes their interference in protein sorptions at relatively low coverages.

The 25° and 37°C adsorption isotherms of γ-GLB on the glass powder adsorbent are displayed in Figs. 2 and 3, respectively.

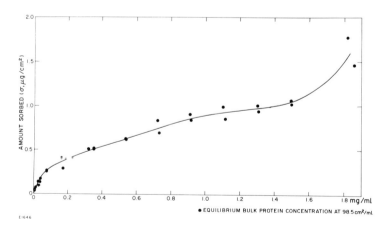

*Fig. 2.  Adsorption Isotherm of Human γ-Globulin on Glass Powder (9.85 $m^2/g$) at 25°C.*

82

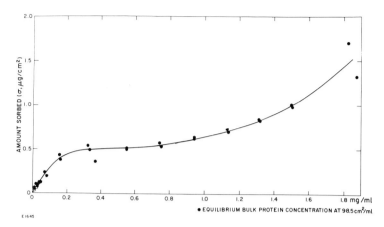

*Fig. 3. Adsorption Isotherm of Human γ-Globulin on Glass Powder (9.85 m²/g) at 37°C.*

Both isotherms indicate multilayer sorption which is similar to that observed, in this work, with the glass powder adsorbent and human fibrinogen[16] or, systems involving γ-GLB with other micro-particulate adsorbents.[15] In the γ-GLB/glass system, the attachment of the native protein has not been observed to reach a true equilibrium but to be irreversible under the conditions employed. When samples of the glass powder, coated with varying quantities of γ-GLB were treated with fresh buffer for 16 hr, no spectroscopically detectable amount of protein was released at either temperature.

In the 25°C isotherm, the completion of the first sorbed monolayer can be estimated to occur at about $\bar{\sigma}_1 (25°) \approx 0.26$ μg/cm². Using the approximate method presented earlier (which is employed since statistical methods of polymer adsorption have not yet been developed to the extent to be applicable to complex proteins), this adsorptivity converts to a computable mean nominal cross-sectional area per sorbed γ-GLB molecule, $\bar{a}_1 (25°) \sim 9,800$ Å² and a mean nominal monolayer thickness, $\bar{\delta}_1 (25°) \sim 31$ Å. In the 37°C isotherm, the completion of the first monolayer is at about $\bar{\sigma}_1 (37°) \approx 0.46$ μg/cm², corresponding to $\bar{a}_1 (37°) \sim 5,300$ Å² and $\bar{\delta}_1$ (37°) $\sim 56$ Å. Although the exact dimensions of the γ-GLB molecule have not been established, hydrodynamic data[17] indicate that the native species is a prolate ellipsoid with a major and minor axes of about 235 Å and 44 Å, respectively. These data would give the "side-on" and "end-on" projections of the undistorted molecule as approx. 6,500 Å² and 1,600 Å², respectively. Based on a comparison involving the $\bar{a}_1$ and $\bar{\delta}_1$ values derived from the 25° isotherm and the dimensions of the native γ-GLB, a substantial degree of adsorption-induced conformational changes in the sorbed protein can be definitely inferred at that temperature.

The $\overline{a}_1(37^O)$ and $\overline{\delta}_1(37^O)$ values indicate either randomized adsorption or a smaller extent of conformational alterations, which do not necessarily exclude each other.

The heats of immersion of the glass powder adsorbent into the standard buffer, $h_I(SLB)$'s are distinctively greater than those measured in water, $h_I(SLW)$'s at either $25^O$ or $37^OC$. In view of the negative electrophoretic mobility of the glass particles, at pH=7.2, this indicates the adsorption of $Na^+$ ions from the buffer.

The heats of immersion of the glass powder adsorbent into the standard buffer containing known amounts of $\gamma$-GLB, $h_I(SLP)$'s, have been determined by direct microcalorimetric measurements at $25^O$ and $37^OC$. To evaluate the extent of interaction between the native $\gamma$-GLB and the glass surface, the net heat of protein sorption at a given temperature, T, has been defined as $\{h_I(SLP)_T - h_I(SLB)_T\}$. In Fig. 4, the net calorimetric heats of $\gamma$-GLB sorption on the glass powder adsorbent at the selected temperatures are shown as a function of both the initial and the resulting equilibrium protein concentrations.

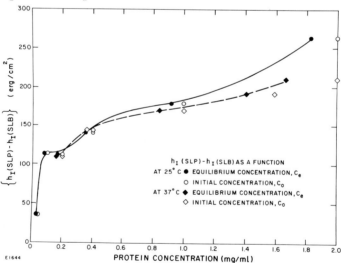

*Fig. 4. Net Calorimetric Heats of Human $\gamma$-Globulin Sorption on Glass Powder (9.85 $m^2/g$).*

The quantity denoted as $h_I(SLP)_T$ is the sum total of enthalpy changes associated with several processes involved in the sorption of a native protein. These include (a) the perturbation and/or displacement of solvent molecules in the interfacial layer; (b) the formation of bonds between the surface and the segments of the initially native protein; and (c) the complex sequence of adsorption-induced conformational changes, if any, entailing the breakage and/or rearrangement of noncovalent intramolecular bonds, which is paralleled by the formation of additional protein-to-surface bonds with the displacement of further solvent. However, the energy term defined as the net heat of protein sorption is related only to the direct protein/surface interaction at the completion of the first monolayer coverage. At this point, adsorbed segments of the protein have displaced most of the solvent, whether the protein remained in its native state or became conformationally altered. The process of solvent displacement has involved an enthalpy change equal to $h_I(SLB)_T$ that is subtracted from $h_I(SLP)_T$. Thus, the net heat of sorption is a quantity characteristic of native protein/foreign surface interactions.

As seen from Fig. 4, the neat heats of $\gamma$-GLB sorption obtained per unit surface area at both temperatures display a sharp rise followed by increases which are almost proportionate to increases in $C_o$. Based on corresponding adsorptivities, the mean net calorimetric heats per mole of sorbed $\gamma$-GLB are shown as a function of the coverage, $\sigma$, in Figs. 5a and 5b for 25° and 37°C, respectively.

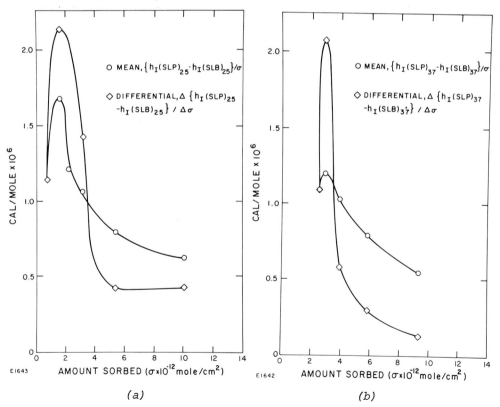

*Fig. 5. Mean and Differential Net Calorimetric Heats of Human $\gamma$-Globulin Sorption on Glass Powder (9.85 m$^2$/g) (a) at 25°C and (b) at 37°C.*

In both cases, the heats released increase sharply with increasing $\sigma$'s and attain maximum values, in the order of $1 \times 10^3$ Kcal/mole of sorbed $\gamma$-GLB, at coverages approximately equal to those which have been derived independently from the $25^\circ$ and $37^\circ$C isotherms as the adsorptivities corresponding to the completion of the first monolayer. The differential net calorimetric heats of sorption, $\Delta\{h_I(SLP)_T - h_I(SLB)_T\}/\Delta\sigma$ are also shown in Figs. 5a and 5b. Although these quantities may involve an average error of about 15-25% because of the uncertainties in determining $\sigma$ at relatively low coverages, the sharp peaks of the differential curves confirm the trend that the relatively strongest interaction occurs in the range of up to the completion of the first monolayer coverage where the protein can become directly attached to the glass surface. Following their maxima, both the mean and differential net heat curves decline to values smaller by an order of magnitude, implying that the attachment of subsequent layers involves an interaction between sorbed and native $\gamma$-GLB molecules. Thus, the calorimetric data are consistent with the adsorption isotherms indicating multilayer sorption and relatively large mean nominal cross-sectional areas for molecules sorbed in the first monolayer, which are attributable to substantial conformational changes.

The maximum of the mean net calorimetric heat of $\gamma$-GLB sorption at $25^\circ$C is about $1.7 \times 10^3$ Kcal/mole. Since the native protein contains about $1.5 \times 10^3$ amino acid residues, the enthalpy change is 1.09 Kcal/mole of residue. The magnitude of this value indicates that not all of the amino acid units of the protein become directly attached to the surface. These values obtained with the $\gamma$-GLB/glass adsorption system at the completion of first monolayer coverage can be compared with that derived for the adsorption of polyethylene glycols (PEG) on Aerosil at $25^\circ$C.[18,19] Adsorbed from benzenic solutions, PEG's having molecular weights of $6 \times 10^3$ and $4 \times 10^4$ gave, respectively, $1.3 \times 10^2$ and $7.2 \times 10^2$ Kcal/mole as their net heats of sorption at the completion of their first monolayer coverage, which converts to about 1.0 Kcal per mole of monomer unit. Should the net heats of sorption of the PEG's scale linearly with their molecular weight, a PEG homopolymer having a molecular weight identical to that of $\gamma$-GLB would yield a net heat of sorption equal to about $2.8 \times 10^3$ Kcal/mole. Since, in contrast to proteins, PEG's are void of any functional groups, the net heats of $\gamma$-GLB sorption appear to be reasonable values.

The structuring, and any potential conformational changes, of $\gamma$-GLB molecules sorbed onto glass have been studied by determining the electrophoretic mobilities of particles coated with varying amounts of that protein. In Figs. 6 and 7, the $25^\circ$ and $37^\circ$C mobilities of these particles are shown, respectively, as a function of the $\gamma$-GLB coverage, $\sigma$, with the values obtained at $37^\circ$C corrected to $25^\circ$C. As indicated, in the range of relatively

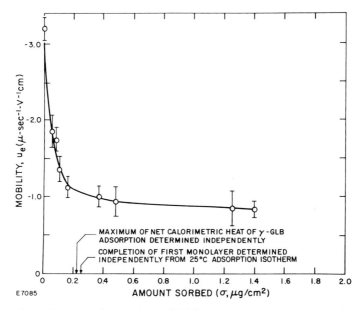

Fig. 6. *Electrophoretic Mobility of Human γ-Globulin Coated Glass Powder Adsorbent Particles at 25°C.*

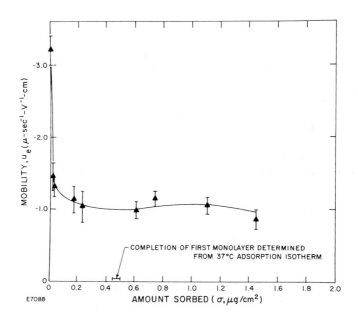

Fig. 7. *Electrophoretic Mobility of Human γ-Globulin Coated Glass Powder Adsorbent Particles at 37°C.*

low adsorbances, the mobilities are shifted quite rapidly in a positive direction at both temperatures. The mobilities reach a steady-state value at coverages reasonably consistent with the adsorptivity which has been determined independently as the first monolayer coverage from the corresponding isotherms. The attachment of additional quantities of sorbed $\gamma$-GLB does not lead to any further changes in the charge of protein-coated particles.

IV. SUMMARY

Consistent with the $25^{\circ}$ and $37^{\circ}$C adsorption isotherms indicating conformational changes in terms of the computable mean nominal cross-sectional area per sorbed $\gamma$-GLB molecule, the mean and differential net calorimetric heats of the protein display maxima in the range of adsorptivities which correspond to near monolayer coverages at the respective temperatures. Both types of heats decline sharply with adsorptivities increasing above that of the monolayer coverage. The changes in the electrophoretic mobility of glass particles coated with sorbed $\gamma$-GLB are consistent with the data derived from the adsorption isotherms and calorimetric measurements inasmuch as the mobilities rapidly decrease with increasing coverages and attain a steady-state value at about the completion of the first monolayer. Thus, the results obtained in all 3 sets of independent experiments are consistent with each other and indicate that (a) the interaction between $\gamma$-GLB molecules sorbed in the first monolayer and the glass surface is relatively the greatest, causing the relatively highest degree of conformational changes, and (b) the sorption of subsequent layers represent interactions between the sorbed and native species of the same protein, involving a different binding mechanism and significantly smaller energies. The results also imply that the net heat of protein sorption is an energy quantity that is characteristic of a particular native protein/foreign surface adsorption system and is a proportionate measure of adsorption-induced conformational changes.

V. REFERENCES

1. Nyilas, E., Proc. 23rd Ann. Conf. Eng. Med. Biol. 12, 147 (1970); and Avço Everett Research Laboratory AMP 327 (1970).
2. Dutton, R.C., Baier, R.E., Dedrick, R.L., and Bowman, R.L., Trans. Amer. Soc. Artif. Int. Organs 14, 57 (1968).
3. Baier, R.E. and Dutton, R.C., J. Biomed. Mat. Res. 3, 191 (1969).
4. Petschek, H., Adamis, D., and Kantrowitz, A.R., Trans. Amer. Soc. Artif. Int. Organs 14, 256 (1968).

5. Schultze, H.E., and Heremans, J.F., "Molecular Biology of Human Proteins," Vol. No. 1, pp. 176-181, Elsevier Publishing Co., New York, 1966.
6. Rowe, D.S., Nature 228, 509 (1970).
7. Nyilas, E., Chiu, T-H., Lederman, D.M., and Micale, F.J., "The State of Water Adjacent to Glass," Paper No. R33 presented at the 50th Colloid and Surface Science Symposium, San Juan, PR, 1976; and in press, "Recent Advances in Colloid and Interface Science," Academic Press, New York.
8. Zettlemoyer, A.C., Young, G.J., Chessick, J.J., and Healy, F.H., J. Phys. Chem. 57, 649 (1953).
9. Skewis, J.D., and Zettlemoyer, A.C., in "Proceedings of the 3rd International Congress on Surface Activity," Vol. No. 2, Sec. B/III/1, p. 401, Cologne, West Germany, 1960.
10. Nyilas, E., Chiu, T-H., Herzlinger, G.A., and Federico, A., "Microcalorimetric Study of the Interaction of Plasma Proteins with Synthetic Surfaces. I. Construction of the Microcalorimeter and Preliminary Results," February, 1974, PB 231 776.[4]
11. Dukhin, S.S., and Derjaguin, B.V., "Electrokinetic Phenomena," in "Surface and Colloid Science" (E. Matijevic. ed.), Vol. No. 7, Wiley-Interscience, New York, 1974.
12. Loveday, D.E.E., and James, E.M., J. Sci. Instr. 34, 97 (1957).
13. Seaman, G.V.F., and Heard, D.H., Blood 18, 599 (1961).
14. Bangham, A.D., Heard, D.H., Flemans, R., and Seaman, G.V.F., Nature 182, 642 (1958).
15. Nyilas, E., Chiu, T-H., Lederman, D.M., and Herzlinger, G.A., "Study of the Interaction of Plasma Proteins with Prosthetic Surfaces," February 1975, PB 242 650.[4]
16. Chiu, T-H., Nyilas, E., and Lederman, D.M., Trans. Amer. Soc. Artif. Int. Organs 22, 497 (1976).
17. Oncley, J.L., Scatchard, G., and Brown, A., J. Phys. Colloid Chem. 51, 134 (1947).
18. Killmann, E., and Eckart, R., Makromol. Chem. 144, 45 (1971).
19. Killmann, E., and Winter, K., Angew. Makromol. Chem. 43, 53 (1975).

---

[4]Public document available from the National Technical Information Service, 5285 Port Royal Road, Springfield, VA 22151.

# THE PARTIAL SPECIFIC VOLUME CHANGES INVOLVED IN THE THERMOTRO-PIC PHASE TRANSITIONS OF PURE AND MIXED LECITHINS.

Peter Laggner and Hans Stabinger

*Institut für Röntgenfeinstrukturforschung der Österreichischen Akademie der Wissenschaften und des Forschungszentrums Graz*

## I. INTRODUCTION

The phase behaviour of natural and synthetic phospholipids has been studied from a variety of different aspects (for review see refs. 1,2). Among the numerous phase transitions of lyotropic and thermotropic nature the thermal transition from a crystalline to a liquid crystalline state is of particular relevance to an understanding of the physical state of biological membranes. From a thermodynamical point of view this transition has been well characterised for some synthetic diacyllecithins and phosphatidylethanolamines with respect to the heats of transition and the respective temperatures (1-7). Another characteristic quantity, equally important to a physical discussion of the transition, the change in partial specific volume has not yet been elucidated in similar detail. So far it is only known from two studies using volumetric dilatometry (8, 9) that the transition is accompanied by a relative volume change in the range of 1 - 4 %, depending on the hydrocarbon chain length of the lecithins.

In the present article we introduce a novel experimental approach to measure the partial specific volumes in dilute solutions as a function of temperature. The method is based upon high-precision density measurement using the oscillator technique (10). We present the results of some experiments on dilute aqueous dispersions of synthetic diacyllecithins, dimyristoyllecithin, dipalmitoyllecithin and distearoyllecithin, and defined binary mixtures thereof. It is shown that the method is capable of resolving the partial specific volume changes involved in liquid-crystalline phase transitions.

## II. EXPERIMENTAL PROCEDURE

### A. The Instrument

The apparatus used in this study is a modification of the commercially available Precision Density Meter DMA 50 (A. Paar K.G., Graz, Austria). It employs two hollow U-shaped glass oscillators of closely matched geometry and mass-to-hollow volume ratio. The two oscillators held in a glass housing filled with hydrogen for good thermal connection are embodied in a copper block which is thermostated by Peltier elements. A temperature control unit with Thermilinear Thermistor Network (Yellow Springs Instrument Co., Ohio) allows to hold and to change the block temperature with a defined gradient in the range from $- 10$ to $+ 65^{\circ}$ C. The construction assures the same temperature for both measuring cells. The period of oscillation for each cell is determined by the mass (involving the density) and spring constant of the glass tube. By using two closely matched oscillators and filling one with the solvent and the other with the solution, and by dividing the two periods of oscillation where the period of one oscillator serves as the time-base for the determination of the period of the other, the thermal influence on spring constant and volume of the solvent are cancelled out and therefore the measured value only reflects the difference in the thermal expansion coefficient between solution and solvent. By calibration of the instrument with air and water in the usual way over the whole temperature range the results can be transformed into actual partial specific volume values (10). Since there is a certain temperature lag between the oscillator and the copper block depending on the heating or cooling rate it is not possible to use the block temperature for exact data evaluation. To this end the period of the solvent filled oscillator is measured simultaneously against a crystal controlled timer and, once temperature calibrated, serves as a thermometer. The necessary sample volume to fill the oscillator is 0,7 ml.

### B. The Samples

1,2-diacyl-L-phosphatidylcholines (dimyristoyl-, dipalmitoyl-, distearoyl ) were purchased from Sigma London and used without further purification since no impurities have been detected by thin layer chromatography. Aqueous dispersions were prepared by sonication for 15 min. in a bath type sonicator (Sonirex, Bandelin Electronics, West Germany) at about $5^{\circ}$ C above the respective transition temperature. For mixtures of two lecithins, appropriate amounts were weighed in, dissolved in chloroform and the solvent evaporated under oil pump vacuum. The dry residue was then dispersed in water as described above. Water was deionised and twice quartz-distilled. Concentrations were determined by phosphorus assay according to the method of Chen et al. (11). The phospholipid concentrations ranged from 5 to 25 mg/ml. Prior to filling into the measuring cell the samples were degassed at the upper limit of the desired temperature range under aspirator vacuum.

## III. RESULTS AND DISCUSSIONS

Figure 1 shows the temperature dependence of the partial specific volumes for the three pure lecithins in dilute aqueous dispersion in the range of the respective transition temperatures. For all three cases two characteristic temperatures

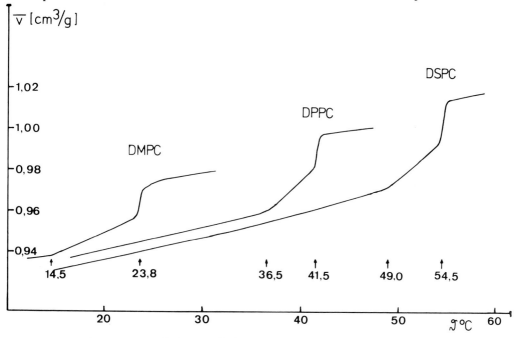

*Fig. 1. Partial specific volume versus temperature curves for dimyristoyllecithin (DMPC), dipalmitoyllecithin (DPPC) and distearoyllecithin (DSPC). The data are mean values from subsequent heating and cooling runs with rates of 0,4° C/min.*

can be defined. For each of the lipids there is an almost discontinuous change in partial specific volume at the temperatures 23,8° C, 41,5° C and 54,5° C, respectively. These values are in good agreement with the crystalline to liquid crystalline transition temperatures measured in these systems by

other techniques (7, 9, 12). In addition to this abrupt thermal expansion there is a marked change in slope at temperatures a few degrees below the main transition. This change in
thermal expansion coefficient coincides with the so-called
pre-transition (3, 7, 13) which is characterised by a heat of
transition by about one order of magnitude lower than the main
transition.

A point of considerable relevance to the structure of biological membranes is the question of the phase behaviour of
mixed phospholipid systems. It has been proposed, that lecithins differing widely in their acyl chain lengths or degree
of saturation may show monotectic behaviour whereas co-crystallisation occurs with chain lengths that differ only by two
carbon atoms (5). Figures 2 and 3 show the temperature depen-
dences of the partial specific volumes of binary mixtures of

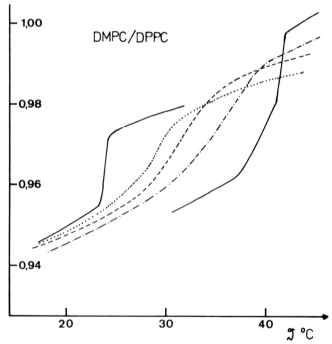

*Figure 2: The temperature dependence of the partial spe-
cific volumes v̄ of binary mixtures of dimyristoyllecithin
(DMPC) and dipalmitoyllecithin (DPPC) with 25 mole-% (·······),
50 mole-% (----), and 75 mole-% (-·-··-·) DPPC.*

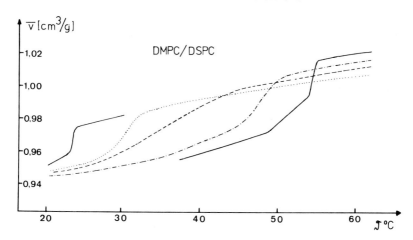

*Figure 3: The temperature dependence of the partial spe-cific volumes v of binary mixtures of dimyristoyllecithin (DMPC) and distearoyllecithin (DSPC) with 25 mole-% (·······), 50 mole-% (----), and 75 mole-% (-··-··-··-) DSPC.*

lecithins differing by two (DMPC/DPPC) and four carbon (DMPC/DSPC) atoms, respectively. In no case the experiments show any indication of a monotectic behaviour which is in good agreement with the results obtained by s in label experiments (12). In the dimyristoyllecithin: dipalmitoyllecithin system the binary mixtures of all proportions studied show little resemblance to the behaviour of the pure lipids. In the case of dimyristoyllecithin/distearoyllecithin.mixtures, however the transitions for the 3:1 and 1:3 mixtures indicate a relative enrichment of the major component in the mixtures that actually undergo the transition. However, also this system shows co-crystallisation as can be judged from the existence of only one broadened transition.

It has been emphasised by several groups of authors (9, 14, 15) that the nature of the phospholipid aggregates, whether they are small single shelled vesicles or multilamellar dispersions, has to be considered in the discussion of physico-chemical data. In the present study, although the dispersions were subjected to sonication, we cannot assure the exclusive presence of one or the other form. In this respect the data have to be regarded as preliminary and a detailed discussion of this problem together with the effects of pressure on the partial specific volume changes will be given elsewhere (P. Laggner and H. Stabinger, in preparation).

In summarising the results it may be said, that the described method provides a useful and precise tool to study the volume changes in liquid-crystalline phase transitions. Preliminary experiments of the same type under different external pressures have shown promising results and should give additional information on the phase behaviour of phospholipids.


## IV. ACKNOWLEDGEMENTS

The authors gratefully acknowledge the support by the Österreichischer Fonds zur Förderung der wissenschaftlichen Forschung.


## V. REFERENCES

1. Chapman, D., in "Biological Membranes" (D. Chapman, and D.F.H. Wallach, Eds.), Vol. 2, p. 91. Academic Press, London and New York, 1973.
2. Chapman, D., in "Form and Function of Pospholipids" (G.B. Ansell, J.N. Hawthorne, and R.M.C. Dawson, Eds.), B.B.A. Library, Vol. 3, p. 117. Elsevier, Amsterdam, London, New York, 1973.
3. Chapman, D., Williams, R.M., and Ladbrooke, B.D., Chem. Phys. Lipids, 1, 445 (1967).
4. Steim, J.M., Adv. Chem. Ser. 84, 259 (1968).
5. Phillips, M.C., Williams, R.M., and Chapman, D., Chem. Phys. Lipids, 3, 234 (1969).
6. Phillips, M.C., Ladbrooke, R.M., and Chapman, D., Biochim. Biophys. Acta, 196, 35 (1970).
7. Hinz, H.J., and Sturtevant, J.M., J. Biol. Chem. 247, 6071 (1972).
8. Träuble, H., and Haynes, D.H., Chem. Phys. Lipids 7, 324 (1971).
9. Melchior, D.L., and Merowitz H.J., Biochemistry 11, 4558 (1972).
10. Kratky, O., Leopold, H., and Stabinger, H., in "Methods in Enzymology" (C.H.W. Hirs and S.N. Timasheff, Eds.) Vol. 27, p. 98, Academic Press, New York and London, 1973.
11. Chen, P.S., Toribara, T.Y., and Warner, H., Anal.Chem. 28, 1756 (1956).
12. Shimshik, E.J. and Mc Connell, H.M., Biochemistry 12, 2351 (1973).
13. Ladbrooke, B.D., Williams, R.M., and Chapman, D., Biochim. Biophys. Acta 150, 333 (1967).
14. Sheetz, M.P., and Chan, S.I., Biochemistry, 11, 4573 (1972).
15. Newman, G.C., and Huang, C., Biochemistry, 14, 3363 (1975).

# A FIRST-ORDER TRANSITION
## IN THE ISOTROPIC PHASE
## OF CHOLESTERYL STEARATE

Paul F. Waters and Lidia Roche Farmer[1,2]

The American University and Gillette Research Institute

The volume coefficient of expansion, $\alpha$, the cohesive energy, $\left(\frac{\delta E}{\delta V}\right)_T$, the compressibility, $\beta$, and the wavelength, $\lambda$, of maximum intensity of reflected light have been measured for cholesteryl stearate above the isotropic/cholesteric transition temperature. The data indicate that the ester undergoes a first-order phase transition in this region.

## I.  INTRODUCTION

While investigating the thermodynamic and reflectance properties of cholesteryl stearate on cooling near 79.5°C, the temperature of the isotropic/cholesteric transition, the sample was observed to exhibit colors in reflection above the transition temperature, in the nominally isotropic phase. This prompted the studies which are reported in this paper.

## II.  EXPERIMENTAL

### A.  Material

Cholesteryl stearate was obtained from the Aldrich Chemical Co.  It was examined by multiple thin layer

---

[1] In partial fulfillment of the requirements for the degree of Doctor of Philosophy.
[2] Gillette Research Fellow, 1972–75.

chromatography, using four different solvent carrier pairs. In all instances only one spot was observed. The sample was therefore free of unreacted cholesterol and/or stearic acid and it was considered to be of satisfactory purity for the measurements contemplated.

B.  Apparatus

1.  Cell

A cell was designed in which thermodynamic and optical measurements could be made simultaneously. It consists of a brass cylinder 1 in. long with a 1.5 in. diameter, in which a 1 in. diameter opening is machined from each base end toward the center, where enough metal remains with a smaller hole bored through to provide a sample holding chamber 0.20 in. deep and 0.19 in.$^2$ in area. Internal channels were machined in the remaining cylinder walls to allow for the circulation of a thermostatic fluid. A clear fused silica optical flat disc is attached with an O-ring (0.065 in. thick) between it and the sample chamber and a similar O-ring between it and a brass cover with which the assembly is held in place by means of 6 brass screws. The movable diaphragm of an air-dielectric capacitor is inserted into the cylinder at the base end opposite the optical flat. An O-ring (0.065 in. thick) is placed between it and the sample chamber and a silicone rubber gasket is positioned between the rigid plate of the capacitor and the brass plate which attaches the capacitor assembly by means of 6 brass screws.

The movable plate of the capacitor is made of 0.001 in. titanium foil. The rigid plate is made of anodized aluminum. The capacitor is a closed system and, as such, the capacitance measured, C, is independent of the dielectric properties of the material in the cell chamber.

Two filling cannulas are soldered to the cell body. One of these is contiguous with the cell chamber and it is fitted with a microsyringe which is held in place with an epoxy cement. The other cannula serves as a conduit for a thermo-couple lead, which is also sealed in with an epoxy cement. All parts of the cell except the silica disc are painted with a flat black paint. The piston of the microsyringe is fitted with a circular vernier and a pointer which, with calibration of the angular displacement of the piston in relation to the volume change in the microsyringe, permits the volume, V, to be read to 0.1 μl. A copper sleeve is placed over the barrel of the microsyringe in order to improve thermal contact with a copper coil which serves as a temperature-control jacket. A 0.25 in. vertical slit in the copper sleeve allows the microsyringe graduations to be read with a cathetometer. The coil is in series with the hydraulic lines which deliver the thermostatic fluid to the cell body. The coil is covered with an insulating cylinder made of asbestos sheet which, in turn, is covered with dull black electrical tape. A schematic of the cell is shown in Fig. 1.

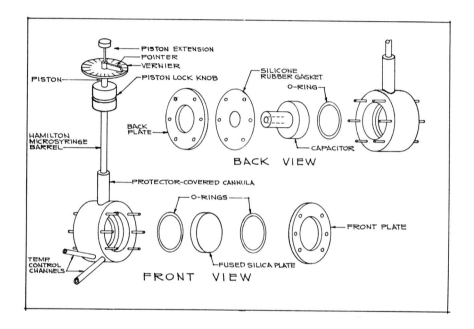

Fig. 1.   Exploded views of cell.

The cell volume was calibrated for the range of temperature of interest at atmospheric pressure with purified mineral oil whose density was measured at each corresponding temperature in the range.  A P vs T calibration was also carried out with the mineral oil and a manometer in a closed-system configuration.  Simultaneous measurements of the capacitance of the capacitor during the determinations allowed the construction  of C vs V and C vs P curves at each temperature examined.

In order to ensure the absence of air from the ester in the cell the sample was degassed in a vacuum oven at 90°C. The cell was then filled in the oven and the brass plates were adjusted to give the cell specific dimensions determined by a previous volume calibration.  A further degassing regime was followed. Only when the cell chamber and the microsyringe were completely free of apparent bubbles was the cell placed in the measuring apparatus to equilibrate before the measurements were taken. The temperature of the cell was maintained constant with water circulated from a bath. It was monitored with a calibrated thermocouple. In no instance did the fluctuations exceed 0.05°K.

2. Reflectometer

Relative intensities of reflected light were measured in a modified Brice-Phoenix light scattering photomultiplier photometer. The mercury lamp and its power supply were replaced with a 12 v tungsten lamp and a dc power supply. The turret filter was replaced with a quartz prism monochrometer. The cell stage was replaced with a cell holder machined from Bakelite. The phototube was housed in a jacket through which cooling fluid could be circulated so that the photocathode could be operated at relatively low noise levels while the cell was heated.

III. RESULTS

A. Volume Coefficient of Expansion

The volume of a sample of cholesteryl stearate was measured as a function of temperature at 1 atm. A cooling regime was followed and the approach to equilibrium was monitored by means of the capacitance. A plot of V vs T appears in Fig. 2.

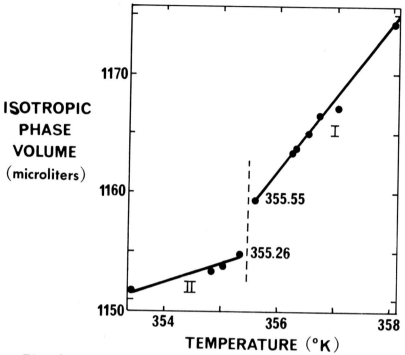

Fig. 2.   Volume of cholesteryl stearate vs temperature in the isotropic phase

The volume coefficient of expansion, $\alpha$, given by:

$$\alpha = \frac{1}{V} \left(\frac{\delta V}{\delta T}\right)_P$$

was determined by a graphical differentiation of Fig. 2.   The data points, except the two immediately adjacent to and on either side of the major discontinuity were subjected to least squares fits and the slopes and standard errors were calculated within 95 per cent confidence limits.   These values of $\alpha$ as well as the magnitude of the errors in each line are shown in Fig. 3.

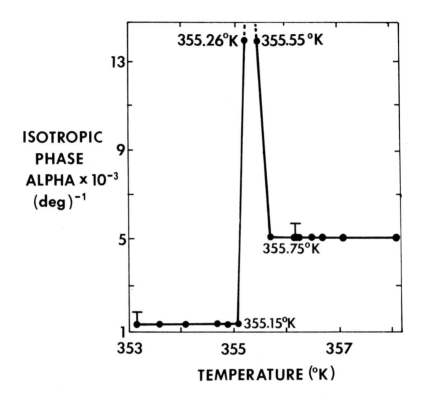

Fig. 3. Volume coefficient of expansion of cholesteryl stearate vs temperature in the isotropic phase.

B. Cohesive Energy Density

The pressure of a sample of cholesteryl stearate was measured as a function of temperature in a closed cell. The pressure was evaluated from the capacitance vs pressure calibration curve. The points were established only after equilibrium was reached at each temperature in a cooling mode as indicated by the constancy of the capacitance. The plot of P vs T is given in Fig. 4.

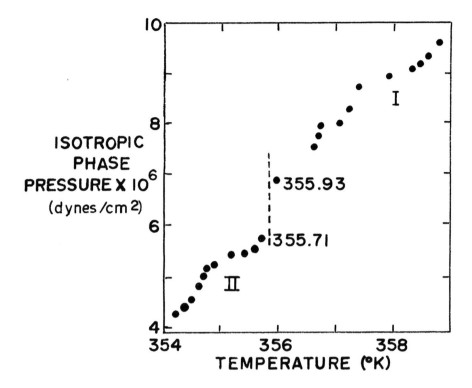

Figure 4. Pressure of cholesteryl stearate vs temperature in the isotropic phase.

The cohesive energy, $\left(\frac{\delta E}{\delta V}\right)_T$, given by:

$$\left(\frac{\delta E}{\delta V}\right)_T = \left(\frac{\delta P}{\delta T}\right)_V - P$$

was evaluated by graphical differentiation of Fig. 4. The isochoric condition was substantially met in these measurements because they were carried out at such a pressure that the diaphragm of the capacitor was extended to the point that increases in the capacitance over the range of variables examined were accompanied by a total volume change that was not detectable, i.e., it was less than 0.1 μl. For convenience in calculating the cohesive energy and in order to emphasize the vertical discontinuity in P between 355.71°K and 355.93°K the data points below 355.71°K and those above 355.93°K were subjected to least squares fits and the slopes and errors were calculated within 95 per cent confidence limits. These values of the cohesive energy as well as the magnitudes of the errors in each line are presented in Fig. 5.

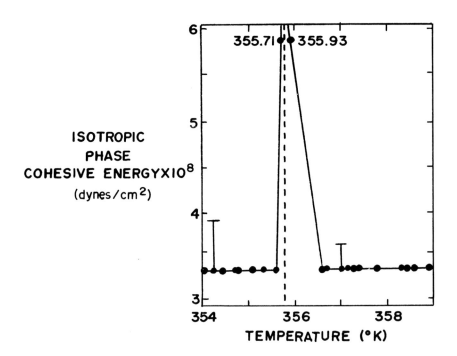

Fig. 5. Cohesive energy of cholesteryl stearate vs temperature in the isotropic phase.

C. Compressibility

The pressure of a sample of cholesteryl stearate was measured as a function of volume. Although, in principle, these measurements might have been made simultaneously with those of P vs T, lengthy periods of time were needed to reach the equilibrium state in each instance. It was convenient to measure them independently. A plot of V vs P is shown in Fig. 6.

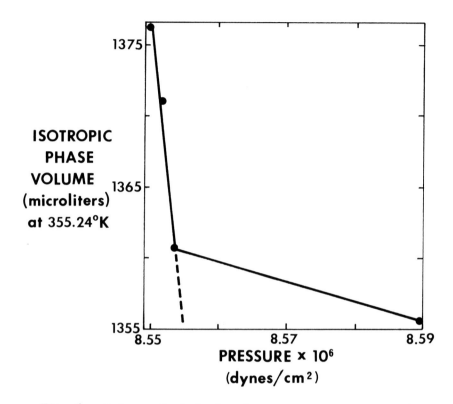

Fig. 6. Volume of cholesteryl stearate vs pressure in the isotropic phase.

The compressibility, $\beta$, is given by:

$$\beta = -\frac{1}{V} \left(\frac{\delta V}{\delta P}\right)_T$$

The partial derivative was evaluated by graphical differentiation of Fig. 6. The error in the term is estimated to be $\pm$ 10%. A plot of $\beta$ vs P is given in Fig. 7.

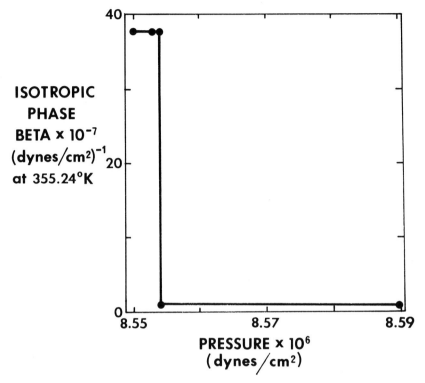

Fig. 7. Compressibility of cholesteryl stearate vs pressure in the isotropic phase.


D.  Reflectance

Since different reflectance colors were exhibited by the samples in the cell as the temperature was varied, an optical scanning was performed after each measurement of V vs T and P vs T.  The cell was fixed in its holder and the phototube was set in the photometer so that the angle of incidence and the angle of reflection were approximately 22.2°.  The intensity of reflected light was recorded as a function of the wavelength of incident light at each equilibrium point. Plots of the kind determined are illustrated in Fig. 8.

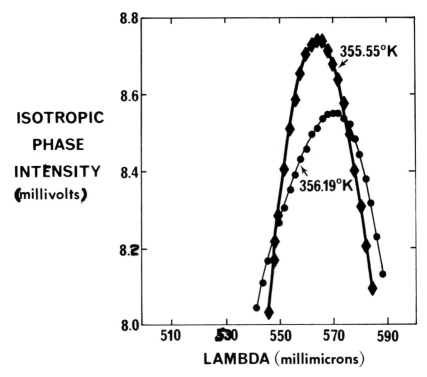

Fig. 8. Intensity of light reflected by a sample of cholesteryl stearate in the isotropic phase vs the wavelength of incident light.

The wavelength of maximum intensity was determined at each temperature at atmospheric pressure and a plot of $\lambda$ vs T was constructed. From this plot the partial derivative, $\left(\frac{\delta\lambda}{\delta T}\right)_P$, was evaluated by graphical differentiation. The derivative is shown plotted vs T in Fig. 9.

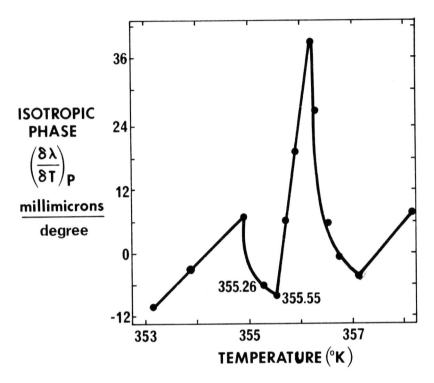

Fig. 9. The temperature derivative of the wavelength of maximum intensity of light reflected by cholesteryl stearate in the isotropic phase vs temperature.

IV. CONCLUSIONS

The vertical discontinuities evident in the thermodynamic properties of cholesteryl stearate indicate that this ester undergoes a first-order transition above the reported iso-tropic/cholesteric transition temperature, i.e., in that temperature region described as the optically isotropic region. At first we were somewhat skeptical of our findings but sub-sequent experiments with fresh loadings of the cell confirmed the original observations.

The cell was designed in the first instance to measure thermodynamic and optical reflectance changes at the isotropic/cholesteric and cholesteric/smectic transition temperatures in as accurate a manner as possible. The capacitor was incorpo-rated into the cell in order to monitor the approach to mechanical equilibrium. The concept underlying the capacitor measurement is this: whereas thermal equilibrium, as measured by a thermocouple embedded in the cell wall, would be reached within a few minutes for small changes of temperature, the mechanical equilibrium of the molecules themselves, with its requirement of more extensive alignment of asymmetric molecules in the more ordered mesophases, would be expected to exhibit a significant time-dependence for measurements recorded in a cooling regime.

Indeed, we have recorded instances wherein cholesteryl stearate requires up to 24 hours to reach a mechanical steady state at constant temperature in the cholesteric mesophase.

We believe that the first-order transition reported here escaped observation heretofore because the experimental techniques used, by their very nature, involve much more rapid temperature scanning than we were able to carry out due to the constraint of attaining constant capacitance at each temperature.

Our measurements confirm a portion of the theory of De Gennes (1), who predicted that nematic mesogens, due to short-range order, should exhibit first-order transitions in the nominally isotropic region.

V.  REFERENCES

1.  De Gennes, P. G., Phys. Lett. 30A, 454 (1969).

# STEADY STATE AND NANOSECOND TIME
# RESOLVED FLUORESCENCE OF DANSYL
# n-OCTADECYL AMINE IN BILAYER LIPOSOMES

Alejandro Romero, Junzo Sunamoto
and Janos H. Fendler
*Texas A&M University*

*Fluorescence spectra and lifetimes of dansyl n-octadecyl amine (DSOA) have been determined in degassed solutions of methanol, ethanol, chloroform, benzene and n-hexane. Correlations have been obtained between the microscopic solvent polarity parameter $E_T(30)$ and the emission maxima and lifetimes of DSOA in these solvents. Fluorescence spectra and lifetimes of this amine have been also determined in single bilayer liposomes under a variety of experimental conditions. The obtained correlations between the fluorescence parameters of the probes in neat solvents afforded information on the microenvironment of DSOA in the liposomes. Lack of quenching by hexadecylpyridinium chloride indicates that DSOA binds to sites which are well shielded from this quencher.*

The utility of fluorescence probes in membrane research is well recognized and amply documented.[1-5] Information is deduced on the apparent microenvironment of the probe in terms of viscosity, polarity, rigidity and proximity within membranes. Assumptions are generally made or tacitly implied on the location of the probe. 1-Anilino-8-naphthalene sulfonate, ANS[-], has been one of the most extensively used fluorescence probes.[6-9] Since it is a relatively small polar molecule, ANS[-] predominantly interacts with the hydrophilic parts of membrane and membrane models. Information on the less polar sites are also needed. We have synthetized fluorescence probes which contain different hydrophobic moieties and report here the results of our experiments with dansyl n-octadecyl amine, DSOA, in organic solvents and in single compartment bilayer liposomes.

## I. EXPERIMENTAL SECTION

Dansyl n-octadecyl amine, DSOA, was prepared by refluxing octadecyl amine (137.6 mg) and dansyl chloride (136.9 mg) in 10 ml chloroform in the presence of a few drops of triethylamine for 8 hours. The resulting product was purified on a basic aluminum oxide column (Woelm) using chloroform as the eluent. 196 mg of the crystals, melting at 69.5-71.0°, were obtained upon evaporation of the solvent *in vacuo*. The crystals were identified as DSOA by satisfactory elementary analysis, and by ir and mass spectrometry.

Synthetic dipalmitoyl-DL-α-phosphatidylcholine (Grade I) was used as received from Sigma Chemical Company, St. Louis, Missouri. It was found to be pure on thin-layer chromatography using pre-coated Silica Gel 60 F-254 TLC sheets (Brinkman Instruments, Inc., Des Plaines, Illinois) and chloroform : methanol : water = 65 : 25 : 4 (v/v) as the eluent. Cholesterol (MCB Co.), stearylamine, (octadecylamine, Aldrich Chemical Company) were recrystallized from ethanol. Dipalmitoyl-DL-α-phosphatidylcholine (choline methyl-C-14) was used as received from New England Nuclear Corp. All other chemicals were reagent grade and used without purification. Water was twice distilled in an all glass still.

Multicomponent dipalmitoyl-DL-α-phosphatidylcholine liposomes were prepared by evaporating 5.0 ml chloroform solutions of phospholipid, cholesterol and stearylamine (in the weight ratios of 6.0 mg, 2.0 mg, and 1.4 mg, respectively) to dryness in a flask using a rotatory evaporator. Liposome entrapped DSOA was prepared by adding 0.5 mg DSOA to the chloroform solution. The thin film remaining on the flask subsequent to evaporation and overnight drying in a desiccator was dispersed by adding 2.5 ml aqueous solution of 5.5 mM phosphate ($KH_2PO_4$ and $K_2HPO_4$) buffer and 0.10 M sodium chloride. The dispersion was carried out at 50° by shaking on a Vortex mixer until foaming ceased. The milky suspension consisted mostly of multilayer liposomes. Single bilayer liposomes were prepared from the multilayer liposomes by sonication under nitrogen in a Braunsonic 1510 sonifier at 70 watts and 50° for 5 minutes at 30 second intervals. The cleared sonified liposome solution was passed through a Sephadex G-50 column in order to separate the free DSOA from that entrapped in the liposome. Liposomes appeared in the void volume and they were recognized by their turbidity.

The liposome solution was centrifuged 1 hour at 45,000rpm in a Beckman L3-50 ultracentrifuge. The transparent supernatant was used for fluorescence determinations.

The concentration of lecithin subsequent to sonication and ultracentrifugation was calculated by measuring the amount of C-14 labelled phospholipid. Radioactivity was determined by means of a Beckman LS-100 liquid scintillation counter. The counting cocktail was prepared by dissolving 60 g of naphthalene, 4 g of POP, 0.2 g of POPOP and 20 ml ethylene glycol in 1.0 liter dioxane.

Absorption spectra was determined on a Cary 118C spectrophotometer. Steady state corrected luminescence spectra were obtained on a SPEX Fluorolog, using the E/R mode. Fluorescence lifetimes were determined on a single photon counting nanosecond spectrofluorometer, based on Ortec components.[10] Fluorescence lifetimes were calculated either graphically[11] or by an IBM 360 computer using the method of moments program.[10]

## II.  RESULTS AND DISCUSSION

The fluorescence spectra of 2-aminonaphthalene-6-sulfonate[12] and 1-(dimethylamino)-5-naphthalanesulfonic acid[13] are solvent sensitive.  Emission maxima, quantum yields and lifetimes are affected.  Regardless of the photophysical origin of these solvent effects, if correlations can be found between the fluorescence and solvent polarity parameters, they can be utilized as rulers to deduce apparent solvent polarities of their environments in membranes and membrane models.  Emission maxima and intensity of DSOA is also highly solvent dependent (Fig. 1).  Fig. 2 includes plots of

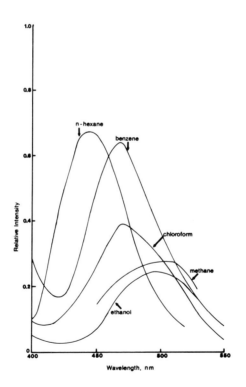

*Fig 1.  Emission spectra of $1.0 \times 10^{-5}$ M DSOA in different solvents.  Excited at 322 nm.  Excitation and emission bandpaths: 10, 10 nm.  Excitation and emission slits: 2.5, 2.5 mm. Determined by SPEX Fluorolog, using Rhodamin-B quantum counter (E/R mode).  Degassed solutions.*

relative fluorescence intensities and emission maxima against the microscopic solvent polarity parameter $E_T(30)$.[14] The emission maximum and relative intensity indicate the apparent environment of DSOA in single compartment liposomes to be somewhat less polar than that in ethanol.

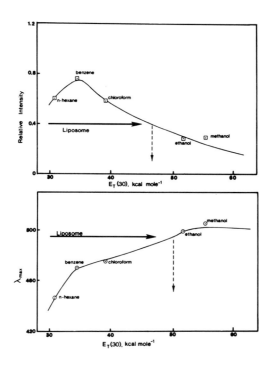

*Fig. 2. Plots of relative fluorescence intensities and emission maxima of DSOA in different solvents versus solvent polarity parameter $E_T(30)$. Non-degassed solutions.*

Concentrations of liposomes were determined by means of measuring the activities of the C-14 labelled lecithin at different stages of the preparation. Based on taking the initial lecithin concentration (6.0 mg) to be 100%, there were 60.8% and 15.8% lecithin subsequent to gel filtration and centrifugation, respectively. Assuming that the external diameter of single compartment liposomes is approximately 250 A, and that each liposome contains approximately 3000 phospholipid molecules[15] the number of vesicles per $\mu$mole of phospholipid was calculated to be $2 \times 10^{14}$. Combining this value with those calculated for the loss of lecithin during gel filtration, sonication and ultracentrifugation, we estimate the concentration of single compartment liposomes to be $8.5 \times 10^{-8}$ M. Concentrations of DSOA in the liposomes were calculated to be $4.7 \times 10^{-6}$ M by assaying the liposome entrapped DSOA in alcohol and by using appropriate spectroscopic calibrations. These concentrations establish that each lecithin vesicle contains 55 molecules of DSOA.

Fluorescence lifetimes of DSOA have also been determined in different solvents and in single compartment liposomes. Typical plots are given in Fig. 3 and the data are collected in Table I. The data distinctly fall into two sets, those in

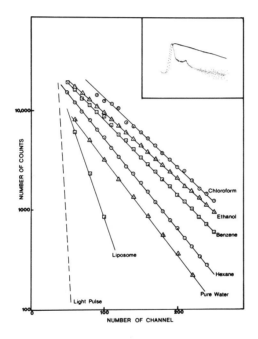

*Fig. 3. Fluorescence decay of DSOA in different solvents. Plotted are the relative intensity versus channel number in the Ortec multichannel analyzer. Each channel, with the exception of the decay of DSOA in liposome, corresponds to 0.33 nanosecond. For the decomposition of DSOA in liposome each channel corresponds to 0.184 nanosecond. The insert shows traces of the light pulse and of a fluorescence decay as they appear on the screen of the multichannel analyzer. The axes are the same as those in the main figure.*

polar solvents and those in nonpolar solvents. This behavior
is analogous to that reported for the solvent effects on
aminonaphthalanes. The plot of fluorescence lifetime *versus*
$E_T(30)$ is hyperbolic (not shown). Lifetimes cannot, therefore, be unequivocally used to establish the microscopic
polarity of DSOA in liposomes.

The available data indicate that the hydrophobic part of
DSOA is lined up along the phospholipid backbones and that
the naphthyl headgroup is in the vicinity of the polar
interior and exterior of the liposome bilayer. Fig. 4 is a
naive representation of the proposed mode of interaction.

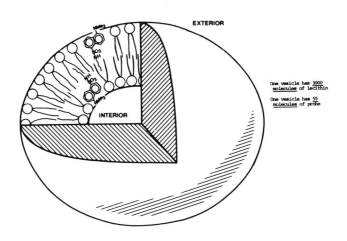

*Fig. 4. Schematic representation of location of
fluorescent probes.*

It is interesting to observe that hexadecylpyridinium
chloride does not quench the fluorescence of DSOA (Table I).
This and other pyridinium halides are efficient quenchers of
DSOA fluorescence in neat solvents. Apparently hexadecylpyridinium chloride is not taken up by the liposome into
regions which are close approximaty to DSOA. This behavior
is different than that observed for liposome entrapped ANS.[7-10]
The hydrocarbon tail of DSOA imparts, therefore, subtle differences to the behavior of naphthylamines as fluorescence

## TABLE I

## FLUORESCENCE LIFETIMES OF DSOA[a]

| Solvent | $E_T,(30)$ Kcal mole$^{-1}$ | $\tau$, nanosecond[b] | |
|---------|------------------|--------------------|---|
| n-hexane | 30.9 | 7.4 | (16.3) |
| benzene | 34.5 | 11.5 | (19.3) |
| chloroform | 39.1 | 14.2 | (21.0) |
| ethanol | 46.0 | 13.2 | (22.0) |
| methanol | 51.9 | 12.5 | (20.8) |
| water | 63.5 | 7.9 | (15.2) |
| liposome | | 5.1 | (5.6[c],6.7[d], 6.0[e],5.2[e]) |

[a] At 25.0°, unless stated otherwise.

[b] Values in parenthesis are obtained in solutions degassed by repeated freese-pump-thaw cycles on a high vacuum line.

[c] Bubbled with $N_2$.

[d] Degassed by freeze-pump-thaw technique on a high vacuum line using dry ice.

[e] Containing 0.5 μmole hexadecylpyridinium chloride.

probes. Clearly due care is required in the experimental design of fluorescence probes in membrane models and in the interpretation of the obtained data.


## III. ACKNOWLEDGEMENTS

We are grateful to the National Science Foundation, the U. S. Energy Research and Development Administration, and the Robert A. Welch Foundation for support of these investigations.


IV. REFERENCES

1. Brand, L., and Witholt, W., in "Methods in Enzymology" (C. H. W. Hirs, Ed.), Vol. 11, p. 776. Academic Press, Inc., New York, 1967.
2. Edelman, G. M., and McClure, W. O., Accounts Chem. Res., 1, 65 (1968).
3. Radda, G. K., in "Current Topics in Bioenergetics", p. 81, Academic Press, New York, 1971.
4. Brand, L., and Gohlke, J. R., Ann. Rev. Biochem., 41, 843 (1972).
5. J. Yguerabide, in "Methods in Enzymology", (C. H. W. Hirs, Ed.), Vol. 26C, p. 498, Academic Press, Inc., New York, 1972.
6. Stryer, L., J. Mol. Biol., 13, 482 (1965).
7. Turner, D. G., and Brand, L., Biochemistry, 7, 3381 (1968).
8. Radda, G. K., and Vanderkooi, J., Biochim. Biophys. Acta, 265, 509 (1972).
9. D. H. Haynes, and H. Staerk, J. Membrane Biol., 17, 313 (1974); D. H. Haynes, J. Membrane Biol., 17, 341 (1974).
10. Pong Sheih, Dissertation, Texas A&M University, May,1976.
11. Shaver, L. A., and Love, L. J. C., Applied Spectroscopy, 29, 485 (1975).
12. Seliskar, C. J., and Brand, L., J. Amer. Chem. Soc., 93, 5414 (1971).
13. Li, Y. H., Chan, L. M., Tyer, L., Moody, R. T., Himmel, and Hercules, D. M., J. Amer. Chem. Soc., 97, 3118 (1975).
14. Reichardt, C., Angew. Chem., lut. Ed., 4, 29 (1965); E. M. Kosower, "An Introduction to Physical Organic Chemistry", Wiley, New York, 1968.
15. Johnson, S. M., Bangham, A. D., Hill, M. W., and Koru, E. D., Biochim. Biophys. Acta, 233, 820 (1971); Huang, C. H., Biochemistry, 8, 344 (1969).

# SELECTIVE TRANSPORT OF IONS ACROSS LIQUID MEMBRANES

D. Fennell Evans, M.E. Duffey,
K.H. Lee and E.L. Cussler
Carnegie-Mellon University

## I. INTRODUCTION

Liquid membranes are potentially useful devices for the removal of contaminates from waste water streams and effluent gases and for the concentration of valuable heavy metals in mining processes. Their use in the removal of metabolic wastes such as urea or potassium ions via the gastrointestinal tract has also been suggested.

Such liquid membranes bear a superficial resemblance to the vesicles used as models of living membranes, which are the focus of much of this symposium. They are fluid; they frequently exhibit a low permittivity to ions and small molar solutes; they can be made selective to the transport of a given solute; and they can utilize some of the mechanisms believed responsible for transport across biological membranes. However, liquid membranes differ from vesicles in two major ways. First, the solvent molecules forming the liquid membrane are generally much simpler chemically than the lipids that are the structural units of biological models. Consequently, such membranes are not self-structuring. Secondly, the flux of solutes across the membrane must be $10^4$ to $10^6$ times more rapid than is generally observed in biological membranes in order to be useful in industrial separations.

In this paper, we will first describe how liquid membranes can be made to selectively concentrate ions and then discuss the membrane geometries that permit such a selective separation to be made rapidly.

## II. CONCENTRATING SOLUTES USING LIQUID MEMBRANES

The simplest type of liquid membrane consists of an organic liquid separating two aqueous phases. Generally, such a system is of limited value, since selectivity is primarily determined by the relative solubilities of the solutes in the organic phase. When the solutes are similar in size and composition, their solubilities are often similar. However, if a carrier molecule that preferentially complexes a particular solute is added to the membrane fluid, selective transport across the membrane can be obtained. In addition, if the carrier is chemically coupled to an energy source, the solute can be concentrated against its concentration gradient.

One system that illustrates such behavior employs a carboxylic acid as the carrier and can be used to concentrate sodium and other alkali metal ions(1). Two carriers that we have studied are shown in Fig. 1. They are insoluble in water, but soluble in the membrane solvent in both their acid and salt forms.

A schematic drawing of the apparatus used in studying these systems is shown in Fig. 2. The top compartment contains a 0.1N NaOH solution and the bottom compartment contains 0.1N NaCl and 0.1N HCl. Both solutions are stirred during the course of the experiment. The liquid membrane separating these two solutions is an octanol solution of the mobile carrier held between two pieces of dialysis paper.

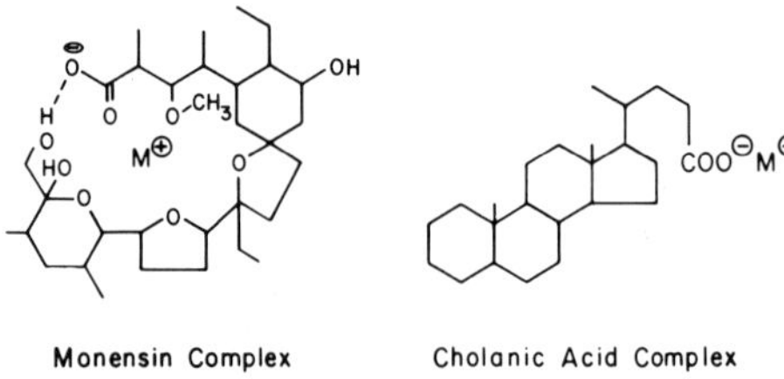

Fig. 1. Chemical structures of the carrier molecules monensin and cholanic acid.

Monensin Complex          Cholanic Acid Complex

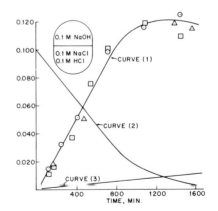

Fig. 2. Sodium transport with monensin. Curve 1 and the open circles represent the sodium transported with monensin; curve 2 is the acid transported with monensin; and curve 3 represents the sodium transported without monensin.

During the course of the experiment, the sodium ion concentration difference across the membrane rises from its initial value of zero to 0.12M(Curve 1, Fig. 2). Simultaneously,
the acid concentration difference drops to zero (Curve 2,
Fig. 2). As the acid concentration difference becomes small,
the sodium ion concentration difference reaches a maximum.
These effects are much smaller when the carrier is absent
(Curve 3, Fig. 2).

The mechanism that is most consistent with these results
in shown in Fig. 3. Carrier molecules located near the interface on the basic side of the membrane undergo the rapid
reaction shown in step 1. The complex diffuses slowly down
its concentration gradient towards the acid side of the membrane (step 2). At this second interface, the sodium ion is
replaced by a proton (step 3). The protonated carrier then
diffuses down its concentration gradient (step 4) to undergo
the reaction shown in step 1. The net result, commonly called
counter transport, is that the sodium ion is moved from left
to right against its concentration difference across the membrane while protons are moved from right to left. The chemical reaction that provides the energy for this process is
the neutralization of the acid.

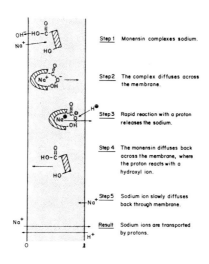

**Fig. 3.** Mechanism for sodium ion transport

We have derived a theoretical equation for describing the flux of sodium ion across the membrane(2). If one assumes that the reactions are rapid compared to the diffusion process and that the diffusion coefficient for the carrier in its two forms are equal, the following expression can be obtained:

$$J_{Na^+} = \frac{Dk_{Na^+}}{\ell}(c_{Na^+_\beta} - c_{Na^+_\alpha}) \qquad \text{(ordinary diffusion)}$$

$$+ \frac{Dk_{Na^+}}{\ell}[R(1 + k_{H^+}K_{H^+}\bar{c}_{H^+})](c_{Na^+_\beta} - c_{Na^+_\alpha})$$

(facilitated diffusion)

$$- \frac{Dk_{Na^+}}{\ell}[R\, k_{H^+}K_{H^+}\bar{c}_{Na^+}](c_{H^+_\beta} - c_{H^+_\alpha})$$

(coupled transport) [1]

where subscripts $\alpha$ and $\beta$ represent constituents on the acid and basic sides of the membrane, respectively; D is the diffusion coefficient of all species; $k_i$ is the distribution coefficient between aqueous solution and membrane; $K_i$ is the equilibrium constant for the carrier reaction; $\ell$ is the membrane thickness; and $\bar{c}_i$ is the average concentration of species "i". The quantity R is equal to:

$$K_{Na^+}\bar{c}/[(1 + k_{Na^+}K_{Na^+}c_{Na^+_\alpha} + k_{H^+}K_{H^+}c_{H^+_\alpha}) \cdot$$
$$(1 + k_{Na^+}K_{Na^+}c_{Na^+_\beta} + k_{H^+}K_{H^+}c_{H^+_\beta})]$$

where $\bar{c}$ is the average carrier concentration in the membrane. At the beginning of the experiment, the sodium flux comes not from its own concentration gradient, but from the third term in this equation. As this term goes to zero, concentrating of sodium ion ceases.

A more exact form of this equation takes into account the fact that the solvent octanol can also serve as a carrier (Fig. 2, Curve 3) and that the HCl, NaOH and NaCl are sparingly soluble in the organic solvent(1). The results of this analysis predict that a plot of $1/J_{Na^+}$ vs. $1/c_{Na^+}c_{OH^-}$ should be linear (Fig. 4). The slope and the intercept depend on the quantities given in Eq. [1]. We have independently determined these quantities and find that we can predict the slope and intercept wi___ n experimental error.

If the experiment shown in Fig. 2 is repeated using equal concentrations of sodium and potassium on each side of the membrane, a selectivity of 3:1 for $Na^+$ over $K^+$ is observed when monensin is used as the carrier. However, no selectivity is observed when cholanic acid is used as the carrier. In the former case, the interaction between monensin and the ion involves a preferential fit into the polar hole in the carrier molecule, while

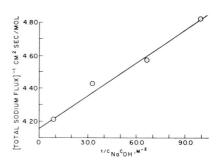

Fig. 4. Test of transport theory with the mobile carrier cholanic acid.

that between cholanic acid and the ion is mainly coulombic.

This sodium ion transport system illustrates how liquid membranes can be used to selectively concentrate ions. These results can be generalized in a number of ways. Other energy sources including oxidation-reduction reaction(3), precipitation, and common-ion effects(4) have been employed. Other types of reactions lead to co-transport(5). Other carriers allow either cations, anions or ion pairs to be moved across the membrane. On this basis, we have developed approximately twenty different liquid membrane carrier systems(5). The reason for specific focus on the sodium system is that the stoichiometry of the reactions are simple and their rates rapid so that a theoretical analysis of the problem is easy.

A system that is more complex and potentially more useful as a separation process employs the polymeric carriers, LIX64N or LIX65N (General Mills International Chemicals Co.). The chemically reactive centers in these low molecular weight polymers are either benzophenoxime or $\alpha$-hydroxy aliphatic oxime which complex with such metal ions as $Cu^{++}$ and $Ni^{++}$. The kinetics in this separation are not known, but are consistent with the following reactions based on equilibrium measurements. On the basic side of the membrane, an ammonium complex reacts with the carrier analogous to step 2 in Fig. 3:

123

$$Cu(NH_3)_4^{++} + 2OH^- + 2H_2O + \underline{2RNOH} \xrightleftharpoons{K_\beta} \underline{(RNO)_2Cu} + 4NH_4OH$$

[2]

On the acid side of the membrane, a similar reaction occurs:

$$2H^+ + \underline{(RNO)_2Cu} \xrightleftharpoons{K_\alpha} \underline{2RNOH} + Cu^{++}$$ [3]

The underlined species represent the carrier in the various forms contained within the membrane; the other species in Eq. [2] and [3] are in aqueous solution outside of the membrane. When this system is studied with the diffusion cell shown in Fig. 2 and with initially equal molar mixtures of $Cu^{++}$ and $Ni^{++}$, $Cu^{++}$ is preferentially concentrated over $Ni^{++}$ by a factor of 40 to 1. Consequently, this system could be used for the simultaneous concentration and separation of $Ni^{++}$ and $Cu^{++}$. However, the time required to achieve a separation is hours.

III. MAKING LIQUID MEMBRANE SEPARATIONS RAPID

Two membrane geometries that appear to be promising candidates for making separations faster are liquid surfactant membranes(LSM)(6) and porous polymer films (PPF). Both produce membranes which are thin and provide a large surface area per volume.

A. Liquid Surfactant Membranes are easy to prepare but difficult to quantify. We used the following preparation in the results reported here. The membrane phase consisted of Span 80, used as a membrane stabilizer; varying concentrations of LIX64N as the carrier; and one of the mixtures of the organic compounds shown in Table I. A water-in-oil emulsion was made by adding nitric or sulfuric acid to the membrane liquid while stirring the mixture for 15 minutes at 1800 rpm in a 250 ml beaker with a 1/70 h.p. motor. The stirring impeller was 1.5 inches in diameter and had 3 vanes with a 5° pitch. The acid to oil ratio was 3:1. Forty ml of the resulting emulsion was added to a 600 ml beaker containing copper and/ or nickel ions in 100 ml of the aqueous ammonia solution. This mixture was stirred at 200 rpm with a 2-inch, 3 vane impeller with a 50° pitch. When the stirring was slowed, the emulsion phases and the aqueous ammonia phase separated in a few seconds, and a sample could be withdrawn for analysis by atomic absorption spectrophotometry.

TABLE I.

THE STABILITY OF LIQUID SURFACTANT MEMBRANES

### Membrane Composition (% by wt.)

| Run | PO* | TE* | TR* | PA* | HE* | DE* | LX* | SP* |
|-----|------|------|-----|------|------|------|-----|-----|
| 1 | 46.7 | - | - | - | - | 46.7 | 4.7 | 1.9 |
| 2 | 46.7 | - | - | - | - | 46.7 | 4.7 | 1.9 |
| 3 | 62.1 | - | - | - | - | 31.1 | 4.9 | 1.9 |
| 4 | 84.5 | 8.9 | - | - | - | - | 4.7 | 1.9 |
| 5 | 79.3 | 14.5 | - | • - | - | - | 4.4 | 1.8 |
| 6 | 72.6 | 21.7 | - | - | - | - | 4.0 | 1.7 |
| 7 | 83.3 | - | 9.8 | - | - | - | 4.9 | . 2.0 |
| 8 | 79.4 | 9.8 | 4.9 | - | - | - | 3.9 | 2.0 |
| 9 | - | - | - | 80.0 | 17.5 | - | 0.5 | 2.0 |
| 10 | 93.1 | - | - | - | - | - | 4.9 | 2.0 |

### Stability (% breakup with time)

| Run | 1(min) | 3(min) | 5 | 10 | 20 | 30 | 40 |
|-----|--------|--------|------|------|------|-----|------|
| 1 | | 10.5 | 12.6 | 15.0 | 21.4 | | 31.8 |
| 2 | | 20.0 | 22.0 | 25.4 | 33.0 | | |
| 3 | 17.0 | 22.0 | 25.0 | 27.0 | | | |
| 4 | 0.0 | 0.0 | 0.0 | 0.0 | | | |
| 5 | 0.0 | 0.0 | 0.0 | 0.0 | | 0.0 | |
| 6 | 0.0 | 0.0 | 0.0 | 0.0 | | 0.0 | |
| 7 | 0.0 | 0.0 | 0.0 | 0.0 | | 0.0 | |
| 8 | 0.0 | 0.0 | 0.0 | 0.0 | | 0.0 | |
| 9 | 0.0 | | | | | 0.0 | |
| 10 | 0.0 | | 0.0 | 0.0 | | | |

*PO = Polybutene; TE = Tetrachloroethane; TR = Trimethyl-
pentane; PA = Paraffin oil; DE = Decyl alcohol; LX = LIX64N;
SP = Span 80; HE = Hexachloro-1,3-butadiene

The resulting system is effectively a water-in-oil-in-water emulsion. Inspection of this system under a microscope shows that one has both single isolated drops and clusters of drops. Such a system is geometrically similar to the phospholipid vesicles used as biological membrane models. However, the stabilization implicit in the ampholytic nature of the lipids is achieved by use of a surface active agent (Span 80). The separation rate for $Cu^{++}$ shown in Fig. 5 using liquid surfactant membranes is much greater than the rate obtained with the configuration shown in Fig. 2. As a result, we decided to study this system quantitatively to determine the transport mechanism and the key parameters responsible for its operation.

We first looked at the question of membrane stability using the HLB surfactant screening method(7). The stability of the organic mixtures shown in Table I was evaluated by replacing the acid solution inside the membrane bubbles by potassium dichromate. Since this salt is insoluble in the membrane fluid, any found in the outer phase must result from break-up of the bubbles. As can be seen, many solvent combinations resulted in no detectable membrane break-up. Span 80 worked best as a surfactant; polybutene or mineral oil were the best membrane thickeners. A more dramatic illustration of stability has been given by Li and coworkers(8) who fed LSM containing lethal doses of sodium cyanide to rats without fatality.

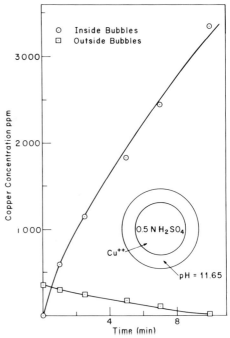

Fig. 5. Concentrating copper ions with liquid surfactant membranes

In order to study the effect of varying the concentration of carrier and chemical constituents on each side of the membrane, we first derived a theory analogous to that shown in Eq. [1] for the sodium system. We assumed that (1) a quasi-steady state concentration profile exists in the membrane at all times; (2) uncomplexed solutes are not soluble in the membrane; (3) diffusion of the complex is rate controlling; (4) the system can be idealized as single emulsion droplets; (5) the aqueous solutions on both sides of the membrane are well mixed; and (6) that concentration polarization can be ignored. On this basis of the stoichiometry shown in Eqs. [2] and [3], the flux is given by:

$$J_C = j_c A = \frac{DA}{L} \frac{\bar{c}}{2} \left[ \left( 1 - \frac{2(\sqrt{x+1} - 1)}{x} \right) - \left( 1 - \frac{2(\sqrt{y+1}-1)}{y} \right) \right]$$

$$\text{where} \quad x = \frac{8K_\beta (Cu(NH_3)_4^{++})_\beta (OH^-)_\beta^2 \bar{c}}{(NH_4OH)_\beta^4}$$

$$y = \frac{8(Cu^{++})_\alpha \bar{c}}{K_\alpha (H^+)_\alpha^2} \qquad [4]$$

The subscripts $\alpha$ and $\beta$ represent constituents on the acid and basic sides of the membrane, respectively; $\bar{c}$ is the total carrier concentration, D the diffusion coefficient for the carrier, and K and $K_\beta$ the equilibrium constants (Eqs. [3] and [2]). The$^\alpha$ total area and effective thickness of the membrane are represented by A and $\ell$, respectively. At the beginning of our experiment, $(Cu^{++})$ is zero and the second term on the right hand side of Eq. $^\alpha$[4] can be ignored. Moreover, at sufficiently high concentrations of $Cu(NH_3)_4^{++}$ and $OH^-$, the flux becomes independent of these concentrations and equals $J = DA\bar{c}/2L$.

Data providing a partial test of Eq. [4] are shown in Fig. 6. The flux does become constant at high $Cu(NH_3)_4^{++}$ and $OH^-$ concentrations. However, this flux increases by only 1.6 when the carrier concentration increases by a factor of 4, which is not consistent with Eq. [4].

There are several possible explanations for this inconsistency. The transport rate may be dominated by the kinetics of the chemical reaction at the membrane surface rather than by the diffusion of the carrier across the membrane; concentration polarization may become important; or the thickness and the surface area per volume of the membranes may change. Although the first two possibilities cannot be ruled out,

127

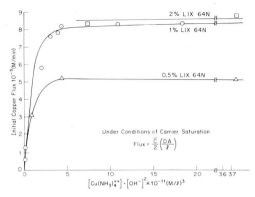

**Fig. 6.** Dependence of initial copper flux in the liquid surfactant membrane system as a function of carrier $Cu(NH_3)_4^{++}$ and $OH^-$ concentrations

direct observation indeed does show that the size and aggregation of the dispersed droplets increases considerably as the carrier concentration increases. This behavior may be general. Carriers for ions will commonly possess ampholytic properties. They must have a non-polar region in order to be solubilized in the organic mixture of the membrane and a polar region which reacts with ions. Thus they can function like emulsion stabilizers and in extreme cases will dominate the physical behavior of the LSM.

The size and aggregation of liquid membranes are also dependent on other factors. We obtained changes in droplet size with changes in the pH on either side of the membrane or with variation of the $Cu^{++}$ or $Ni^{++}$ concentrations outside the membranes. In some cases, these changes destabilized the water-in-oil-in-water emulsion. For example, with the membrane compositions shown in run 9 of Table I, an increase in the acid concentration inside the membrane from 0.5 to 2N $H_2SO_4$ resulted in a 50% decrease in the average LSM droplet size. As a result, the initial flux increased by a factor of two. Decreasing the pH outside of the membrane caused an increase in the size of droplets. As a second example, for a membrane solution consisting of 32.2% polybutene, 63.4% hexachloro-1,3-butadiene, 1.6% Span 80, and 0.8% LIX64N, and an aqueous phase of 0.5N $H_2SO_4$ inside and $12 \times 10^{-3}$ M/ℓ $Cu^{++}$ outside of the membrane, the initial flux of $Cu^{++}$ was $12 \times 10^{-5}$ M/min. When we changed only the aqueous outer solution to $6 \times 10^{-3}$ M/ℓ $Cu^{++}$ and $6 \times 10^{-3}$ M/ℓ $Ni^{++}$ the whole system became a W/O emulsion in 3 minutes. In some extreme cases, we could not even get the same size droplets using identical procedures. Details of one such case are given in Table II where the initial flux varies by a factor of 3 in identical experiments.

TABLE II

INITIAL FLUX MEASUREMENTS

| Membrane Composition | $NH_3$ Concentration | Initial pH Basic | Acid |
|---|---|---|---|
| 2% Span 80 93.1% Polybutene 4.9% LIX64N | 1.57 M/$\ell$ | 11.35 | 2.5 |

| Run | Time (min) | $Ni^{++}$ Conc. $(10^{-3}$ M/$\ell$) | Initial Flux $(10^{-6}$ M $Ni^{++}$/min) |
|---|---|---|---|
| 1 | 0 | 1.703 | 34.951 |
|   | 0.5 | 1.533 | |
|   | 1.5 | 1.226 | |
|   | 3 | 1.056 | |
|   | 6 | 0.818 | |
| 2 | 0 | 1.703 | 19.009 |
|   | 0.5 | 1.618 | |
|   | 1.5 | 1.431 | |
|   | 3 | 1.277 | |
|   | 6 | 1.022 | |
|   | 10 | 0.698 | |
| 3 | 0 | 1.703 | 12.826 |
|   | 1.5 | 1.533 | |
|   | 3 | 1.295 | |
|   | 6 | 1.030 | |
|   | 10 | 0.766 | |
|   | 15 | 0.511 | |
| 4 | 0 | 1.703 | 16.863 |
|   | 1.5 | 1.601 | |
|   | 3 | 1.277 | |
|   | 6 | 0.920 | |
|   | 10 | 0.511 | |

For the reasons cited above, we found it impossible to quantitatively study the behavior of the LIX64N $Cu^{++}$ liquid surfactant membrane system. Although this system may be an extreme example of some of the difficulties inherent in this geometry, it is difficult to envision how any system which simultaneously possesses elements of both a W/O and an O/W emulsion can be free of some of these disadvantages.

B. Porous Polymer Films are the other membrane geometry which will permit rapid concentration of $Ni^{++}$ and/or $Cu^{++}$ ions. We have studied most extensively polypropylene films which are 25 microns thick and contain pores with an average size of 0.04 microns ("Celgard", Celanese Corp.). The pores occupy approximately 45% of the film's volume. This type of material is not wetted by water or any other liquid with a surface tension greater than 35 dynes/cm$^2$. Consequently, the membrane can be filled with the organic membrane fluid and stabilized by surface tension.

In our experiments, the pores in a 200 cm$^2$ membrane sheet were filled by placing the membrane in an evacuated container and allowing the membrane fluid to flow onto it under pressure. The membrane solution consisted of varying concentrations of LIX65N in octanol saturated with water. After excess membrane solution was wiped off the surface of the polymer film, the membrane sheets were clamped between Technicon teflon dialyzer plates. The dialyzer configuration provided counter-current flow of the aqueous phases past the membrane. The area of membrane actually in contact with the aqueous phase was 30 cm$^2$ and the volume of the aqueous phase on each side of the membrane was 2.0 cc. Both the $H_2SO_4$ and the copper-ammonium hydroxide solution were saturated with octanol. Identical flow rates on both sides of the membrane were effected by a Harvard infusion pump.

Some preliminary results for this system are shown in Fig. 7. The flux was determined by collecting aliquots of the acid solution as it left the dialysis unit. The $Cu^{++}$ ion concentration was determined by atomic absorption. When the aqueous fluid velocity is greater than 0.5 cm/sec the flux becomes independent of fluid flow indicating that the transport rate is dominated by properties of the membrane. At lower fluid flow, our results are consistent with a linear dependence on fluid velocity to the 1/2 power, consistent with that expected for mass transfer in the bulk. However, more data is required to verify this behavior.

For the three lowest carrier concentrations (Curves 2,3 and 4 in Fig. 7), the flux is proportional to the carrier

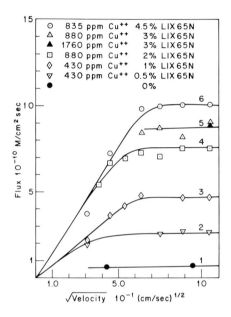

**Fig. 7.** Dependence of copper flux in the porous polymer film membrane system as a function of carrier concentration and fluid flow.

concentration when a correction for transport in the absence of carrier (Curve 1) is made. However, at the higher concentrations (Curves 5 and 6), the increase in flux is considerably less than expected. This behavior is superficially similar to that observed with the LSM and may reflect either concentration polarization or a change in mechanism.

IV.  CONCLUSIONS

We have described how carrier containing liquid membranes can selectively concentrate ions. We then compared two liquid membrane geometries, the liquid surfactant membrane and the porous polymer film.

Liquid surfactant membranes give very rapid separations and are capable of selectively concentrating specific ions. They are most easily used as a semi-batch process. However, their exact geometry can be ambiguous, making a quantitative understanding difficult.

Porous polymer films can also rapidly and selectively concentrate specific solutes. Their geometry is much more easily defined, and they can be directly used in a continuous process. Whether this geometry is superior to liquid membranes (LSM) depends on the relative speed. From the data presented here, one cannot compare the relative speed of these two systems, since the $A/\ell$ value cannot be directly obtained for the LSM.[1] However, one can compare the flux per volume of feed. For liquid surfactant membranes, this is $8 \times 10^{-7}$ moles/min $cm^3$ in our experiments. For the porous polymer films, it is $6 \times 10^{-7}$ moles/min $cm^3$. Thus the normalized flux is comparable for the two systems. Deciding between them must depend on specific applications.

V.  REFERENCES

1.  Choy, E.M., Evans, D.F., and Cussler, E.L., J. Am. Chem. Soc. 96, 7085 (1974).

2.  Cussler, E.L., Evans, D.F., and Matesich, M.A., Science 172, 377 (1971).

3.  Bdzil, J., Carlier, C.C., Frisch, H.L., Ward, W.J., and Breiter, M.W., J. Phys. Chem. 77, 896 (1973).

4.  Caracciolo, F., Cussler, E.L., and Evans, D.F., AIChE J. 21, 160 (1975).

5.  Schiffer, D.K., Choy, E.M., Evans, D.F., and Cussler, E.L., AIChE Symp. Series, Water-1974, 70, 150 (1974).

6.  Li, N.N., AIChE J. 17, 459 (1971); Ind. Eng. Chem. Process Des. Dev. 10, 215 (1971).

7.  Becher, P.,"Emulsions: Theory and Practice," 2nd Ed., Reinhold Publishing Co., New York, 1965, p 232.

8.  Science 193, 134 (1976).

---

[1] R.D. Steele and J.E. Halligan have directly determined the thickness of the LSM and found them to be orders of magnitude thicker than had been originally suggested (Separation Science 9, 299 (1974) and 10, 461 (1975)).

# THE NATURE OF PORES IN A PARTICULAR ASYMMETRIC ULTRAFILTRATION MEMBRANE AS DETERMINED BY MEASUREMENT OF ELECTROKINETIC PHENOMENA AND ELECTRON MICROSCOPY.

Mark P. Freeman
Dorr-Oliver Incorporated

## I.  INTRODUCTION

The asymmetric ultrafiltration membrane is well-known and indeed has enjoyed considerable commercial success in recent years.  The method of casting such a membrane was developed by Loeb and Sourirajan (1) and considerable serious effort has gone into understanding the thermodynamic condi- tions required for the formation of this kind of membrane (2). Nonetheless, the nature of the pores within the "skin", those pores that perform the actual size rejection, is something of an enigma as they have thus far defied definition.

Tanny and coworkers (3) have devoted considerable atten- tion to characterizing hyperfiltration membranes by streaming potential measurements, but in this case well defined tubules do not form - penetration is more of a diffusion process - and it is not clear that the same methods and theoretical framework are applicable.

It is the intent here to examine in some detail the mem- brane used in the commercial device shown in Fig. 1, by a two pronged approach.  On the one hand the membrane, which is known to reject diameters greater than 50A, is perfused with silver nitrate and sodium chloride in that order so as to deposit silver chloride in the pores.  Electron microscopy on thin sections of skin - especially when viewed in 3 dimen- sions - then does much to reveal the pore pattern within the skin.  On the other hand, we measure electroretardation (4) and first electroosmosis for solutions of a carbonate buffer. It is found that combining these measurements generate infor- mation about:  * pore size
                * pore shape (some)
                * zeta potential
                * pore frequency

133

*Fig 1. A commercial 20 ft$^2$ ultrafiltration cartridge (50A rejection) showing the leaves and plastic collection header.*

When these results are combined with information obtained from the photographs a picture of the pores starts to emerge with some clarity.

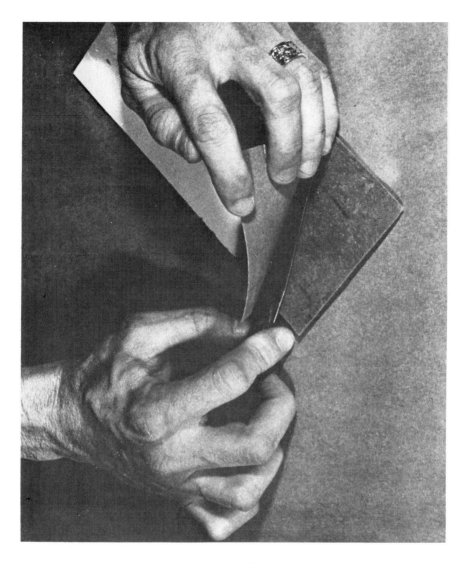

Fig. 2.  Delaminating the substrate of an ultrafiltration leaf so as to provide a test sample for the ultrafiltration cell.


II.  EXPERIMENTAL

The proprietary membrane is cast by the non-solvent method (1) from a terpolymer consisting of polyvinyl chloride, polyvinylidine chloride and polyacrylonitrile.  Samples for the electrokinetic studies were always taken from discarded commercial membranes; however, the electron micrographs were taken of membranes especially cast on a Vyon substrate.

"Staining" of the pores with silver chloride was achieved as described above and critical-point freeze dried by an outside vendor who developed the technique for the purposes of this study. Sections for transmission electron microscopy were 2-3 microns thick.

Fig. 3. Apparatus used to perform electrokinetic measurements.

The cell used for making the electrokinetic measurements was adapted from a 3" Amicon batch cell. The reservoir was fitted with a stainless steel cathode. The filtrate side was accommodated with a specially coated titanium anode and a second fluid port so that a back-up buffer solution nearly equivalent to that being filtered can be pumped through the filtrate cavity. This precludes chemical moieties generated at the anode from interfering with the membrane processes.

A special gas collection chamber allows correcting the volume of the filtrate for oxygen gas generated at the anode.

The buffer solutions were equimolar sodium carbonate-sodium bicarbonate solutions and concentrations are given in terms of $p(Na^+) = -\log_{10} [Na^+]$.

Electroretardation measurements were accomplished by averaging, say, six runs in which water was passed through the ultrafiltration membrane and six with the buffer. The datum was the differential in flux vs the water flux. Water or buffer were changed every second measurement. Electro-osmosis measurements were made in pairs by passing currents from 8 ma to 600 ma through the membrane while measuring flux of buffer solutions and alternately filtering without electricity. Polarity was such that the flux decreased when the current was on (negative zeta potential). It was verified that this reverse pumping rate was independent of filtration pressure and was taken to be a true measure of first electroosmosis.

Filtration pressure differentials were 5" Hg for all data reported here.


III.  THEORY

A.  Characterization by Electrokinetic Phenomena

The capillaries in the skin are small and the zeta potential of this membrane material in aqueous solution is appreciable. Thus conditions are optimum for large and easily measurable electrokinetic effects. The two considered here, electroretardation (4) and electroosmosis (5), require least sophistication in measurement and interpretation. Combined they will be seen to provide, at least formally, considerable information about the pores.

The most convenient entry into the analysis is through the formulation of irreversible thermodynamics (6) expressing the individual pore electrical and material currents, $\hat{I}$ and $\hat{J}$, in terms of the two generalized forces, voltage drop and pressure differential, $\Delta\phi$ and $\Delta P$:

$$\hat{I} = L_{11} \Delta\Phi + L_{12} \Delta P \tag{1}$$

$$\hat{J} = L_{21} \Delta\phi + L_{22} \Delta P \tag{2}$$

In addition we will need the reciprocity relation (8)

$$L_{12} = L_{21} \tag{3}$$

and the relationship between gegenion charge, $Q$, per unit length, $\ell$, of cylindrical capillary and the zeta potential, $\zeta$,

$$Q/\ell = -2\varepsilon\zeta \tag{4}$$

where $\varepsilon$ is the dielectric constant for water. From Poiseuille's law we know for a composite cylindrical capillary of varying radii $a_i$ and lengths $\ell_i$

$$L_{22} = \frac{\pi}{8\mu\Sigma_i \ell_i/a_i^4} \tag{5}$$

where $\mu$ is viscosity of the solution. From Ohm's law:

$$L_{11} = \frac{\pi}{\Sigma_i \ell_i/(a_i^2\lambda_i)} \tag{6}$$

where

$$\lambda_i = C_{Bi}\Lambda_{Bi} + C_{gi}\Lambda_{gi}^+ \tag{6a}$$

Note that the conductivity of the material in the pores, $\lambda$, has been expressed in terms of the bulk normal concentration, $C_B$, and equivalent conductivity $\Lambda_B$, and the pore gegenion normal concentration, $C_g$, and ionic equivalent conductivity, $\Lambda^+$. This concentration forms the most convenient connection between the microscopic processes and the phenomenological equations. It is obvious from the definition of streaming current (5) and Eqs. (1) and (2):

$$SC = \frac{I}{J}\Big|_{\Delta\phi=0} = \frac{Q^+}{\pi a^2\ell} = C_g F = \frac{L_{12}}{L_{22}} \tag{7}$$

where $F$ is the number of esu in an equivalent. This relation, perfectly consistent with other microscopic treatments of the same phenomenon, is clearly based on the model of a uniform distribution of the gegenion atmosphere throughout the capillary. This is strictly true under static equilibrium conditions only when the Debye length is much longer than the capillary radius, however the rotational character of the superimposed laminar flow will work to homogenize the contents of the capillaries so that this assumption is not as restrictive as it first appears. Furthermore, it is clear that if the capillary is not of uniform diameter, then the streaming current will be heavily weighted to reflect conditions at the "exit" (see below).

Electroretardation (4), a non-linear and less well-known electrokinetic effect, is due to the streaming potential (5)

$$\Delta\phi_s = \frac{\Delta\phi}{\Delta P}\bigg|_{I=0} \Delta P = -\frac{L_{12}}{L_{11}} \Delta P \tag{8}$$

acting on the body charges in the fluid to retard the flow. Note that a streaming potential will develop not only at the exit of the capillary but at every diameter change. Insofar as such a potential will only be effective over a distance of the order of a Debye length, the resulting electroretardation will be expected to be a composite of that occurring in every region of size change.

Substituting Eq. (8) in Eq. (2) we obtain the result

$$\hat{J} = L_{21}(\Delta\phi - \Delta\phi_s) + L_{22}(1 - Ae)\Delta P \tag{9}$$

and

$$Ae = \frac{L_{12} L_{21}}{L_{11} L_{22}} \tag{10}$$

Electroretardation for pure water may be shown to be negligible. Thus Ae may be determined for an electrolyte solution simply by comparing the solution flux through a membrane to that of pure water

$$Ae = 1 - J_s/J_w \tag{11}$$

The other measurable we will find interesting is first electroosmosis (5)

$$EO = \frac{J}{I}\bigg|_{\Delta P=0} = \frac{\hat{J}}{\hat{I}}\bigg|_{\Delta P=0} = \frac{L_{21}}{L_{11}} \tag{12}$$

We see immediately from Eqs. (7), (10) and (12) that the ratio of Ae to EO is formally equivalent to the streaming current, which should be just the concentration of the gegenion atmosphere near the "exit" of the pores. In fact, however, because Ae is a measured composite over all diameter changes we would expect the thus derived gegenion concentration to be some sort of composite

$$C_g = \frac{1}{F} \frac{Q^+}{\pi a^2 \ell} = \frac{1}{F} \frac{Ae}{J/I} \Big|_{\Delta P=0} \tag{13}$$

where $F$ is the number of esu in an equivalent. The pore average conductivity $\lambda$ now follows from Eq. (6). Note that if we take the ratio of Ae to $EO^2$ than all cross coefficients drop out using Eq. 3:

$$\Gamma = \frac{Ae}{J/I} \Big|_{\Delta P=0}^2 = \frac{L_{22}}{L_{11}} = \frac{\Sigma_i \frac{\ell_i}{a_i^2 \lambda_i}}{8 \mu \Sigma_i \frac{\ell_i}{a_i^4}} \approx \frac{a^2}{8 \mu \lambda} \tag{14}$$

If d is the weighted average (Eq. 14) <u>diameter</u> of a cylindrical capillary then

$$d = \left(\frac{32 \mu \lambda}{\Gamma}\right)^{\frac{1}{2}} \tag{15}$$

The concentration found in expression (13) may be used with the thus derived capillary dimensions to establish $Q/\ell$ and hence from (4) the $\zeta$ potential.

Finally, insofar as we have the membrane permeability to water we can again use the pore dimension to get a measure of the number of pores per unit area, n:

$$\frac{Kn}{\Delta P} = L_{22} \cdot n = \frac{\pi a^4}{8 \mu} \frac{n}{\ell} \tag{16}$$

Note that it is necessary to use a pore length estimated microscopically to get an estimate of n.

The pore "exit" must somehow be defined by the Debye length. When a pore diameter gets to be some multiple of a Debye length we will expect that the center will not have any charge and there will thus be no electroretardation. Thus in a uniformly tapered pore one might expect the electrokinetically measured diameter to decrease as the Debye length decreases with increase in ionic strength because the pore exit occurs at a progressively smaller diameter while the residual effect is due to a weighted average of all smaller diameters. The Debye length for aqueous solutions is conveniently given by (7):

$$D(A) = \frac{3.08}{I^{\frac{1}{2}}} \tag{17}$$

where the ionic strength is given by

$$I = \sum_i C_i Z_i^2 \tag{18}$$

and $Z_i$ is the number of elementary charges on each ionic moiety in solution. The smallest diameter of a capillary will clearly determine its rejection. As the Debye length gets small compared to that diameter we might well expect a radically different relationship in the behavior of electroretardation as a function of concentration.


IV.  RESULTS

A.  Electrokinetic Measurements

1.  *Electroretardation*
     Fig. 4 shows a typical set of measurements of the electroretardation coefficient. The line "faired" through the data is shaped with more confidence that *these* data appear to justify insofar as it is an easily and reliably repeatable phenomenon and has been observed many times. For clarity it was necessary to omit error bars from the figure, however typical standard deviations are shown in Table I.

2.  *Electroosmosis*
     The electroosmosis data show considerably more scatter and the points were merely averaged for purposes of computation. The crudeness of these data probably reflect needed improvements in the apparatus. Note that the deviations from a smooth curve are probably real insofar as the data at $p(Na^+)=$ 2.3 are probably the most precise.

3.  *Membrane-Pore Evaluation*
     In Table I the selected and averaged data are used to evaluate pore diameter, $\zeta$ potential and pore density. As expected the measured radii show as a general trend a decrease with increasing concentration reflecting the heterogeneous nature of the capillaries. The electrokinetically measured pore diameters are shown plotted semi-logarithmically in

Fig. 5 together with the computed Debye length for the bulk
solution.  A line representing 5.4 Debye lengths has been
drawn through the measured pore diameters and would appear
to fit at least within the precision of the data.

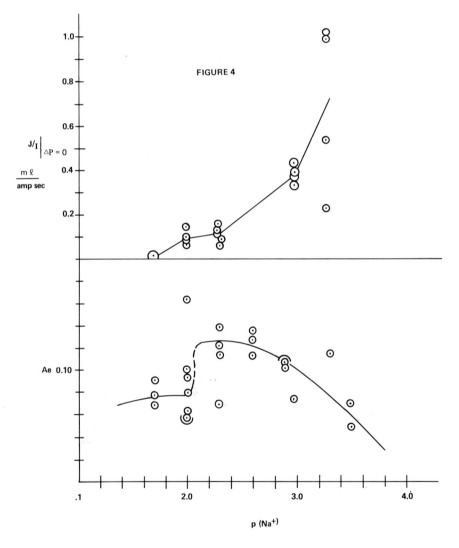

Fig. 4.  *Typical plots of Ae and EO vs $p(Na^+)$ showing
the lines used to select data for Table 1.*

TABLE 1

### Electrokinetically derived parameters
### Sodium Carbonate Buffer Solutions

| p(Na$^+$) | 1.7 | 2.0 | 2.3 | 3.0 | 3.3 |
|---|---|---|---|---|---|
| Avg. Ae | .078+.016 | .079+.010 | 0.12+.021 | 0.10+.014 | 0.076+.016 |
| Avg. EO (ml/amp sec) | 0.0078+.002 | 0.100+.016 | 0.112+.013 | 0.384+.022 | 0.697+.19 |
| Eq. Cond. | 90.5 | 96 | 103 | 114 | 121 |
| Geg. (Na$^+$) | 10.372(-5) | .819(-5) | 1.11(-5) | .27(-5) | .113(-5) |
| Pore Cond. (mho/cm) | .00710 | .00138 | .00108 | .00025 | .00012 |
| d(A) | 40.6 | 228 | 183 | 331 | 478 |
| Debye 1. (A) | 13.35 | 18.9 | 26.6 | 59.6 | 84.2 |
| ζ pot. (mv) | 7.48 | 18.65 | 16.28 | 12.96 | 11.3 |

Typical water permeability $= 0.0112 \times 10^{-6}$ cm/sec//dyne/cm$^2$

n/l (assuming d = 305A) $= 0.5 \times 10^{14}$ cm$^{-3}$

estimated n(l = $10^{-5}$ cm) $= 5$ holes/(μ)$^2$

Note that the abrupt change in Ae at p(Na$^+$)=2.05 is at a bulk Debye length of about 20A. This must somehow relate to the minimum capillary dimension assumed to be 50A and thus the membrane rejection. The quantitative connection is not yet clear and must probably await the accumulation of some experience. The "bench" in the measurement down to p(Na$^+$)=2 is probably real and reflects a powerful weighting of the larger capillaries through length as well as diameter (Eq. 14). An average capillary diameter for p(Na$^+$)=2 and above is 305A. One interpretation of the data would be a uniform diameter of ~300A through out most of the membrane "skin".

## B. Electron Micrographs

### 1. Scanning Electron Micrographs

Fig. 6 shows a series of progressively magnified scanning electron micrographs of the same membrane section. Note that the understructure is very typical of this sort of membrane. We are however only interested in the skin, which is much thinner than it first appears insofar as the microtome turns the edge slightly. Looking closely at the higher magnifications, bits of the true edge show up indicating a thickness of the order of 0.1 micron. Note that at the highest magnification the lamellar composition of the "skin" is starting to show.

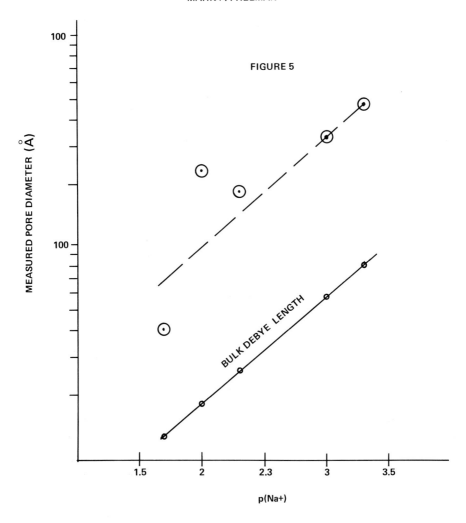

Fig. 5. A plot of electrokinetically measured pore diameter vs p(Na⁺) demonstrating the important influence of the Debye length.

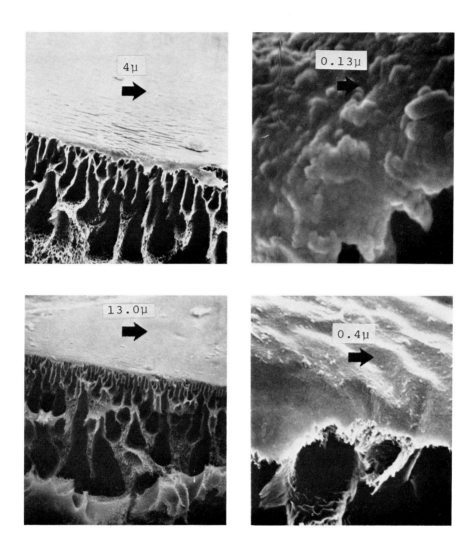

Fig. 6. A series of progressively "closer" SEM photos of the same membrane cross section clearly showing understructure and skin.

## 2. *Transmission Electron Micrographs*

Figs. 7, 8, and 9 are "stained" sections of the skin about 2-3μ thick. They all clearly show the lamellar (though non-crystalline) lumps of which the membrane is composed.

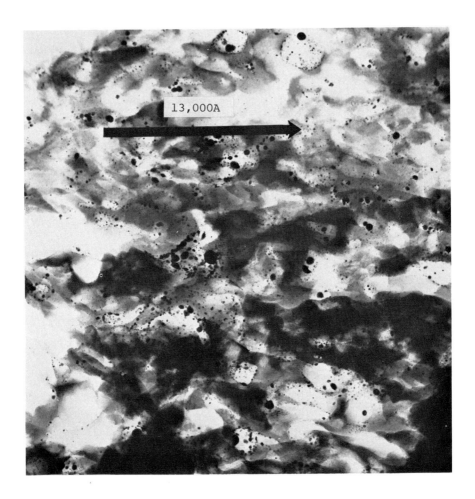
13,000A

*Fig. 7. A low power TFM showing the multiplicity of water passages through the membrane and the heterogeneous nature of the plastic skin.*

Recall that Table 1 indicated a passage density of 5 holes/micron$^2$ or 5 for each half micron of Fig. 7. This would seem to be quite reasonable.

Fig. 8 is taken at high magnification deep within the skin and shows several water passages of diameter about 300A. Indeed a number of passages of about this size may be seen in Fig. 7 as well. It is thus probably the preferred size for most of the distance across the skin which is consistent with the interpretation of the electrokinetic data.

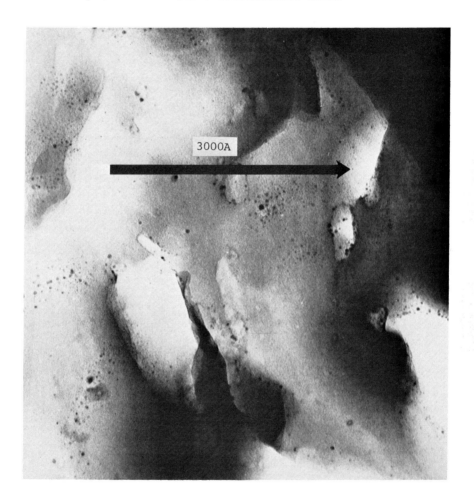

*Fig. 8. A high magnification TEM from deep within the skin showing several 300A water passages.*

Fig. 9 is perhaps the most interesting. It is taken near the surface and shows a few channels 50-70A in diameter but mainly wide bands of these "canonical" channels 1500-2000A wide. It is interesting to note that such a composite "slot"

50A thick and 1500A wide would have the same area as a 300A
cylindrical capillary.  Thus we may propose two hypothetical
explanations which may in fact be equivalent for the rejection of these membranes.  On the one hand, the membrane is
crossed by 300A tubules which flatten to 50A wide slots near
the true interface.  On the other, a multiplicity of 50A
tubes penetrate the surface (mostly in "bands") and then
coalesce into 300A tubules.

Fig. 9 has been presented as a stereo pair so as to
illustrate a ubiquitous truth.  Namely, the water passages
of whatever size never penetrate the lamellae, but rather
follow the curved spaces between the lamellae.  This figure
may be seen in three dimensions without a machine if an 8"
card is placed between your eyes and between the pair.  A 3D
visualization will appear about 30 seconds after you accommodate your eyes so as to make the pictures merge.

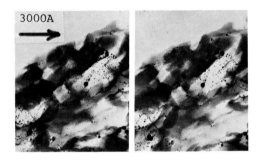

*Fig. 9.  A stereo pair of TEM's showing the multiplicity
of five channels near the surfaces and how they follow the
lamella interstices.*

V.   DISCUSSION AND SUMMARY

The pore image now emerging is that a multiplicity of
water channels form near the surface.  They often run parallel
to form a sort of slot, but sometimes separately, and they
always follow the interstices between the lamellae of which
the membrane is composed.  These capillaries are between 50
and 70A in diameter, reflecting the rejection of the membrane.
As they go deeper in the membrane they would appear to merge
in channels about 300A in diameter that occur at a frequency
of about 5 pores/$\mu^2$ (which implies the smaller channels must
have been present at an average frequency of about 125 pores/
$\mu^2$).

It is clear that more work needs to be done in coming to grips with the complex averaging used in interpreting the electrokinetic data. Nevertheless, these electrokinetic data are highly reproducible as such measurements go. Especially on these membranes reproducibility of this order in any measurement is difficult because of the high probability of plugging of some of the pores by, e.g., bacteria. It seems likely that use of these combined measurements to characterize membranes will develop as a potent diagnostic tool.

VI.  REFERENCES

1.  S. Loeb and S. Sourirajan, Advan. in Chem. ser. 38, 117 (1962).

2.  H. Strathmann, K. Kock, P. Amar and R. W. Baker, Desalination 16, 179 (1975).

3.  G. Tanny and O. Kedem, J. Colloid Interface Sci. 51, 177 (1975).

4.  M. Dosoudil, "Influence of the Electrical Double Layer on the Filtration Process", 10th Magdeburg Conference on Mechanical Separation, 1972.

5.  D. G. Miller, "Thermodynamics of Irreversible Processes", in Chemical Reviews, 60, 15 (1960).

6.  G. N. Lewis, M. Randall, K. S. Pitzer and L. Burns, "Thermodynamics". McGraw Hill, New York, 1961.

7.  R. F. Fowler and E. A. Guggenheim, "Statistical Thermodynamics", p. 393 Cambridge Press, Cambridge, (3rd Ed.) 1952.

8.  D. G. Miller, "Transport Phenomena in Fluids", H. J. Hanley, ed., p. 426 Marcel Dekkar Inc., New York, 1969.

# THE ATPASES ACTIVITY ON IONIC TRANSPORT IN ARTIFICIAL BIOMEMBRANES

by M. DELMOTTE and J. CHANU
*Laboratoire de Thermodynamique des Milieux Ioniques et Biologiques – Université Paris VII – Tour 33-43 – E2 – Equipe de Recherche Associée au C.N.R.S. N° 370*

## I. THERMODYNAMICS RECALLS

When we describe phenomena such as the transport of bio-chemical or chemical components through biological membranes, two ways are adopted :

*a.* At a microscopic* level, we look in priority at the structure, the configuration and the conformation of all par-ticules from ions up to very complicated molecules such as macromolecules or membrane proteins.

*b.* As long as the transport processus through the mem-branes never affect only one *isolated* molecule but involve a large number of particules, Thermodynamics plays here an es-sential role.

Clearly, these transport processus are due to cooperative phenomena specifically of thermodynamic character.

It is well known that all transport phenomena belong es-pecially to non-equilibrium thermodynamics (1). Furthermore, biological systems are always *open* ones and the corresponding laws cannot figure within the framework of classic thermody-namics.

In this paper, we shall focus on the analysis of membra-nes taken as physical chemical systems exhibiting *chemical reactions and diffusion.* For each component (subscript $i$) the evolution equation can be written as

$$\frac{\partial C_i}{\partial t} = F_i (\{C_j\}) - \nabla J_i \qquad (1)$$

$(i,j = 1,\ldots N)$ if $N$ is the number of the implicated components.

---

*Microscopic level means atom or simple molecule level.

In the divergence term, the flow $J_i$ figures and is given by the so-called Fick diffusion law

$$J_i = - D_{ii} \nabla C_i \qquad (2)$$

so that in the concentration gradient range when coefficient $D_{ii}$ is constant we can write (2)

$$\frac{\partial C_i}{\partial t} = F_i \left( \{ C_j \} \right) + D_{ii} \nabla^2 C_i \qquad (3)$$

On the other hand, the mathematical structure of the source term $F_i$ is not a linear function of concentration $C_j$. Indeed, $F_i$ represents the overall rate of change of $i$ as a result of the chemical reactions[*]

In living systems, we observe most of the time, stationary states characterized by a null left hand side of the equation 3. Such stationary states are found in resolving the non linear equation 3 in this case, as

$$0 = F_i \left( \{ C_j \} \right) + D_{ii} \nabla^2 C_i \qquad (3')$$

Therefore, we obtain most of the times multi-solutions to which correspond multi-stationary states. The solution multiplicity of such equation is one of the main character of situations belonging to far from equilibrium thermodynamics.

To avoid misunderstanding, it should be stressed that the equilibrium state is a very particular steady state, obtained when both right hand side terms of equation 3' are null separately. The corresponding solution is thus unique.

Another fundamental result has been established in advance thermodynamics which concern the stability of stationary states. All stationary states under the *critical point* which ends up the thermodynamic branch, are stable with respect to small pertubation. Beyond this threshold, instability can occur. A general criteria of the above range of matter stability has been found by the Brussels school of Prof. I. Prigogine. In the present case without convective movement it has been deduced from entropy production given by

$$P = \sum_\alpha J_\alpha X_\alpha \qquad (4)$$

---

[*] In general reaction rates are complicated polynomials or rational fractions of component concentrations.

(where $J_\alpha$ is flow or chemical rate and $X_\alpha$ the conjugated
forces gradient or chemical affinities)
that the stability of steady states is assured if, and only if,

$$P\ (\delta S)\ =\ \Sigma_\alpha\ \delta J_\alpha\ \delta X_\alpha > 0\ . \qquad (5)$$

The critical threshold is reached when

$$\Sigma_\alpha\ \delta J_\alpha\ \delta X_\alpha\ =\ 0\ . \qquad (5')$$

When we analyse the membrane behavior from the point of view
of the evolution equation 1 the stability test of each stationary states is given by condition 5 .


## II. THEORITICAL APPROACH OF EXPERIMENTS

A few years ago, one of us M.D. has been able to explain
some specific phenomena taking place in a sac membrane of
outer rod segment of retina cells. He found three stationary
states made of two different stable states separated by an
unstable one (3). Other works lead to similar results (4).

A very large programme was thus elaborated with the
ultim goal to study *in situ* the living membrane behavior. Of
course, such programme suppose many intermediate steps to
analyse in great details the delicate mechanisms of transport.

The present work is the first try of this programme and
reproduce the fundamental physico-chemical conditions characterizing the living membrane activity. In fact, experiments
analyse chemical kinetic transport of ion species when ATPase
specific enzymes are placed in a permanent concentration gradient in such a way to obtain real steady non-equilibrium
conditions.

To realize such experiments we reticulate or absorb ATP-
ase enzymes inside a support-membrane. This latter was placed
in diffusion reaction cells in such a way to limit the concentration gradient zone. To counterbalance the permanent flows
of ionic components driving by Fick forces, a very sophistica-
ted device was elaborated to maintain concentration gradient
clamp. As initial information, we obtain directly the transport flows of different components and precise data on chemical activity of the. membrane.


## III. EXPERIMENTAL APPARATUS

We only give here a general scheme of the. apparatus for

the detailed description is given elsewhere (5) .

The diffusion reaction cells was a bicompartments altu-glassmade (roughly 280 cm³). The active artificial membrane loaded with ATPase enzymes separates the top from the bottom compartments. In each of them a set of very sensitive detectors, controlling parameters (temperature, salt concentration, pH, conductivity, membrane potential,...) drove the servo-mechanisms (see Fig. 1.)

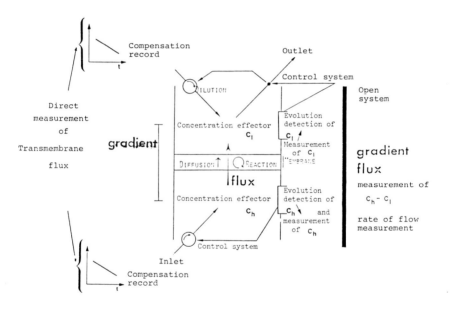

*Fig. 1. Theoritical scheme of the experimental device.*

These servo-mechanisms maintain the physico-chemical properties of both media *fixed in advance* and *clamped* in each compartment all through experiments. Notice the temperature was strictly kept *isothermal* everywhere, less $5.10^{-3}$°C. Moreover, a magnetic transmission stirring system assumed a real homogeneization of both solutions. It worked at constant, adjustable and controlled speed. The goal of homogeneization was both :

- to optimize the work of servo-mechanisms
- to standardize with accuracy the geometry of the concentration gradient in which the experimented membrane and its two limit layers were.

For ionic concentrations the detectors driving servo-mechanisms were specific crystalline-type electrodes, sensitive to $Na^+$, $K^+$ and $Ca^{++}$. The signals given by null indicator and after amplification drove volumetric pumps injecting a higher concentrated solution in the high concentration compartment and a more dilute solution in the other one. Whatever the spontaneous fluctuations, such a device assumes a compensation of the normal evolution of both solutions communicating through the membrane. With consumption measured in the reservoirs of servo-mechanisms two direct estimates of flux were possible.

One of the most delicate points of the experiment was the membrane preparation (6). Many kinds of support-membranes were used according to the nature of ATPase enzymes. For example, the experimented enzyme active on $Na^+$ and $K^+$ transport was entrapped into an inert cellulose nitrate membrane. Such membranes have no osmotic properties.


IV. RESULTS

The main part of the results presented in this paper is devoted to the ATPase enzyme activity on $Na^+$ and $K^+$ ion transport.

Previously, it was necessary to check up the expected capacities of the apparatus. It was made using a free enzyme membrane. As expected, when constant clamped gradient exists, we observe a strictly constant flow through it. In other words, in the entire range of explored concentration gradient, Onsager-type linear relation between flows and forces was extensively verified.

The first experimented enzyme chosen was an ATPase extracted from pork brain and known to be active on sodium and potassium ions. The concentrated and dilute solution in close contact with the membrane contained buffered ATP at constant concentration ($2.5$ $m$ $mol/l$  $Na_3H$ $ATP$ + $2.5$ $m$ $mol/l$ $Na_4$ $ATP$) so that $pH$ was about $6.8 - 6.9$. The concentration gradient only concerns $Na^+$, concentration $K^+$ was kept constant.

Fig. 2 and 3 exhibit the characteristic nature of the answers obtained through the experiment.

a. It is quite striking that the transport phenomena with respect to ionic gradient is not linear : the Fick law is

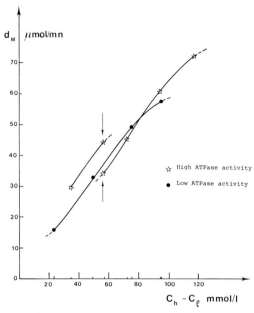

Fig. 2. *Total apparent fluxes concentration gradient dependent. Notice the different dynamic permeabilities of both zones.*

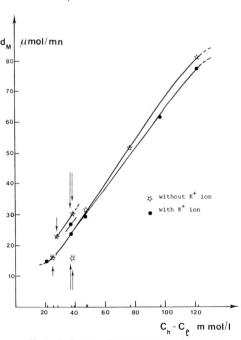

Fig. 3. *Total apparent fluxes concentration gradient dependent. Notice the high dynamic permeability for low gradients.*

thus no longer applicable.

*b*. The modification of the flux aspect with regard to pure diffusion fluxes, leads to conclude to the enzyme interference from which a transport of chemical nature.

*c*. The ATP enzyme activity increases so much at times that it is an additional prouf the enzyme "proceeds" in a chemical way during the transport.

*d*. Experimental data in which the flow is plotted against the concentration gradients are particularly striking in the membrane behavior in presence of chemical transport. We notice in particular in Fig. 2. two different steady states corresponding to a gradient around $60m$ $mol/l$.


V. DISCUSSION (7)

The preceding results and the experimental works in progress lead to new outlooks.

Clearly, transport phenomena as seen before, show we cannot use traditional theoretical laws. For, these flows are more or less linear function of concentration gradient alone or potential and concentration gradient both, like in Fick equation or Nernst-Planck equation respectively. In fact, the experiments have obviously demonstrated the chemical interference of enzymes on ionic transport through the artificial membrane in stationary states. Beyond the primary organic relationship from equation 3' we get another relationship as phenomenological law.

$$J = k\nabla C + k'\nabla\phi + G(\{C_j\}) \qquad (6)$$

where $k$ and $k'$ are constant easily accessible and $G(\{C_j\})$ describes the chemical contribution. This latter expression is never a linear function of $C_j$ and $\nabla C_j$.

The previous experimental work done with great accuracy has clearly demonstrated the traditional concepts of permeability outcoming from Fick or Nernst-Planck equation is insufficient. If the permeability concept is preserved, it is necessary to stress on its chemical reaction and concentration gradient dependent character. In other words, we can propose in the mean time the term "dynamic permeability" which describes the real behavior of diffusion reaction interference transports more realistically.

VI.   REFERENCES

1.  Glansdorff, P.,and Prigogine, I., "Structure, Stabilité
    et Fluctuations", Masson, Paris, 1971. English Ed. Wiley,
    New-York, 1971.
2.  Nicolis, G. in "Advances in Chemical Physics" (G. Niolis
    and R. Lefever Eds.) Vol. 19, p. 29. (1975).
    Prigogine, I. and Lefever, R. in "Advances in Chemical
    Physics" (G. Niolis and R. Lefever Eds) Vol. 19, p. 1,
    (1975)
3.  Delmotte, M. Vision Res., 10, 671 (1970).
4.  Blumenthal, R., Lefever, R. and Changeux, J.P. J. Memb.
    Biol. 2, 351 ( 1970).
5.  Delmotte, M. Thèse d'Etat - Université de Paris 1975.
6.  Thomas, D., Broun, G., and Sélégny, E. Biochimie, 54, 229
    (1972).
7.  Delmotte, M. and Chanu, J. in "Proceeding of 3rd Symposium
    on Bioelectrochemistry", Jülich 1975 (to appear in 1976)

# INTERACTIONS OF A PHYSICAL MODEL OF THE MEMBRANE/CYTOPLASM INTERFACE WITH COCARCINOGENS

A.W. Horton, C.K. Butts and A.R. Schuff
*University of Oregon Health Sciences Center*

*The interface between buffered aqueous KCl and a solution of selected membrane lipids in decahydronaphthalene provides a useful model for gaining an understanding of the interactions of cocarcinogenic hydrocarbons, alcohols, and esters with biological membranes. We measure the rate of transfer of an anionic dyestuff probe, sodium p-(p-anilinophenylazo)-benzene sulfonate (Orange IV) from the aqueous phase of this system to the lipid micelles in the organic phase under conditions of emulsification.*

*The kinetics, using phosphatidyl cholines as the primary lipid component, are dependent on the fluidity of the micelles. Using egg phosphatidyl choline, (2 double bonds/molecule) at 21°, a comparatively high rate of transfer of the dyestuff probe is observed, equilibrium being reached in approximately 10 minutes with 70% of the probe in the bulk phase micelles. In contrast, the kinetics are drastically retarded by inclusion of lysolecithins in the organic phase.*

*The combination of cholesterol with these 2 phospholipids produces a micelle of intermediate fluidity that is selectively and quantitatively sensitive (i.e., the kinetics of transfer of the dyestuff probe) to those n-alkanes that are promoters of carcinogenesis.*

In the past few months increasing recognition has been given to the contribution of cofactors in the progression of latent cancer (1). We are all exposed to chemical and physical carcinogens, even in the most remote wilderness areas. Cancer epidemiologists are now stressing the importance of personal habits of diet, smoking, and drinking in determining which of us are at risk from these environmental factors (2). Fortunately, at least for men, they haven't indited sex or song yet as cocarcinogens.

By definition a cocarcinogen is any factor unable to induce cancer itself, that increases the incidence of tumors, shortens the latent period, or reduces the dosage of a carcin-

ogen required to induce cancer. In this paper we shall focus
our attention on non-carcinogenic hydrocarbons that meet all
of these criteria. Human encounters with these promoters have
resulted in skin and scrotal cancers in industry (3,4) and
respiratory cancers in smokers (5,6). Currently concern about
the presence of cocarcinogenic n-alkanes in petroleum-based
single-cell protein is holding up approval for manufacture in
Europe (7).

A standard protocol will illustrate the biological results of exposure to these cocarcinogenic hydrocarbons (4).
Two out of three groups of mice are given a single initiating
topical dose of a classical polycyclic aromatic carcinogen.
Starting two weeks later, one of these two groups and the
third, previously unexposed, group are treated repeatedly
(twice each week for 30-40 weeks) with the promoting hydrocarbon.

In Figure 1 are shown the resulting tumor incidences in
such a test of the promoting activity of n-dodecane. Exposure to 80 applications of dodecane alone resulted in no
tumors at all. The single application of carcinogen resulted
in tumors in less than 10% of the mice, but, as evidenced by
the results on the 3rd group, the single dose had initiated
potential neoplasms in all the mice. All that was required
to complete their progression to cancer was chronic exposure
to the promoting effects of the long chain hydrocarbon.

Figure 1
INFLUENCE OF LONG-CHAIN CATALYSTS
ON INCIDENCE OF TUMORS

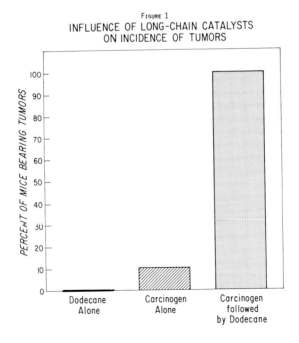

Among the n-alkanes this cocarcinogenic activity begins with the $C_{10}$ homologue. As shown in Figure 2 it increases with increasing chain length to $C_{20}$. Little information is available on the higher paraffins, partly due to their low solubility. The $n-C_{28}H_{58}$ is about as active a promoter as $n-C_{20}H_{42}$ on a equimolar basis (8).

Figure 2

Cocarcinogenic Activity of n-Alkanes versus Their
Effects on Phospholipid-Cholesterol Micelles

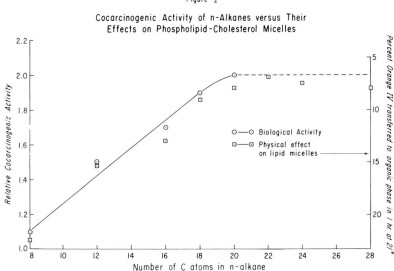

Higher cocarcinogenic activity is shown by alicyclic and aromatic hydrocarbons of similar molecular length. Peak activity is reached at 18-20 carbon atoms in most series (4,9).

The oil/water partition coefficient is so high for all of these compounds that it is apparent that they would be localized in the lipoprotein membranes of the cells into which they are absorbed. The major lipid components of these membranes are phospholipids and cholesterol. Hence research on the mechanism of the cellular action of cocarcinogenic hydrocarbons has focussed on their physical interactions with these components.

In 1970 it was found that the effects of n-alkanes on the physical properties of phospholipid micelles in decalin solution could be probed by the kinetics of uptake of a water-soluble anionic dyestuff from an adjacent aqueous phase (10). A useful relationship developed from the observation that the rate of transfer to egg phosphatidyl choline micelles was retarded strongly by those alkanes known to be cocarcinogenic. Predictions of biological activity for certain hydrocarbons not adequately tested previously have been borne out fully by subsequent animal experiments (8).

In 1975 we began to investigate the possible cocarcinogenic activity of isoprenoid hydrocarbons present in various
fossil fuels, cigarette smoke, etc., using this physical
model system as a screening tool.  The purified egg phosphatidyl choline behaved quite differently from that obtained
from the same supplier 5 years previously.  It became apparent that a single spot on thin layer chromatography on silica
gel is not an adequate criterion for reproducibility.

Changes in the acyl groups have altered micellar structure in hydrocarbon solution in a direction such that the rate
of transfer of the dyestuff probe is so rapid that equilibrium
is reached in approximately 10 minutes, as compared to 2 hours
for the previous lecithin. Only those n-alkanes above $C_{20}$ now
affect the rate significantly.

Attempts to find a synthetic lecithin with the properties
of the 1970 egg lecithin have not met with any success.  None
has been found to date with comparable sensitivity to promoting hydrocarbons.

Research in the interim on the effects of cholesterol and
other steroids on the kinetics of transfer of the dyestuff
probe provided a clue that has lead to a solution to the problem (11).  Following the suggestion of Chapman that the role
of cholesterol in biological membranes may be to control the
fluidity that would vary widely with changes in phospholipid
acyl groups (12), we have developed a mixed lipid micelle that
seems to be quantitatively sensitive to those alkanes that are
cocarcinogenic.


MATERIALS AND METHODS

The solvent cis- and trans-decahydronaphthalene (decalin),
practical grade; (Matheson, Coleman and Bell, Norwood, Ohio)
was purified by chromatography on activated silica gel, as
were the following alkanes: n-dodecane, n-hexadecane, n-octadecane, n-eicosane, n-docosane, n-tetracosane, and n-octaco-
sane (Humphrey Chemical Co., North Haven, Conn.), and pristane
(2,6,10,14-tetramethylpentadecane; Aldrich Chemical Co., Mil-
waukee, Wis.).  This procedure eliminated any hydroperoxide or
aromatic impurities.  Tert-dodecylbenzene (gift from Oronite
Chemical Co., Richmond, Calif.) was used as received.

Phospholipids included egg phosphatidyl choline (GIB-EL)
(Grand Island Biologicals, Grand Island, N.Y.) dried in a
vacuum desiccator for at least three days; egg lecithin
(Schwarz/Mann, Orangeburg, N.Y.), and lysolecithin (Grand
Island Biologicals, by hydrolysis of GIB-EL with the lipase
from snake venom) used as received.  Both of the Grand Island
Biologicals materials produced a single spot on thin layer
chromatography on silica gel, developed with 65:25:4 chloro-

form:methanol:water. USP cholesterol (Merck, Rahway, New Jersey) was purified by triple recrystallization from methanol.

The crude egg lecithin from Schwarz/Mann (SM-EL) contained lysolecithin and sphingomyelin components as well as a variety of other phospholipids. The material was chromatographed on activated silica gel, using chloroform and methanol as developing solvents (1:1 v:v). A fraction 8-11 was prepared containing the lecithin, lysolecithin, and sphingomyelin components (approximately 3:4:2 parts), plus approximately 10% of methyl esters formed on the column. A sample of purified egg phosphatidyl choline, Sample C (one spot on thin layer chromatography) was prepared from the SM-EL by chromatography on activated alumina using methanol:water (95:5 v:v), followed by chromatography on activated silica gel using chloroform: methanol (1:1 v:v).

The "C" lecithin was hydrolyzed with porcine pancreatic lipase (steapsin; ICN Pharmaceuticals, Cleveland, Ohio) according to the method of Walker et al (13), with the following modifications: Powdered enzyme, equal in weight to the "C" used, was added to the reaction mixture, half the lipase being added before sonication, and half added between sonication and shaking at 37°. Sodium deoxycholate (Difco Laboratories, Detroit, Michigan) was substituted for the ox bile extract. One drop of 30% bovine albumin (Hyland, Costa Mesa, Calif.) was added for each 25 mg lecithin used. The crude hydrolysate was chromatographed over activated silica gel using chloroform and methanol.

By this procedure, fraction "LL-1", containing lysolecithin as the major component with lecithin as the minor, was prepared by partial hydrolysis of C. Fraction "LL-4", a lysolecithin that yields only 1 spot on TLC, was also prepared from C.

Unsaturation of the various lipid fractions was determined by bromination by the procedure of Trappe (14), but carried out in the dark to avoid substitution reactions. The standard used was cholesterol, having one double bond/molecule.

The physical model of the membrane/cytoplasm interface was set up by a procedure previously described (10). A buffered aqueous solution, ionic strength 0.15, of the recrystallized (2 times from water and once from 95% ethanol) dyestuff, sodium p-(p-anilinophenylazo)-benzene sulfonate (Matheson, Coleman and Bell), Orange IV, was mixed with an equal volume of a decalin solution of the phospholipids with or without cholesterol. (With cholesterol alone in the organic phase, adsorption of Orange IV to the interface occurs but no transfer to the cholesterol micelles in the bulk phase.) After selected periods of time the emulsion was broken by centrifugation and distribution of dyestuff determined by absorption spectrophotometry, the absorbance at $\lambda_{max}$ at approximately

430nm used to analyze the clarified phases.


RESULTS

The following experiments were carried out in the course
of development of a lipid mixture with specificity for and
quantitative sensitivity to those alkanes that are cocarcinogenic. The active hydrocarbons retard the initial rate of
transfer of the ionic probe to the lipid micelles, but do not
change the ultimate equilibrium distribution. Hence the oper-
ational criterion was followed that the time required to reach
equilibrium should be several times that required for manipulation of two-phase systems, including the time for breaking
the emulsion formed during mixing.

Egg phosphatidyl cholines from various sources show wide
variations in the degree of unsaturation of their acyl groups.
For example GIB-EL showed 1.7 double bonds per molecule,
whereas sample C only 1.1 double bonds/molecule.

Neither the GIB lysolecithin nor sample LL-4 added any
bromine at all. Apparently both enzymes, the lipase from
snake venom used commercially and the pancreatic lipase in
our experiments, were selective for hydrolysis of unsaturated
acyl groups. It was found that the pancreatic lipase readily
hydrolyzed sample C, but had no effect on the GIB-EL under
these experimental conditions.

It was found previously that the rate of transfer of the
dyestuff by the GIB-EL could be retarded markedly by the inclusion of cholesterol in the lipid mixture (11). However
these mixed micelles did not prove to be sensitive specifically to the cocarcinogenic alkanes.

Similarly lysolecithin added to the lecithin micelles is
found to retard the rate of transfer of the dyestuff probe in
this system. The percent of Orange IV transferred to the
organic phase micelles in 1 hour at $21°$ by GIB-EL lecithin
plus the GIB lysolecithin (2:1 molar ratio of lecithin:lyso-
lecithin) was reduced to 4%, compared to 70% with the lecithin
alone. The lysolecithin, LL-4, produced a parallel response
but a much higher concentration of this lysolecithin was required for the same reduction in the rate of transfer of the
dyestuff probe.

In contrast to its retarding effect when added to lecithin micelles cholesterol has the opposite effect when added
to mixed micelles of lecithin and lysolecithin. Addition of
cholesterol to the GIB lysolecithin-lecithin mixture approximately doubled the rate of transfer of the dyestuff probe.
In the case of the mixture with the lysolecithin LL-4 the rate
was approximately tripled.

The comparative control of the micellar structure by the GIB lysolecithin (produced by the lipase from snake venom) was notable. Even a 100/1 molar ratio of cholesterol to phospholipid in the organic phase was unable to alter the rates to a very large degree on an absolute basis. (Note that it is not inferred that the ratio of cholesterol to phospholipid in the mixed micelle is 100/1. The bulk decalin solution contains cholesterol micelles as well as phospholipid micelles.)

Although these mixed micelles containing purified lysolecithins did not prove to be quantitatively satisfactory for estimating the activity of cocarcinogenic alkanes added to the organic phase, the greater effect of cholesterol on those containing LL-4 pointed towards a successful development.

The effects of the two cocarcinogens n-dodecane and n-octadecane were used in the following experiment for a quantitative comparison of sensitivity of different lipid mixtures for screening purposes. As shown in Figure 2 and in previous publications (8) octadecane is a more active cocarcinogen than the $C_{12}$ homologue. Thus in mixtures with decalin the same activity is demonstrated biologically by 20% (by volume) of the $C_{18}$ homologue as for 40% of the $C_{12}$.

Satisfactory mixed lipid micelles to make this distinction are produced by the combination of the highly unsaturated GIB-EL, cholesterol, and the fraction LL-1. Transport experiments were carried out using a mixture of these phospholipids at 0.225 mg/ml total, containing 40% (by weight) of the fraction LL-1, plus 30mM cholesterol in the decalin-alkane phase. This produces a mixed micelle that is quantitatively sensitive to the action of dodecane and octadecane at various dilutions. Essentially the same effects of these alkanes were observed using an alternate combination of phospholipids provided by 75% (by weight) fraction 8-11 plus 25% GIB-EL.

The results for the latter combination of phospholipids plus 30mM cholesterol are shown in Figure 3 (lower curve). In contrast it is seen that the addition of a lower concentration of cholesterol (10mM) to the 8-11 plus GIB-EL mixture did not produce sufficient control of micellar behavior. A random pattern was obtained with the various dilutions of the normal alkanes. (See next page for Figure 3.)

This difference in the effects of 10 and 30mM cholesterol was noted previously (11) in equilibrium experiments using 0.3mM GIB-EL (21 hours mixing). A transition was observed at 22mM cholesterol. Below this concentration the emulsion formed was easily broken and almost all of the dyestuff probe was found in the lipid micelles. However, at higher concentrations, an abrupt decrease in the transfer of interfacially adsorbed dye to the clarified organic phase was seen.

Figure 3

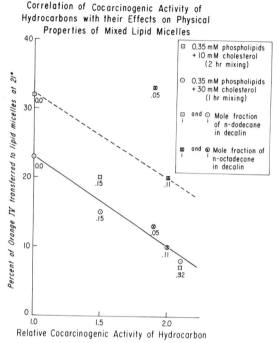

Correlation of Cocarcinogenic Activity of
Hydrocarbons with their Effects on Physical
Properties of Mixed Lipid Micelles

Relative Cocarcinogenic Activity of Hydrocarbon

The fact that better sensitivity is obtained with the chromatographic fractions, LL-1 or 8-11, containing a series of related phospholipids, than with the mixtures of purified lecithin and lysolecithin suggests that the micellar structure for best sensitivity requires intermediate components difficult to separate from the major lecithin and lysolecithin components. Since the isolated pure lysolecithin proved to be completely saturated it seems possible that the minor components essential for this purpose are unsaturated lysolecithins present in much lower concentrations.

Finally the application of such mixed lipid micelles to the screening of a variety of normal alkanes is shown in Figure 2. In this case the lecithin-lysolecithin used was fraction 0 11. A total phospholipid concentration of 0.225 mg/ml (75% fraction 8-11 plus 25% GIB-EL) was combined with 30mM cholesterol. The various alkanes indicated by the squares in Figure 2 were tested at 20% (by volume) for their effects on the rate of transfer of Orange IV to the organic phase micelles at 21°. The correlation offers further confirmation of the previous indication that the $C_{28}$ homologue is about as active as the $C_{18}$ and $C_{20}$ alkanes.

It is notable that the lecithin-lysolecithin fractions
that produce the most selective interaction with the promoting hydrocarbons are those whose response to cholesterol produced the same rate of transfer of the dyestuff probe as was
obtained when cholesterol was added to the original purified
egg lecithin itself, approximately 20-23% transferred in
1 hour at 21°. The lecithin-lysolecithin combinations in
which cholesterol was less effective in raising the respon-
siveness to this ionic probe were not as quantitatively sensitive to the promoting alkanes.

The same system using fraction 8-11 has also been shown
to be sensitive to the active aromatic cocarcinogen, tert-
dodecylbenzene (DDB). Compared to a control rate (0% DDB in
the organic phase) of transfer of the probe of 20% in 1 hour,
the inclusion of 10,20,30 and 40% DDB yielded 13,12,8, and
7% respectively.

An example of the application to isoprenoid hydrocarbons
not as yet tested biologically were the results using 2,6,10,-
14-tetramethylpentadecane (pristane). At a concentration of
20% by volume, pristane retarded the Orange IV transfer from
18% (control) to 6% in 1 hour at 21°.


DISCUSSION

The lipoprotein membranes of cells are conceived of as
semi-permeable barriers (low permeability to salts but
moderately permeable to water). The region of the membrane
within 10 Å of the aqueous interface is an anisotropic mesophase. The four ringed structure of cholesterol plays an important part in the structure of this region of the cell envelope and the nuclear membrane, acting to control their
fluidity (12).

The physical model of membrane/cytoplasm interface, described in this paper responds to the moderating effects of
cholesterol similarly. The very fluid micelles of egg phosphatidyl choline are condensed by cholesterol, whereas the
steroid alters the mixed micelles of lecithin plus lysolecithin in the opposite direction, i.e. towards greater fluidity.

The action of promoting hydrocarbons may be looked upon
as a competition interfering with this control by cholesterol.
The conformation of proteins that are either adsorbed on or
are essential components of these biological membranes may
thereby be significantly altered by the absorption of cocarcinogenic hydrocarbons.

Membrane-bound enzymes that may be significantly affected
by such hydrocarbon interactions include the important adenylate cyclase responsible for the production of cyclic AMP from
ATP at the inner surface of the cell envelope. It has been

shown that the specific interactions of the cell envelope with hormones that stimulate the adenylate cyclase enzyme are dependent upon the presence of the phospholipids of the membrane (15).

Other important surface interactions with serum proteins of the immunosurveillance systems may be affected by this competition of promoting hydrocarbons with cholesterol in the cell membrane. Important interactions of the membrane with serum complement and the proteins responsible for its activation may be affected adversely.

ACKNOWLEDGMENT

We wish to acknowledge the valued contribution of W.H. Perman in the work on the mixed phospholipid micelles. This research was supported in part by Chesebrough-Pond's, Inc., through the Medical Research Foundation of Oregon.

REFERENCES

1. Weisburger, J., Chem. & Eng. News, p.18, March 22, 1976.
2. Hammond, E.C., in American Cancer Society Science Writers Seminar, St. Petersburg, Florida, March 29, 1976.
3. Hendricks, N.V., Berry, C.M., Lione, J.G., and Thorpe, J.J., Arch. Ind. Hlth. 19, 524 (1959).
4. Horton, A.W., Denman, D.T., Trosset, R.P., Cancer Res. 17, 758 (1957).
5. Doll, R., Cancer Res. 23, 1613 (1963).
6. Gelpi, E., and Oro, J., J. Chrom. Sci. 8, 210 (1970).
7. "Petro-protein - Stalled on Purity Issue", reported in Chem. Week, p.79, April 28, 1976.
8. Horton, A.W., Eshleman, D.N., Schuff, A.R., and Perman, W.H., J. Natl. Cancer Inst. 56, 387 (1976).
9. Saffiotti, U., Shubik, P., Natl. Cancer Inst. Monogr. 10, 489 (1963).
10. Horton, A.W., McClure, D.W., Biochim. Biophys. Acta 225, 248 (1971).
11. Horton, A.W., Schuff, A.R. and McClure, D.W., Biochim. Biophys. Acta 318, 225 (1973).
12. Chapman, D., in "Biological Membranes; Physical Fact and Function" (Chapman, D., Ed.), Vol. 1, p.151. Academic Press, London, 1968.
13. Walker, R.W., Barakat, H. and Hung, J.G.C., Lipids 5, 684 (1970).
14. Trappe, W., Biochem. Zeit. 296, 180 (1938).
15. Pfeuffer, T., Helmreich, E.J., in "Cell Surface Receptors" (Nicolson, G.L., Raftery, M.A., Rodbell, M., Fox, C.F., Eds.), Vol. 8. Alan R. Liss, Inc., New York, 1976.

# HINDERED DIFFUSION OF MACROMOLECULES
## IN TRACK-ETCHED MEMBRANES

Jaime Wong H. and John A. Quinn
*Department of Chemical and Biochemical Engineering*
*University of Pennsylvania*
*Philadelphia, Pennsylvania 19174*

*Hindered diffusion in porous membrane transport refers to the composite effect of two separate phenomena, both of which retard the movement of molecules through pores of comparable dimension. The two retarding mechanisms are: (1) solute molecules entering a pore encounter steric restrictions which, effectively, decrease the pore concentration, and (2) the translating molecule within a pore is subject to enhanced viscous drag forces as a result of the proximity of the pore walls. We have made independent measurement of these two effects using bovine serum albumin (BSA) as diffusing solute in uniform micropores formed (in mica membranes) by the track-etch process. Steric partitioning is determined by measuring pore conductivity to small electrolytes in the presence and absence of BSA; the decrease in pore conductance can be related to the concentration of non-conducting particles, i.e. BSA molecules, in the pore. Net transport of BSA through the pores is measured by using a fluorescent tag (fluorescamine) to monitor the steady-state flux of BSA through a membrane of known pore characteristics. Transport measurements are compared with theoretical predictions based on hydrodynamic models of the diffusion process. Preliminary measurements have shown that BSA adsorbs as a monolayer that reduces the effective pore radius by 70Å. At λ=0.19, BSA shows a hindrance effect twice that predicted by the centerline approximation given by the Renkin equation.*

## I.  INTRODUCTION

### Background

Hindered diffusion in porous membranes refers to the restricted mobility encountered in permeation through fluid-
filled pores wherein the size of the translating solute molecule is some non-negligible fraction of the pore diameter.
The hindrance is the composite effect of two, separate pore-
permeant interactions:  one "thermodynamic" and the other "hydrodynamic" in origin.  The thermodynamic effect refers to the
equilibrium interaction between the permeating species and the
pore wall; for rigid, non-polar molecules this effect is equivalent to a geometric partitioning arising from the "excluded"
volume, i.e. the solute-free region near the pore wall – that
region inaccessible by virtue of the finite dimension of the
solute.  Its measure is the partition coefficient between the
exterior phase and the pore interior.  The hydrodynamic factor is accounted for by the increased drag (increased with
respect to the unconfined fluid) experienced by the permeating
molecule – its movement is hindered by the proximity of the
pore wall.

The theoretical basis for interpreting hindered diffusion in pores is examined in Bean (1) and in Anderson and
Quinn (2).  The latter paper reviews the literature through
1973 and presents the hydrodynamic models considered in this
work.  For reference, we list below (and display in Figure 1)
the analytical relations which characterize various aspects of
the hindered diffusion phenomenon.  For a spherical molecule
(radius $a_s$) in a cylindrical pore (open radius $r_o$) the steric
partition coefficient is simply:

$$\Phi = (1-\lambda)^2 \ , \ \lambda = a_s/r_o \tag{1}$$

The local ratio of pore-to-bulk solute diffusivity, F, for a
spherical molecule moving along the centerline of a cylindrical pore is given by the hydrodynamic expression (3):

$$F(\lambda,0) \cong 1 - 2.104\lambda + 2.089\lambda^2 - 0.948\lambda^5 \tag{2}$$

The combination of Eqns. (1) and (2) leads to an equation frequently referred to as the Renkin equation:

$$R(\lambda,0) = (1-\lambda)^2 \cdot R(\lambda,0) \tag{3}$$

where $R(\lambda,0)$ is the hindered diffusivity ratio using the centerline drag expression.  For Brownian particles, all avail-

able radial positions are equally probable and, therefore, the appropriate particle drag should reflect an average value of $F(\lambda,\beta)$ where $\beta$ is the dimensionless particle position, $r/r_o$, and r is the distance from the pore centerline to the center of the particle. A general expression for $F(\lambda,\beta)$ is not available and Anderson and Quinn utilize available hydrodynamic results to calculate the appropriate averaged value of $R(\lambda,\beta)$ valid for $\lambda<0.1$. This relation is also shown in Figure 1.

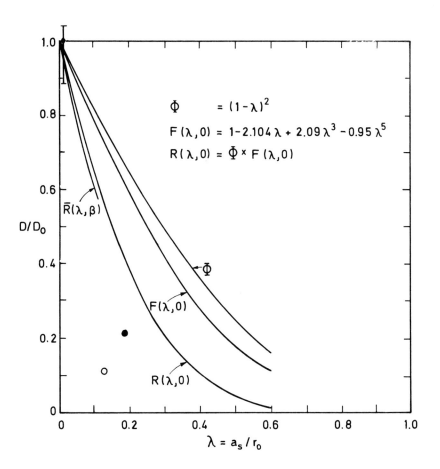

Fig. 1.  Restricted diffusion as a function of the ratio of solute to pore size.  $R(\lambda,0)$ is the centerline approximation given by the Renkin equation; $R(\lambda,\beta)$ is its radially-averaged counterpart. Pore diffusivities at $\lambda{\sim}0$ and $0.2$ are shown by (●). The open circle (○) represents the hindered diffusion value obtained if BSA adsorption is ignored.

## Previous Work

A major experimental advance which has opened the way to a critical evaluation of the hydrodynamic theory of hindered diffusion is the track-etch process for the fabrication of model membranes with uniform pores of near molecular dimension. The experimental techniques involved in the preparation and characterization of these membranes is summarized in Bean et al. (4,5) and Quinn et al. (6).

Track-etched membranes were first used in hindered dif- fusion studies by Beck and Schultz (7,8). They examined a wide variety of solutes, extending from solute-solvent size ratios near unity to macromolecules, the largest of which (ri- bonuclease) has a radius of 21.6Å. Their results show rea- sonable agreement with the Renkin equation for $\lambda$ as large as 0.40. Their results are somewhat obscured by the fact that they observed a pore-size distribution with their track-etched membranes. Also, they did not measure solute partitioning and there is evidence that irreversible adsorption was occuring in their measurements with ribonuclease (9).

In addition to Beck and Schultz's work with track-etched membranes, there have been several other investigations of hindered diffusion using different porous materials as model pores. Two recent examples are the papers of Satterfield, Colton and Pitcher (9) and Colton, Satterfield and Lai (10). They have measured restricted diffusion in silica-alumina bead catalysts (median pore diameter 32Å) and in finely porous glass (pore radii of 25 to 476Å). In the former work they found no evidence for steric exclusion and they claim no agree- ment with the hydrodynamic models summarized above. In their 1975 paper (porous glass) they report results which are in marked contrast with their earlier findings – here they ob- served partition coefficients less than unity and in general agreement with those predicted on the basis of steric exclu- sion. Although there is some ambiguity in their determination of average pore size, their limited data on protein compare favorably with the Renkin equation when the pore radius is reduced by an amount equivalent to that of an assumed adsorbed monolayer.

## Experimental Plan

As a critical test for the hydrodynamic theory of hin- dered diffusion the present study was aimed at obtaining data on both diffusion and equilibrium partitioning using a well-characterized solute macromolecule in an ideal, isoporous membrane. To satisfy these conditions we studied the diffusion of bovine serum albumin (BSA) through track-etched mica membranes. Experimental details and preliminary results are presented in this paper.

## II.  EXPERIMENTS

### Track-Etched Membranes

As noted above, details on membrane preparation and characterization are well established (4,5,6). Here we mention particular features which enter into the hindered diffusion study. The membranes are made from sheets of muscovite mica 7μ thick. Discrete pores are created by collimated irradiation with massive fission fragments emitted from a $Cf^{252}$ source. The damage tracks produced by the fission fragments are susceptible to etching by aqueous hydrofluoric acid. The pores are rhombic (60°) in cross-section and all pore axes are within 16° of perpendicularity with the membrane face.

Pore number, length, and size are closely controlled and precisely measured. Number density is determined by irradiation time and calibrated by etching pores to micron-size and counting under a light microscope. Pore length is equal to membrane thickness within an average factor of 1.02, the spread reflecting the collimation angle. Uniformity of individual pores has been confirmed by scanning and transmission electron microscopy (4,11). Pore size is determined by the etching process (5). Pore radius measurement is discussed below; all pore radii are expressed in terms of a circle of equivalent area. The smallest pores that we have attempted to work with have an equivalent diameter of approximately 50Å.

### Materials and Procedure

Bovine serum albumin (BSA) was selected as the permeating species because it met our requirements for a compact, rigid solute "particle" of known structure and diffusivity. The molecule resembles a prolate ellipsoid with "hydrodynamic radius" of 36Å.[1]  For BSA diffusing between two reservoirs at

[1]The hydrodynamic radius for BSA was derived from its diffusivity at infinite dilution, using the Stokes-Einstein equation. As a basis for comparison, assuming the intrinsic viscosity for BSA is 3.7 $cm^3/g$ [see Table 23-1 in Tanford (12)], the calculated BSA radius is 34Å.

25 g BSA/liter against pure solvent, the diffusion coefficient at 25°C and pH 8, ionic strength >0.05 is estimated to be $D_o$ = $6.9 \pm .1 \times 10^{-7}$ cm$^2$/sec (13,14,15). The various effects of pH, protein concentration, and ionic strength are reported by Raj and Flygare (14). Protein solutions were prepared by dissolving bovine serum albumin in 0.05M phosphate buffer solutions at pH 8, to which 0.005% sodium ethylmercurithiosalicylate (Thimerosal) was added to suppress bacterial and microbial growth. (This concoction will be referred to as the solvent for BSA). The protein was obtained from Pentex (Research Products, Miles Laboratories, Inc., Elkhart, IN) as a lyophilized protein monomer standard which had been reacted with L-cystine to block free sulfhydryl groups and stabilize against dimerization (16), and then separated from its oligomeric fraction by gel permeation chromatography (17,18).

A recently introduced fluorescent tag for peptides, proteins and primary amines was used to analyze simply and precisely small samples of 1-20 ppm BSA. Fluorescamine or 4-phenylspiro [furan-2(3H), 1'-phtalan]-3,3' dione (Roche Diagnostics, Hoffman LaRoche Inc., Nutley, N.J.) is a small organic molecule that reacts with the primary amino groups in proteins to form a strongly fluorescent tagged molecule (19). Excess reagent is quenched in aqueous media, and both fluorescamine and its hydrolysis products are non-fluorescent. The reaction time for tagging is 10-100 times faster than the rate of destruction of tag by hydrolysis (20).

BSA sample solutions are assayed for their protein content by reacting 10 ml of the sample with an aliquot of the fluorescent dye at a molar ratio of tag to protein of at least 150:1. The tagged sample exhibits maximum fluorescence at pH 8, with the excitation wavelength at 3900Å and the emission wavelength above 4750Å. Analysis was performed with an Aminco Fluoro-Colorimeter (American Instrument Co., Silver Spring, Md.).

The mica samples used had an irradiated circular spot 1.27 cm across and were clamped between two Lucite half-cells, each with 6 ml fluid capacity. Once in place, the membranes were never exposed to an air-liquid interface to prevent surfactants from depositing onto and into the pores. Instead, liquid solutions were changed and samples taken by positive displacement from each half-cell with several volumes of the desired new solution, using a syringe and Teflon tubing, as represented in Figure 2. The inlet solutions were filtered through sterile Millipore syringe filters (Millipore Corporation, Bedford, MA). These filters are made from thin cellulose ester polymer films - type TF, surfactant free - with

174

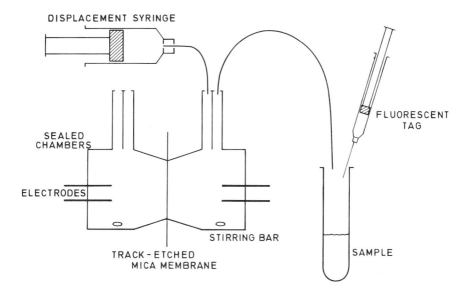

*Fig. 2.   Schematic representations of the diffusion cell showing the sampling and fluorescence tagging procedure.*

0.22μ pores.  The half-cells were stirred with Teflon-coated 7x2 mm magnetic spin-bars, spinning at 400-500 RPM.  A pair of platinum/platinum black wire electrodes was mounted in each half-cell for *in situ* monitoring of the electrolyte concentration surrounding each side of the membrane.

## Calibration Experiments

Three types of experiments are necessary to fully describe hindered diffusion, and three general classes of membranes are required for these experiments.  The different experiment-membrane combinations are depicted in Table 1.  Class II and III membranes have porosities $f \sim 10^{-3}$, differing only in the range of pore sizes.  Class I membranes have small pores and porosities in the order of $f \sim 10^{-6}$.  (Here, the porosity, f, is defined as the fraction open area per unit of irradiated membrane area; i.e. membrane pore density multiplied by the cross-sectional area of each pore.)

The use of f = 0.001 porosity membranes was dictated by the desire to maximize fluxes through the membranes and to minimize the effects of pore overlap and convective boundary layer resistances without sacrificing strength of the mica substrate.  At 0.001 porosity, over 99.5% of the pores are

TABLE 1

*Different Experiments Considered, and the Three Classes of Membranes Used*

| Membranes | Experiments | | |
|---|---|---|---|
| | pore conductance | electrolyte diffusion | BSA diffusion |
| I.<br>$f=10^{-6}$<br>$50<r_o<1000\text{Å}$ | pore size, adsorption, partitioning | pore size, adsorption | |
| II.<br>$f=10^{-3}$<br>$100<r_o<500\text{Å}$ | (pore size) | pore size, adsorption, boundary layer resistance | hindered diffusion |
| III.<br>$f=10^{-3}$<br>$r_o=5000\text{Å}$ | (pore size) | pore size, boundary layer resistance | free diffusion |

singlets, and the membrane comprises 90 to 95% of the resistance to mass transfer, the boundary layer and the pore mouth effects making up the remainder (21,22).

In addition to net protein diffusion, used to obtain hindered and free pore diffusivities for BSA, two other types of experiments are called for: net electrolyte (KCl) diffusion serves to characterize the membrane pore size at $f>10^{-4}$, and the extramembrane resistance of the boundary films; and membrane conductance defines pore size, adsorption, and partitioning when membranes are less porous than $f=10^{-4}$. This conductimetric method is explained in detail by Quinn et al.(6), who derived a simple expression relating the pore specific conductance of an electrolyte solution to its bulk conductivity and the number and size of those pores in a heterogeneous membrane.

Net protein diffusion experiments were carried out in class II membranes having about $10^8$ pores/cm$^2$ in the 100–500Å range, and in class III membranes with larger pores (5000Å) and lower pore densities ($10^5$ pores/cm$^2$). Fluxes obtained were on the order of $10^{-11}$–$10^{-13}$ mol/hr for each g/l concentration difference of protein across the membrane. Results obtained from class II membranes provide the hindered diffusion data, while class III membranes produced information on the free diffusivity of protein. The equivalent pore radii in these membranes were determined by electrolyte diffusion using a 1.0M KCl solution diffusing into 0.1M KCl. The integral (time-space averaged) diffusivity at 25°C was taken as 1.86 x $10^{-5}$ $cm^2$/sec (23). Diffusion was monitored by measuring the conductivity of the downstream solution at 3 minute intervals.

To measure steric partitioning in class I membranes, the pore walls were pre-coated with either high molecular weight polysaccharides (dextrans) or cetyltrimethyl ammonium chloride to preclude protein adsorption. Then, pore conductivity to small electrolytes was measured in the presence and absence of BSA. The decrease in pore conductance can be related to the concentration of non-conducting particles (BSA molecules) in the pore, and hence a measure of the distribution of protein between pore and bulk is obtained. Adsorption of BSA can be measured either by electrolyte diffusion in class II membranes or by the conductimetric method with class I membranes. The conductimetric method repeats the procedure used to measure partitioning, but this time, the pore walls were not pre-coated. BSA adsorbed onto the mica walls, and the reduced pore conductivity was a measure of the combined effects of adsorption and partitioning.

Electrolyte diffusion was also used to measure BSA adsorption onto the pore walls, and the added resistance of the unstirred film at the mouth of each pore. Using a membrane with known pore radius, one can predict the expected solute flux under given conditions. The difference between this expected value and the actual experimental flux was attributed to resistance to diffusion outside the membrane pores (21,22). As mentioned before, this resistance was less than 5-10% of the total resistance to diffusion and is accounted for in the analysis. Osmotic effects due to the near semipermeable nature of these micropores leads to coupled retarding effects (back flow) which are no more than 6-10% of the total flux (private communication from J.L. Anderson, October 1975). Streaming potential arises from net charges on the surface of the protein molecules, at pH conditions other than isoelectric, being transported across the membrane. This electrically coupled flow is negligible if the ionic strength is above 0.1M, for the conditions of our experiments (24).

## III.   RESULTS AND DISCUSSION

Partitioning measurements with BSA in pre-coated micro-pores show close agreement (11) with the geometric exclusion predicted by Eqn. 1. Measurements of protein diffusion in large pores verified the literature value for unrestricted diffusivity of BSA as shown by the data point at $D/D_o = 1$ in Figure 1. The rather large amount of scatter in the free diffusivity data represents what is believed to be convective transport through these large pores. The data for hindered diffusion was more precise because solution streaming, related to $r_o^4$ by the Poiseuille equation, is greatly reduced in the small pores. Figure 3 represents the sequence of experiments performed with a single (class II) membrane. The results of protein diffusion may be compared directly with those of electrolyte diffusion and membrane pore conductance on a single diagram such as Figure 3 showing predicted pore radii from each experiment. The pore radii predicted from BSA flux data assumes the unrestricted diffusivity applies in the pores. Electrical resistance measurements, R, to characterize the pore radii by the conductimetric method, were not dependable because the membrane resistance was too small (about the same magnitude as

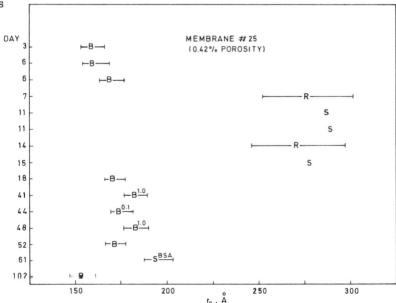

*Fig. 3.   Equivalent pore radii predicted for membrane #25.   R represents results from conductimetric data.   Pore radii predicted from electrolyte diffusion in the presence and absence of BSA are shown by $S^{BSA}$ and S.   B represents net protein diffusion and the superscripts 0.1 and 1.0 indicate the addition of 0.1M and 1.0M KCl.*

the bulk solution in each half-cell). (The symbols R, S, and B are explained in Figure 3.) Electrolyte diffusion, S, on the other hand, predicted very precisely the actual pore radius. By adding BSA to the salt solutions, $S^{BSA}$, the effective pore radius reduced by a monolayer of adsorbed protein, was determined. This layer reduced the effective pore radius by 70Å.[2]

Protein diffusion, B, was carried out both in buffer solutions and with added support electrolyte (KCl) at different concentrations. $B^{0.1}$ and $B^{1.0}$ used 0.1M and 1.0M KCl respectively to increase the ionic strength. No difference was observed in the net flux of protein. The closed circle in Figure 1 near $D/D_o$ = 0.19 represents a measure of hindered diffusion in a pore with an adsorbed monolayer of BSA on the pore walls. The open circle was computed from the protein flux data assuming no adsorption; that is, the pore radius as given by S alone was used. It serves to point out the magnitude of the error that can be introduced by neglecting to consider solute adsorption effects.


Conclusion

Track-etch mica membranes are ideally suited for studies of various phenomena associated with transport through molecular-sized pores. The parameters which govern transport are amenable to analysis by several complementary calibrating techniques. As a check on accuracy, free diffusion of BSA was measured in large pores and the results show agreement within 10% of literature values. For accurate measurement of hindered diffusion, precautions must be taken to account for adsorption, extramembrane resistance, electrostatic effects, osmosis and convective transport. Preliminary measurements have shown that BSA adsorbs as a monolayer that reduces the effective pore radius by 70Å. At $\lambda$ = 0.19, BSA shows a hindrance effect twice that predicted by the centerline approximation given by the Renkin equation.


*Acknowledgements* - This work was supported in part by grants from the National Science Foundation and the General Electric Foundation.

---

[2]A single layer of close-packed hard spheres 36Å in radius gives an excluded volume equivalent to a hard shell 60Å thick.

## IV. REFERENCES

1. Bean, C.P., in "Membranes: a Series of Advances" (G. Eisenman, Ed.), Vol. 1. Marcel Dekker, Inc., New York, 1972.
2. Anderson, J.L., and Quinn, J.A., Biophys. J. 14, 130 (1974).
3. Happel, J., and Brenner, H., "Low Reynolds Number Hydrodynamics," Chapter 7. Prentice-Hall, Inc., New Jersey, 1965.
4. Bean, C.P., "Characterization of Cellulose Acetate Membranes and Ultrathin Films for Reverse Osmosis," Office of Saline Water, Research and Development Progress Report No. 465.
5. Bean, C.P., Doyle, M.V., and Entine, G., J. Appl. Phys. 41, 1454 (1970).
6. Quinn, J.A., Anderson, J.L., Ho, W.S., and Petzny, W.J., Biophys. J. 12, 990 (1972).
7. Beck, R.E., and Schultz, J.S., Science 170, 1302 (1970).
8. Beck, R.E., and Schultz, J.S., Biochim. Biophys. Acta 255, 273 (1972).
9. Satterfield, C.N., Colton, C.K., and Pitcher, W.H., Jr., AIChE J. 19, 628 (1973).
10. Colton, C.K., Satterfield, C.N., and Lai, C.J., AIChE J. 21, 289 (1975).
11. White, I., and Quinn, J.A., unpublished results, 1974.
12. Tanford, C., "Physical Chemistry of Macromolecules," John Wiley & Sons, Inc., New York, 1961.
13. Keller, K.H., Canales, E.R., and Yum, S.I., J. Phys. Chem. 75, 379 (1971).
14. Raj, T., and Flygare, W.H., Biochemistry 13, 3336 (1974).
15. Lin, C.H., Ph.D. thesis, Univ. Illinois, Urbana, 1971.
16. Isles, T.E., and Jocelyn, P.C., Biochem. J. 88, 84 (1963).
17. Andersson, L-O., Biochim. Biophys. Acta 117, 115 (1966).
18. Pedersen, K.O., Arch. Biochem. Biophys., Suppl. 1 157 (1962).
19. Böhlen, P., Stein, S., Dairman, W., and Udenfriend, S., Arch. Biochem. Biophys. 155, 213 (1973).
20. Udenfriend, S., Stein, S., Böhlen, P., Dairman, W., Leimgruber, W., and Weigele, M., Science 178, 871 (1972).
21. Malone, D.M., and Anderson, J.L., "Boundary Layer Resistance for Heterogeneous Membranes," submitted for publication, 1975.
22. Wong H., J., M.S. Thesis, Univ. Pennsylvania, Philadelphia 1974.
23. Robinson, R.A., and Stokes, R.H., "Electrolyte Solutions," 2nd ed., rev. Butterworths & Co. Ltd., 1959.
24. Anderson, J.L., and Quinn, J.A., J. Colloid Interface Sci. 40, 273 (1972).

# A SEDIMENTATION TECHNIQUE
## FOR THE STUDY OF
## DEMINERALIZATION AND REMINERALIZATION
## OF HARD TISSUES IN VITRO

D.B. Boyer and F.R. Eirich
University of Iowa, Polytechnic Institute of New York

The density changes of particles of bovine dentin were determined by way of finding their frictional coefficients during sedimentation experiments in columns filled with EDTA or water. Stokes' law was obeyed by particles smaller than 500 microns. The relative decrease in sedimentation velocity which can now be related quantitatively to the relative decrease in the density thus becomes a reliable indicator of progressing demineralization or remineralization.

The frictional coefficients were found to increase with particle size at constant particle density, and to decrease with decreasing density at constant particle size for particles larger than about 500 microns. Mathematical techniques were developed to determine the frictional coefficient as a function of velocity, so that the densities of large particles could also be quantitatively determined. This was accomplished by the use of two dimensionless parameters, the drag coefficient and the Reynolds number.

It was hypothesized that the divergence of the frictional coefficients of large particles of high and low density is caused by the more rapid development of turbulance around the denser, faster falling particles.

## I. INTRODUCTION

Chan and Eirich (1) introduced a new method of studying the rates of demineralization and remineralization of hard tissue in which changes in density are continually measured by monitoring the velocity of fall of a particle through a column filled with a decalcifying agent or a calcifying solution. They used this technique to study the decalcification of bovine cortical bone in EDTA and lactic acid and demonstrated a high correlation between the mineral content of a bone particle and its velocity of fall. The calcification of rachitic rat cartilage was also investigated.

Evaluating their data, the authors noted an inconsistency which appeared important to resolve if the technique is to be used to determine the particle density quantitatively from its sedimentation velocity. If the velocity changes were functions of the changes in particle density only, the relative decrease in particle velocities would have to be exactly equal to the relative decreases in densities. Instead discrepencies of up to 20% were found. The possibility that a change in the frictional coefficient of the falling particle with change in density may be responsible for the discrepancy lead us to study the frictional coefficient as a function of of particle size and density. To this end, several modifications in the procedure and data treatment have been employed in the present study.

## II. MATERIALS AND METHODS

The apparatus and its application are the same as is described by Chan and Eirich (1). The apparatus consists of glass columns about two feet long, and a stand for mounting and rotating them in an air thermostat. The glass columns were made by ACE Glass Inc., Vineland, N. J. Near one end, each column has a restricted channel to center the fall of the particle within the column. The solution in the column is stirred by a micro stirring bar while readings are not being taken. The bar is held in place by an external magnet during readings.

The decalcification experiments were conducted by filling the column with 220 ml of 0.5 M EDTA adjusted to pH 7.4 and equilibrating in the thermostat to 37°C. The hard tissue sample was introduced into the column through a teflon valve with a disposable pipet. The time of fall of the particle between bench marks 30.5 cm apart was periodically measured with a stopwatch. The columns were airfree and airtight and did not have to be opened for sampling. Calcification experiments were conducted in the same manner by filling the columns with calcifying solution.

Bovine dentin was selected for use throughout the investigation. The lower jaws of calves were obtained from an abattoir the day of slaughter. The teeth were removed immediately, cleaned of soft tissue, and frozen until use. The roots were the source of dentin.

The dentin was found by chemical analysis to contain 24.4% calcium, 11.1% phosphorus, and 3.52% carbonate. The Ca/P ratio was 1.71 when uncorrected for calcium bound by carbonate and phosphate bound by protein. The densities of wet dentin and demineralized dentin (collagen) were found by pyknometry to be 1.90 and 1.10 respectively. The density of demineralized dentin in 0.5 M EDTA was 1.15. The density of dentin could not be measured in EDTA but was calculated to be 1.92 from the volume fraction composition of dentin in terms of the solution, mineral, and organic portions.

Samples of bovine root dentin for the sedimentation studies of demineralization and remineralization were cut with a scalpel under a dissection microscope into the shape of small cubes. The dimensions of the cubes were measured to the nearest 5 microns with a planimeter by methodically rotating the sample so that the height and width of each face were measured at least twice. All the values for one sample were combined into an average edge length, d. The density determined from the measured sample weight and the calculated volume $(d^3)$ and the density measured by pyknometry correlated well.

The availability of size data makes possible the determination of the frictional coefficients of the falling particles, as well as the expression of rate data in terms of the geometric surface area. The particle diameters were generally about 0.03 cm. because this proved to be the smallest size particle to cut and to observe. Small sized particles were desirable in order to shorten the time required for demineralization or remineralization.


III.  RESULTS

Samples of dentin of various sizes were demineralized in the sedimentation columns in 0.5 M EDTA. Typical data are shown in Figure 1. The initial velocity and the final velocity, when no further demineralization took place, were used to calculate the frictional coefficients in EDTA. Knowing the density of a particle, $\rho$, its volume, V, its velocity, u, and the density of the fluid, $\rho_s$, the frictional coefficient, f, can be calculated using Equation 1. The gravitational constant is g. This equation was derived by setting the sum of the forces acting on the particle equal to zero for the case of constant velocity.

$$f = \frac{gV}{u} (\rho - \rho_s) \qquad (1)$$

The frictional coefficients of fully mineralized dentin and demineralized dentin (collagen matrix) in EDTA are listed in Table 1 and are plotted as a function of particle size in

Table 1.
Frictional Coefficients of Dentin in EDTA

| | Dentin | | | | Demineralized dentin | | | |
|---|---|---|---|---|---|---|---|---|
| d | Velocity | f | $C_D$ | $R_e$ | Velocity | f | $C_D$ | $R_e$ |
| .0357 | 2.75 | .0133 | 6.90 | 8.84 | .362 | .0133 | 52.5 | 1.16 |
| .0402 | 3.27 | .0160 | 5.50 | 11.8 | .439 | .0157 | 40.2 | 1.59 |
| .0427 | 3.97 | .0158 | 3.96 | 15.3 | .490 | .0169 | 34.2 | 1.88 |
| .0473 | 4.31 | .0198 | 3.72 | 18.4 | .559 | .0201 | 29.2 | 2.38 |
| .0580 | 5.45 | .0288 | 2.86 | 28.5 | .770 | .0268 | 18.8 | 4.02 |
| .0663 | 6.10 | .0384 | 2.60 | 36.4 | .978 | .0315 | 13.4 | 5.84 |
| .0920 | 7.09 | .0625 | 2.38 | 52.4 | 1.26 | .0463 | 9.95 | 9.31 |
| .107 | 10.7 | .0917 | 1.37 | 103. | 1.96 | .0659 | 5.36 | 19.0 |
| .125 | 12.2 | .129 | 1.23 | 138. | 2.23 | .0926 | 4.84 | 25.0 |
| .160 | 14.2 | .232 | 1.16 | 205. | 2.82 | .154 | 3.88 | 40.7 |

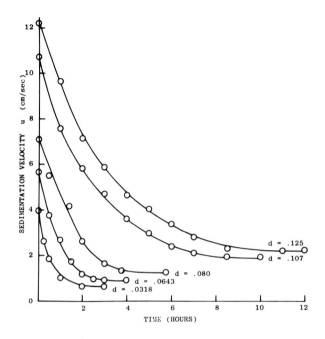

Fig. 1. Demineralization of bovine dentin. Effect of particle size (d = edge length in cm) on the sedimentation velocity in 0.5M EDTA, pH 7.4, 37°C.

Figure 2. The velocities of the particles were also measured in water before and after demineralization in EDTA. The density of the calcifying solution is close to that of water, so water was used as the calibrating medium for remineralization experiments. These data are shown in Table 2 and Figure 3.

IV. DISCUSSION

The frictional coefficients increase with increasing particle size at constant density. For particles greater than about 500 microns, the frictional coefficient changes with density at constant size. The variability of the coefficient for larger particles explains the discrepancy between the relative decrease in velocity and the relative decrease in density. The data also reveal that for small particles, Stokes' law is obeyed; i.e., f is constant for constant size and fluid viscosity. Stokes' law, Equation 2, is generalized

Table 2.
Frictional Coefficients of Dentin in Water

| d | Dentin | | | | Demineralized dentin | | | |
|---|---|---|---|---|---|---|---|---|
| | Velocity | f | $C_D$ | $R_e$ | Velocity | f | $C_D$ | $R_e$ |
| .0289 | 2.89 | .00744 | 6.23 | 11.9 | .572 | .00744 | 31.5 | 2.36 |
| .0325 | 3.63 | .00846 | 4.44 | 16.9 | .700 | .00867 | 23.7 | 3.25 |
| .0365 | 4.45 | .00973 | 3.32 | 23.2 | .840 | .0102 | 18.5 | 4.39 |
| .0427 | 5.62 | .0124 | 2.44 | 34.3 | 1.21 | .0114 | 10.4 | 7.39 |
| .0473 | 6.63 | .0143 | 1.94 | 44.9 | 1.28 | .0146 | 10.3 | 8.66 |
| .0580 | 8.03 | .0217 | 1.62 | 66.6 | 1.71 | .0201 | 7.08 | 14.2 |
| .0663 | 8.97 | .0290 | 1.48 | 85.1 | 2.03 | .0253 | 5.74 | 19.2 |
| .0820 | 9.53 | .0516 | 1.63 | 112 | 2.46 | .0395 | 4.84 | 28.9 |
| .125 | 13.1 | .133 | 1.31 | 234 | 3.94 | .0874 | 2.88 | 70.5 |
| .160 | 14.5 | .252 | 1.37 | 332 | 4.74 | .153 | 2.54 | 108 |

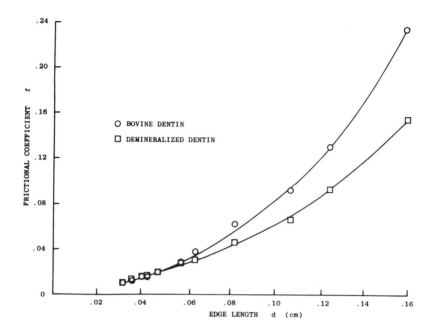

Fig. 2. Dependence of frictional coefficient, f, on particle size of bovine dentin and demineralized dentin in 0.5M EDTA, pH 7.4, 37°C.

here to include particles of any shape:

$$f = kd\eta \qquad (2)$$

where k is a number depending on particle shape, d, is a linear dimension, and $\eta$, is the viscosity of the fluid. If (2) holds, the density of a particle at any degree of demineralization can be found quite simply from Equation 1. The fraction of a particle mineralized, $1-\xi$, at any time is then related to the velocity at that time, $u_t$, and the initial and final velocities, $u_i$ and $u_\infty$.

$$1-\xi = \frac{\rho_t - \rho_C}{\rho_D \quad \rho_C} = \frac{u_t - u_\infty}{u_i - u_\infty}$$

The subscripts, D and C, stand for dentin and collagen.

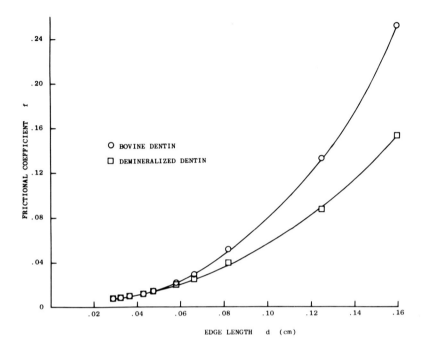

Fig. 3. Dependence of frictional coefficient, f, on particle size of bovine dentin and demineralized dentin in water at 37°C.

With larger particles, where the frictional coefficients vary, the density and extent of mineralization can be determined only after the coefficient has been found as a function of velocity. This was accomplished by rewriting the equations in terms of a dimensionless drag coefficient, $C_D$, and studying its behavior as a function of the Reynolds number, Re, which depends on the velocity. See Albertson, et al (3). The general drag equation is:

$$F_f - \frac{C_D A \rho_s u^2}{2} = tu \qquad (4)$$

where, $F_f$, is the frictional force and, A, is the cross-sectional area of the particle. Substituting for f from Equation (1)

$$C_D = \frac{2gV}{Au^2} \cdot \frac{\rho - \rho_s}{\rho_s} \tag{5}$$

which for a cubic particle can be written

$$C_D = \frac{2gd}{u^2} \cdot \frac{\rho - \rho_s}{\rho_s} \tag{6}$$

The Reynolds number, Re, also dimensionless is

$$Re = \frac{ud\rho_s}{\eta} = \frac{ud}{\nu} \tag{7}$$

where $\nu = \eta/\rho_s$ is the kinematic viscosity in stokes ($cm^2$/sec) and represents the ratio of the inertial to viscous forces.

The viscosity of the EDTA solution was measured at 37°C in a viscometer and found to be 1.222 centipoise. The kinematic viscosity of EDTA is $0.01222/1.098 = 1.11 \times 10^{-2}$ stokes.

$C_D$ is calculated from the data by means of equation (6), see Tables ] and 2, and is plotted against Re on a log-log scale in Figures 4 and 5. The drag coefficient decreases with increasing velocity and becomes constant at high Reynolds numbers. The value of $C_D$ at high Re is 1.3 in water, compared with 1.1 cited by Albertson for cubes falling in water. At low Re, the curves become linear indicating the region of laminar flow where Stokes' law is obeyed. In this region, a given particle will have a single value of f for both dentin and collagen as must be if the velocity changes introduced into (3) are to be correctly translated into calcification changes.

At higher Reynolds numbers the drag coefficients begin to curve away from a straight line, i.e., friction becomes higher than ideal due to the onset of nonlaminar flow, or turbulence. At Re = 50 in EDTA, and 100 in water, the inflection in the curve corresponds to the onset of turbulence which is visually apparent. The drag coefficient of the calcified dentin, which falls at much higher velocity than the decalcified samples, decreases less rapidly from the ideal values with increasing velocity than it does for demineralized dentin. Since $C_D$ and $\rho$ are inversely related, f increases more rapidly with particle size for dentin than for demineralized dentin. The cause of this behavior is probably increased turbulence around the faster falling dentin which

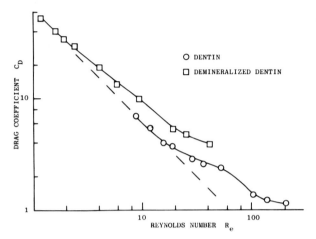

Fig. 4. Dependence of drag coefficient, $C_D$, on Reynolds number, $R_e$, for dentin and demineralized dentin in 0.5M EDTA, pH 7.4, 37°C. Dashed line is Stokes' law.

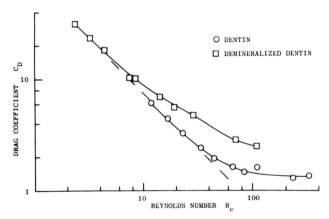

Fig. 5. Dependence of drag coefficient, $C_D$, on the Reynolds number, Re, for dentin and demineralized dentin in water 37°C. Dashed line is Stokes' law.

gives rise to apparently larger frictional coefficients.

The following procedure was adopted for the determination of the density of particles larger than 500 microns from the velocity of fall. The values of $C_D$ and Re for the fully cal-cified and decalcified particle were calculated, plotted on log-log paper, and connected by a straight line. It was assumed that values of $C_D$ and Re for all intermediate densi-ties would fall on this line. For each value of a velocity measured during an experiment, the Reynolds number was calcu-lated. The corresponding value of the drag coefficient was obtained from the curve, or from the equation of the line, $C_D = a/Re^n$, where, a, is the y intercept, and n = log y intercept/log x intercept. The density at any time, $\rho_t$, was calculated from a rearrangement of equation (6)

$$\rho_t = (1 + C_D u^2/2gd) \rho_s \qquad (8)$$

In the studies of demineralization and remineralization, small particles were used because of the simplification in the calculations, except when it was desired to study the effect of variation in particle size on these processes.

There is no fundamental difference in the behavior of the frictional data in water or in EDTA. The curves of Figures 4 and 5 are practically superimposable, since this treatment takes into account the difference in the viscosity of the two solutions. The ratio of the frictional coefficients for small particles in water and EDTA from Tables 1 and 2 should be equal to the ratio of the viscosities of the two solutions, if Stokes' law is obeyed. This is found to be true.

$$\frac{f(dentin, H_2O)}{f(dentin, EDTA)} = \frac{\eta (H_2O)}{\eta (EDTA)} = 0.56 \qquad (9)$$

V.  CONCLUSIONS

Two advantages of the sedimentation method are that the rather slow process of remineralization of hard tissue can be followed with ease and that the solution composition remains constant. Measurement of the change in the particle density rather than of the change in the solution permits the study of a variety of solution constituents which may affect demin-eralization or remineralization. For example, the effect of the topical treatment of hard tissues by fluorides, or by cal-cification inhibitors, can be conveniently followed.

Another advantage is the exclusion of air from the columns, the presence of which hastens the spontaneous precipitation of supersaturated calcifying solution. Thus, remineralization of hard tissues at physiological concentrations of calcium and phosphate salts, a very slow process, may be studied for several weeks.

An advantage of the technique of particular importance with respect to biological materials is the smallness of the samples which facilitates obtaining homogeneous samples and allows many experiments to be conducted with a modest amount of material.

The use of small particles, as shown, has the advantages of reducing the duration of experiments and simplifying the calculations, however, a lower limit is put on the sizes because of the decrease in experimental accuracy. Small particles cannot be cut and measured precisely and the minor density difference between the fluid and the demineralized particle, 1.10 vs. 1.15, causes small dentin particles to move too slowly with poor reproducibility.

A disadvantage of the sedimentation method for some purposes is that the details of the processes in terms of the individual molecular species cannot be determined. For example, the ratio in which calcium and phosphorus are being deposited or removed from the tissue cannot be measured.

The technique is at a disadvantage compared with the rotating disk method for studying demineralization (3) because dentin can be cut to the theoretically required flat, circular shape. However, in the sedimentation method, we cannot cut spheres of dentin, but must use cubes, the hydrodynamic resistance of which has not been rigorously solved. The sedimentation method probably has the most favorable application to the study of remineralization of hard tissue. The results of study of demineralization and remineralization of dentin by the sedimentation technique are reported elsewhere.

VI. REFERENCES

1. Chan, M.S., and Eirich, F.R., Proc. Soc. Exp. Biol. Med. 143, 919-924 (1973).
2. Albertson, M.L., Barton, J.R., and Simons, D.B., "Fluid Mechanices for Engineers," Prentice-Hall, Englewood Cliffs, N.J., 1960.
3. Linge, H.G. and Nancollas, G.H., Calc. Tiss. Res. 12, 193-208 (1973).

# CHEMILUMINESCENT AUTOXIDATION OF UNSATURATED FATTY ACID FILMS. LINOLENIC ACID AND DOCOSAHEXENOIC ACID ON SILICA GEL.

Arthur W. Adamson

University of Southern California

Vida Slawson

Laboratory of Nuclear Medicine

University of California at Los Angeles

*Both linolenic acid and 4,7,10,13,16,19-docosahexenoic acid are found to undergo chemiluminescent autoxidation when present as monolayer and submonolayer films adsorbed on chromatographic type silica gel. The emission is autocatalytic, the intensity increasing over successive heating/cooling cycles and once well initiated, shows a temperature dependence corresponding to about 10 kcal/mole apparent activation energy. The quantum yield is $10^{-11}$ in absolute value for both acids. The emission spectrum contains two components in the case of linolenic acid, one centered around 550 nm and the second, at 600 + nm and may correspond to excited state ketone and singlet oxygen emission, respectively.*

## I. INTRODUCTION

The present investigation is an extension of earlier work on the analytical autoxidation rates of various unsaturated fatty acid esters adsorbed on silica gel. When present at 1% or less of a monolayer coverage, the autoxidation is slow relative to that in homogeneous solution and shows little or no induction period, in contrast to the behavior in solution (1). Later work with coverages approximating a monolayer gave rates of autoxidation more nearly comparable to these in solution and showing the typical induction periods (2).

Autoxidations of organic substances either neat or in solution typically are accompanied by weak chemiluminescence (3-5), and we were interested in whether the surface autoxidations would also be detectably chemiluminescent. If so, a further interest would be in the emission spectrum and implications from it as to type of emitting species present. There would be the possibility of using emission as a rapid means of following chemical reaction. Finally, the change in character of the thermal autoxidation kinetics between very

low and near monolayer coverage might be accompanied by some
significant change in the character of the emission.  If
emission were due to bimolecular processes between high energy intermediates, for example, and such was not possible at
low surface coverage, one would expect that emission, if
observed at all, would be seen only around the monolayer
region of coverage.

The present report is largely on a continuation of preliminary studies with linolenic acid (hereafter designed as
18:3 acid).  These results (6) discovered chemiluminescence
of the 18:3 acid adsorbed on silica gel at around monolayer
coverage.  It was of interest to extend the work to at least
one other compound, and 4,7,10,13,16,19-docosahexenoic acid
(hereafter referred to as 22:6 acid) was chosen.  General
experience has been that the more highly unsaturated fatty
acids are more prone to autoxidation both in solution and in
adsorbed films at around monolayer coverage.  For autoxidations in solution, the induction period increases with
degree of unsaturation (7).  We were interested in whether
chemiluminescence would again be observed and, if so, in what
pattern of behavior.

II.  EXPERIMENTAL

A.  Materials.

Solvents were of reagent grade and were used without
further purification.  The silica gel was the same as before,
J. T. Baker 3405, lot 28074.  The 22:6, 18:0 and 20:0 acids
were of greater than 99% purity, obtained from Nu-Chek Prep.,
Elysian, Minnesota.

B.  Procedures.

The equipment used was that previously described (6).
Figure 1 adds some detail, however, on the "hot finger"
arrangement.  The availble equipment did not allow the thermo-
statting of samples, being designed for measurements at
ambient temperature only.  It was very desirable to make at
least qualitative measurements at higher and at least approximately known temperatures and the mean for doing this is
shown in Fig. 1.  A Dewar test tube was designed to fit the
1" sample well; the insulation was sufficient to reduce the
cooling rate of hot samples to a manageable level.  To provide
further temperature ballasting, a stainless steel round ended
cylinder or "hot finger" was used.  This could be pre-heated
to a desired temperature, determined by placing a thermometer

194

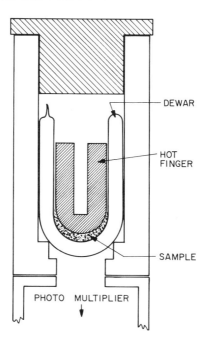

*Fig. 1   Dewar test tube and "hot finger" sample holder.*

in the well shown in the figure. The finger was then placed
firmly in contact with 0.15-0.3 g of fatty acid coated silica
gel contained in the Dewar test tube, and the unit inserted
into the sample well. It was possible periodically to with-
draw the unit and measure the temperature of the finger (and
thus the approximate temperature of the sample itself).

The general procedure for depositing the 22:6 acid was the
same as was used for the 18:3 acid. The silica gel coated
samples were stored over dry ice until just before use. As
before, about 10% by weight of marker saturated acid, in the
present case n-$C_{19}H_{39}COOH$ (or 20:0 acid), was added so that
the deposited film actually consisted of the mixed acids. In
subsequent gas-liquid chromatographic (GLC) analyses, it was
assumed that the 20:0 acid was negligibly oxidized so that
its chromatogramic peak area allowed calculation of the extent
of loss of the 22:6 acid. As a further control, a second
marker species, stearic (18:0) acid, was added to the ether
extract from each sample so that by comparison of its peak
area with the others the absolute amounts of the 22:6 and
20:0 species could be obtained.

## III.  RESULTS

Several types of reference measurements were made prior
to each experiment.  The background count rate for the empty
sample container (either with or without the metal finger)
was about 30 counts/sec.  If the finger was heated to $120^{0}$C
and then placed in the empty Dewar test tube, the unit now
showed a background of around 100 counts/sec., diminishing to
the 30 counts/sec value as the unit cooled.  The sample container with non-heated sample gave the same background as the
empty container.  That is, the fatty acid coated silica gel
showed no chemiluminescence if it had never been heated above
room temperature.

Figure 2 shows the results of three experiments, Runs 1,
2, and 3.  The silica gel, of previously estimated specific
surface area 430 $m^2$/g, had deposited on it sufficient 22:6
(plus marker 20:0) acid to give 1.12, 0.80, and 0.17 of a
complete monolayer of 22:6 (assuming the molecules lie flat),
respectively.  The sample weights used were 0.15, 0.30, and
0.20 g, again respectively.

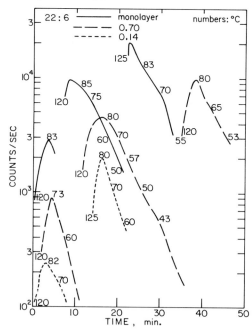

Fig. 2    Count rate over successive heating/cooling cycles of
          22:6 acid adsorbed on silica gel.  Full line: Run 1;
          dashed line: Run 2; dotted line: Run 3.

Each run consisted of two or three heating/cooling cycles. The metal finger, heated to $120^{\circ}C$ or $125^{\circ}C$, was placed on top of the sample, and the measurement begun. After the finger had cooled to around $40^{\circ}C$, it was removed and a new finger, again at $120^{\circ}C$ or $125^{\circ}C$, inserted. In the case of Runs 1 and 2, the procedure was repeated another time. The figure thus shows the time sequence of counting rate as each of the samples was cycled. The numbers at various points along the curves give the approximate temperature at that time.

The curves shown are slightly smoothed around the maxima; it was found that the act of removing the sample container unit in order to make a temperature measurement led to an approximately 10% increase in the count rate. This is presumably because of the improved exposure to oxygen. The curves thus given only an approximate presentation of the emission rate as a function of temperature, but Arrhenius plots of the cooling sections of the second and third cycles of Runs 1 and 2 have slopes corresponding to about the same 10 kcal/mole apparent activation energy as found for films of the 18:3 acid.

Quantum yield estimations were also made. For Runs 1, 2, and 3, the total quanta emitted were $1.9 \times 10^{8}$, $9.8 \times 10^{7}$, and $1.2 \times 10^{7}$, respectively. The figures are obtained from area measurements on the actual chart recordings of emission rate vs. time, using a previously estimated geometry or counting efficiency factor of 0.06. Correction for the background count was less than 10% in all cases. The percent oxidation was 53%, 44%, and 16%, again respectively, corresponding to $2.7 \times 10^{-5}$, $3.4 \times 10^{-5}$, and $1.75 \times 10^{-6}$ moles of 22:6 acid oxidized. The corresponding efficiencies for light production are $1.2 \times 10^{-11}$, $4.8 \times 10^{-12}$, and $1.1 \times 10^{-11}$ moles of photons per mole of initial 22:6 acid reacted. A fourth run, not shown in the figure to avoid excessive overlay of curves, gave very similar results. For Run 4, the surface coverage was 0.48 of a monolayer, 0.20 g of sample were used, $6.7 \times 10^{7}$ quanta were emitted over two cycles of heating/cooling, the percent oxidation was 34%, and the emission efficiency, $1.1 \times 10^{-11}$. We attribute the lower figure of Run 2 to the fact that too large a sample was used; it is likely that only light from the bottom half or two-thirds of the sample was received by the detector.

The 22:6 acid also emitted light when undergoing autoxidation in the neat state. A thin (but macroscopic) film of material ($6 \times 10^{-5}$ mole) deposited in a scintillation vial showed an easily detected emission, although only after first heating the vial to around $100^{\circ}C$. Two heating/cooling cycles gave essentially the same count rate vs. time plots, however.

The emission efficiency was not estimated, although it was determined that about 14% oxidation had occurred.

## IV.  DISCUSSION

The general observation is that the chemiluminescent be-havior of films of 18:3 and 22:6 acids is rather similar. Figure 3 shows one of the heating/cooling sequences for the 18:3 acid.  As with the 22:6 one, the final relatively high level of emission developed only after several heating/cooling cycles.  The surface coverage in this case was 75%.  It does appear, however, that the induction effect is greater with the 18:3 than with the 22:6 acid (compare with Figure 1); the latter samples reached wht 10,000 to 20,000 counts/sec level of emission after fewer and shorter periods of heating.

The existence of an induction effect in a monolayer re-action carries some implications.  First, any such effect is hard to explain except in terms of some type of chain or at least product assisted reaction; that is, surface bimolecular reactions seem firmly to be implicated.  This conclusion in turn implies surface mobility.  The unsaturated fatty acids involved probably form liquid type monolayers (in the Adams classification for spread monolayers on water--see Ref. (8),

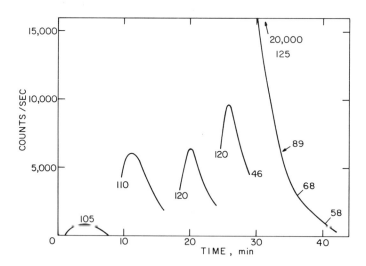

Fig. 3.  *Count rate over successive heating/cooling cycles of 18:3 acid adsorbed on silica gel.*

and such mobility is not unreasonable at our high surface coverages even though the substrate is solid. Improved equipment planned for future work may allow measurements at low surface coverages, and it seems likely that the induction period (and perhaps the emission) will disappear.

A second implication of the type of induction effect observed here is that the intermediate species involved must be relatively stable ones. Thus, previous heating cycles carried out many minutes earlier still leave the system pre- pared for enhanced chemiluminescence when reheated. A possi- bility is that the surface autoxidation leads to high energy products which are stabilized by adsorption and thus survive to take part in reaction on the next heating cycle. The species may be the same as in the homogeneous reaction but it is entirely possible that they are different.

The 18:3 and 22:6 acids are very similar in the tempera- ture dependence of their chemiluminescence. Although the experimental arrangement allowed only for approximate measure- ments, the decay of emission intensity as the samples cooled corresponded in both cases to an apparent activation energy of about 10 kcal/mole. The measurements for the two acids were under comparable conditions, namely about 75% surface coverage and after sufficient heating/cooling cycles to be past the induction period.

The nature of the emitting species is probably the same for the two acids. The spectral measurements made in the case of films of the 18:3 acid led to the approximate emission spectrum shown in Figure 4. The photomultiplier used was not sensitive beyond 650 nm, and the long wavelength shoulder is almost certainly an actual peak in emission intensity. As noted (6), the emission peak at 550 nm could be due to excited state ketone or aldehyde; however, the probable peak between 600 and 650 nm suggests the presence of single oxygen. The second spectrum shown in the figure is that of ozonized isopropyl ether, known to produce singly oxygen (9). The similarity of the emissions in the 600-650 nm region in the two cases at least supports the possibility that singlet oxygen production has occurred in the autoxidation of the 18:3 acid.

Finally, the yield for emission is about $10^{-11}$ for both the 18:3 and the 22:6 acids. The absolute yields are subject to uncertainties in the estimation of the light collection efficiency and could be in error by a factor of three. The order of magnitude is in the range of values typical for

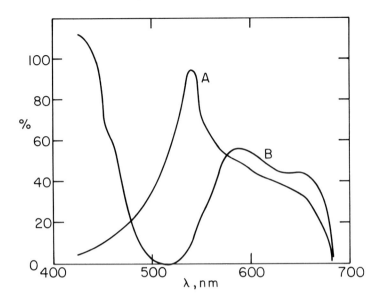

*Fig. 4:    Chemiluminescence emission spectrum.    A.    18:3 acid
            adsorbed on silica gel; B. Ozonized isopropyl ether.*

homogeneous autoxidations, $10^{-8}$ to $10^{-15}$ (3,5,10).  The relative yield for the two unsaturated fatty acids is not so
subject to apparatus uncertainties, and analysis of both sets
of data indicates that the yields for the 18:3 and 22:6 acids
are the same within at least 30%.  Moreover, in the case of
the 22:6 acid, our determinations show that for surface
coverages ranging from a complete monolayer down to 0.17 of a
monolayer, the yield is essentially constant.  This constancy
suggests that no major change in reaction mechanism takes
place.

    In summary, we confirm our discovery of chemiluminescence
in surface autoxidation reactions, finding that the behavior
is quite similar for 18:3 and 22:6 unsaturated fatty acids.
The principle qualitative difference observed that the latter
shows less of an induction effect than does the former.

V.   ACKNOWLEDGMENT

    We gratefully acknowledge the use of quantum counting
equipment in the laboratory of Prof. B. Abbott and the very
helpful assistance of Dr. J. W. Meduski.  The investigations
have supported in part by contract (National Science Founda-
tion--Grant No. MPS74-02766).

## VI.  REFERENCES

1.  Slawson, V., Adamson, A. W., and Mead, J. F., *Lipids*, 8, 129 (1973), see also Honn, F. J., Bezman, I. I., and Daubert, B. F., *J. Amer. Oil Chemists Soc.* 28, 129 (1951); Togashi, H. J., Henick, A. S., and Koch, R. B., *J. Food Sci.* 26, 186 (1961); Porter, W. L., Levasseur, L. A., and Henick, A. S., *Lipids*, 6, 1 (1971); Slawson V., and Mead, J. F., *J. Lipid Res.* 13, 143 (1972); Porter, W. L., Henick, A. S., Levasseur, L. A., *Lipids*, 8, 31 (1973).

2.  Mead, J. F., and Wu, K., *Lipids*, Vol. 1, p. 197-201. (R. Paoletti and G. Jacini, Eds.)  Raven Press, New York, 1975.

3.  Gundermann, K. D., *Angew. Chem. Int. Ed. Engl.* 4, 566 (1965).

4.  Hofert, M., *Angew. Chem.* 76, 826 (1964).

5.  Vasil'ev, R. F., and Vichutinskii, A. A., *Nature*, 194, 1276 (1962).

6.  Slawson, V., and Adamson, A. W., *Lipids*, in press.

7.  Privett, O. S., and Blank, M. L., *J. Amer. Oil Chemists Soc.* 39, 465 (1902).

8.  Adamson, A. W., "Physical Chemistry of Surfaces," 3rd ed., Wiley-Interscience, New York, 1977.

9.  Murray, R. W., Lumma, W. C. Jr., and Lin, J. W. P., *J. Amer. Chem. Soc.* 92, 3205 (1970).

10. Kellogg, R. E., *J. Amer. Chem. Soc.* 91, 5435 (1969).

# WATER VAPOR SORPTION OF THE SUBUNITS OF α-CRYSTALLIN AND THEIR COMPLEXES

Frederick A. Bettelheim and Joseph Finkel
Chemistry Department
Adelphi University

α-crystallin is one of the structural proteins in the lens fibers of the eye which together with the β and γ crystallins provide specific hydrated ultrastructures that account for the transparency of the lens. In order to study the nature of the hydration, water vapor sorption isotherms were obtained gravimetrically on eleven samples: on the four subunits of α-crystallin, $\alpha A_1$, $\alpha A_2$, $\alpha B_1$ and $\alpha B_2$ polypeptides and a number of combination of these four subunits. The results implied that in an aqueous environment the aggregation of different subunits is preferred to that of self-aggregation.

## I INTRODUCTION

Bovine α-crystallin is the most thoroughly studied lens protein. It represents about 20-25% of the total soluble proteins of the lens. It is heterogeneous and its polydispersity is age dependent (1). Only the newly synthesized α-crystallin (NSα) is homogeneous. The weight average molecular weight of this low molecular weight (LMW) α-crystallin is $7x10^5$ daltons (2).

The adult bovine α-crystalline has a molecular weight range of $1-50 \times 10^6$ dalton (1,3). These high molecular weight (HMW) aggregates are essentially composed of four polypeptide chains: $\alpha A_1$, $\alpha A_2$, $\alpha B_1$ and $\alpha B_2$ (4). In contrast the NSα contains only two chains: $\alpha A_2$ and $\alpha B_2$ (5). The amino acid sequence of these two polypeptides have been elucidated. $\alpha A_2$ contains 173 amino acids (6) and the $\alpha A_1$ arises from $\alpha A_2$ by the deamidation of glutamine at position 126 (7). $\alpha B_2$ contains 175 amino acids (8) and it was suggested (5) that $\alpha B_1$ is derived from $\alpha B_2$ by a similar process.

The mode of aggregation is of primary importance because of sufficiently high concentration of high molecular weight (HMW) α-crystallin is present in the lens fibers they may cause lens opacity (9). Li and Spector (10) found that the subunits of α-crystallin can reaggragate from the basic $2x10^4$ dalton

to $4 \times 10^5$ dalton and larger units. However, the size
of the aggregated $\alpha A_1$ was less than that of the LMW
$\alpha$-crystallin. The same was true for $\alpha A_2$ and $\alpha B_2$.
Furthermore, even the urea treated $\alpha$-crystallin[2] (all
four subunits present) aggregated to a size somewhat
smaller than the native LMW $\alpha$-crystallin. Similar
results were obtained by van Kamp et. al. (11).
Both studies implied that some small molecular
weight compounds have been eliminated during the
urea deaggregation process that these small molecular
weight compound (s) enhance the aggregation to high
molecular weight compounds.

In our laboratory we have isolated an $\alpha$-crys-
tallin complex that contained beside $\alpha$-crystallin
some $\beta$-crystallin chains and two small molecular
weight phosphopeptides (12,13). The complex was
crystalline as viewed by X-ray diffraction. However
when the phosphopeptides were eliminated by dialysis
the complex became amorphous. It was evident that
the small molecular weight compounds had some organ-
izational ability in aligning the different subunits
in a crystalline matrix.

**Recent (14) electron** micrographs of LMW $\alpha$-crys-
tallin showed spherical aggregates. On the other
hand, the HMW $\alpha$-crystallin displayed random coil
like arrangements. Bloemendal et. al. (15) showed
that $\alpha$ and $\beta$ hybrid aggregates are rod like structures as opposed to the globular entities of $\alpha$ and
$\beta$-crystallins.

The fact that native crystallins are dissociated into their subunits in urea or in guanidinium HCl
suggested that the subunits aggregate via hydrophobic interactions. If the hydrophobic interaction
is stronger between a combination of different subunits than between the same polypeptide subunits
themselves, this tendency may show up in hydration
studies. In other words a combination of subunits
may project more hydrophylic sites toward incoming
water vapor than an aggregate of a single subunit.
Consequently this study was conducted to investigate
the water sorptive capacities of $\alpha$-crystallin subunits alone and in different combinations as well
as that of the $\alpha$-$\beta$ crystallin complex containing
phosphopetides.

II MATERIALS AND METHODS
$\alpha$-crystallin was isolated from calf lenses by
gel filtration on Bio-Gel-A-5m according to Li and
Spector (16). Peak II from Bio-Gel-A-5m was used as
the LMW $\alpha$-crystallin. The subunits of $\alpha$-crystallin

were isolated by a combination of affinity and ion exchange chromatography according to Stauffer et. al. (17). The α-crystallin was first dissociated into its subunits by dissolving 100-200 mg/ml in 7 M urea that contained 0.005 M Tris at pH 7.6. This buffer was prepared from freshly purified 8 M urea which had been passed through a 3x20 cm column of Amberlite MB-3 ion exchange resin immediately before use (17).

The affinity resin was prepared by covalently linking p-amino-phenylmercuric acetate to Bio-Gel-A-5M using the procedure of Sluyterman and Wijdenes (18). The affinity resin was equilibrated with 0.005 M Tris buffer at pH 7.6 in 7 M urea. The resin was packed in a 1.5x19 cm column. The dissolved α-crystallin was applied to this column and the column was washed with the equilibrating buffer until the non-binding polypeptide chains were eluted. The appearance of proteins in the eluate was monitered at 280 nm in a Carey 15 Spectrophotometer. Two peaks appeared with non-binding polypeptides. Peak I contained about 3.6% of the material and contained $\alpha A_1$ and $\alpha A_2$ chains. Peak II contained about 35% of the α-crystallin and was composed largely of $\alpha B_1$ and $\alpha B_2$ but still substantial amounts of $\alpha A_1$ and $\alpha A_2$ polypeptide chains were also present (17).[1] This was confirmed by gel electrophoresis in cellulose-acetate gel in a buffer of Tris-glycin at pH 8.9 in 7 M urea (19), in a Millipore gel electrophoresis apparatus.

At this point the equilibrating buffer was adjusted to 0.01 M in mercaptoethanol and the polypeptides bound to the affinity resin were eluted. Peak III, thus obtained, contained 60% of the α-crystallin and was composed of $\alpha A_1$ and $\alpha A_2$ (17). Again this was confirmed by gel electrophoresis.

Peaks I and II were pooled and sufficient mercaptoethanol was added to make it 0.01 M. The pooled samples were applied to a 2.2x27.0 cm column of DEAE cellulose (capacity 0.95 meq/g), which had previously been equilibrated with 0.005 M Tris, pH 7.6 containing 7 M urea and 0.01 M mercaptoethanol. Stepwise elution was carried out using first the equilibrating buffer to obtain the first peak and subsequently applying 0.007 M and later 0.015 M phosphate buffers each containing 7 M urea and 0.01 M mercaptoethanol. The peak contained $\alpha B_2$; the fraction eluted with 0.007 M phosphate had two peaks one containing $\alpha B_1$ and the other containing mostly $\alpha A_2$. The third fraction eluted with 0.015 M phosphate contained largely $\alpha A_1$ with minor contaminations (17). These were confirmed by gel electrophoresis. Only the $\alpha B_2$ and $\alpha B_1$ polypeptides were used from this isolation procedure. They were dialyzed against 20ℓ of distilled water for 48 hours with three changes of water. After dialysis they were lyophilized.

Peak III of the affinity column was similarly chromatographed on the DEAE cellulose column and $\alpha A_2$ was isolated by eluting with 0.007 M phosphate and $\alpha A_1$ was obtained by eluting it with 0.015 M phosphate. These polypeptides were dialyzed against distilled water and lyophilized. Thus, the four subunits of α-crystallin were obtained in pure form. Combinations of the subunits were prepared by dissolving weighed samples of the individual polypeptides in water and lyophilizing the mixture. The following mixtures were prepared in proportions representing their natural abundance in α-crystallins (17).; $(\alpha A_1 + \alpha A_2)$ in 22:33; $(\alpha A_2 + \alpha B_2)$ in 33:26 and $(\alpha A_2 + \alpha A_2 + \alpha B_2)$ in 22:33:26.

The original LMW α-crystallin obtained from the Bio-Gel-A-5m gel filtration was also dialyzed and lyophilized. Finally, a complex of α-crystallin containing some β-crystallin subunits and phosphopeptides has been isolated from soluble lens proteins by precipitation at pH 5.1 (12,13). This fraction corresponds to the original designation of α-crystallin by Mörner (20). The rest of the soluble proteins, β and γ crystallins were also obtained by isoelectric precipitation (21). All these were also lyophilized. A total of eleven samples were studied regarding their water vapor sorptive capacities.

Water sorption and desorption isotherms of the samples were obtained gravimetrically at two temperatures (20° and 30°C) in a high vacuum sorption apparatus using quartz springs (22). The B.E.T. (23) and D'Arcy-Watt (24) isotherm parameters were obtained using a CD 3000 computer. The D'Arcy-Watt equation was rearranged to have a linear form with eleven coefficients and the equations were solved by the use of an inverse matrix method.


III RESULTS AND DISCUSSION

All the eleven samples showed water vapor sorption isotherms of the B.E.T. II type. The sorption isotherms of the subunits of α-crystallin at 20°C are given in Fig. 1.

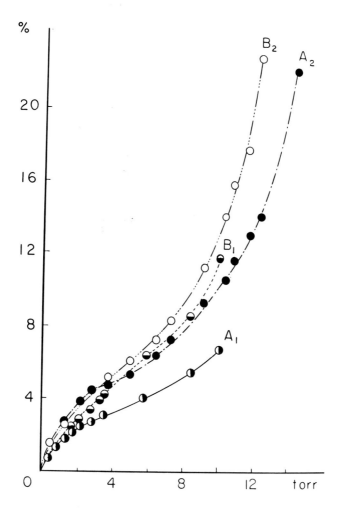

Fig. 1. Water vapor sorption isotherms at 20°C of $\alpha A_1$
⦶——⦶; $\alpha A_2$ ●——●; $\alpha B_1$ ◒——◒ and $\alpha B_2$ O——O polypeptides.

The subunit $\alpha A_1$ has the lowest sorptive capacity.
The subunit $\alpha A_2$ which differ from $\alpha A_1$ by an additional amino group only, has a much higher sorptive
capacity. Similar relationship exists between $\alpha B_1$
and $\alpha B_2$, the latter sorbing more water throughout
the whole water vapor pressure range investigated.

In general, the acidic subunits of α-crystallin
$A_1$ and $A_2$ sorb less water than the basic subunits $B_1$
and $B_2$. The isotherms $\alpha A_2$ and $\alpha B_1$ overlap somewhat.
The $\alpha B_1$ exhibiting lower water uptake at low vapor

pressure and somewhat greater uptake at high vapor pressure than that of α A$_2$. Similar relationship was observed among the isotherms at 30°c.

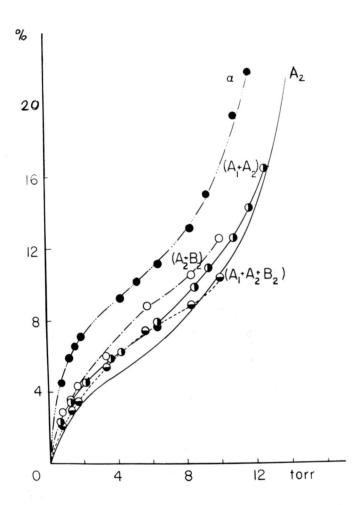

Fig. 2. Water vapor sorption isotherms at 20°C of ——— αA$_2$; ◐——◐ α(A$_1$+A$_2$); ○——○ α(A$_2$+B$_2$); ◕——◕ α(A$_1$ +A$_2$+B$_2$); and ●——● α-crystallin.

The water vapor sorption isotherms of combined subunits are presented in Fig. 2. Since αA$_2$ subunit is the most abundant subunit of the α-crystallin, its isotherm is also presented for comparison. It is interesting to note that a combination of αA$_1$ and αA$_2$ subunits sorb more water than αA$_2$ alone. One would expect the opposite since αA$_1$ had much lower water sorptive capacity than $\alpha A_2$, therefore if two subunits were acting independently in water sorption the combination of $\alpha A_1$ and $\alpha A_2$ as a sorbent should yield a lower water uptake than that of $\alpha A_2$.

This implies that the combination of these subunits is a preferred state in a hydrophylic environment and that the interaction of $\alpha A_1$ and $\alpha A_2$ enables more hydrophobic sites to be buried inside the aggregate than in the cases of $\alpha A_1$ and $\alpha A_2$ aggregates alone.

The same can be said regarding the water vapor sorption of $\alpha A_2 + \alpha B_2$. Again the combination of the subunits provide greater water binding capacity than either subunits alone.

However, the situation is not as simple that one could state that any combination of subunits will increase its water uptake compared to the previous less complex state. The combination of ($\alpha A_1 + \alpha A_2 + \alpha B_2$) illustrates this. Although this combination gives rise to a water sorptive capacity that is higher than those of the subunits alone, it is less than that of ($\alpha A_1 + \alpha A_2$) or ($\alpha A_2 + \alpha B_2$).

With the exception of this combination, however, one can perceive a trend in the combination of the subunits. The $\alpha$-crystallin containing all four subunits sorbs more water than either the subunits alone or the binary and tertiary combinations of subunits used in these experiments. Furthermore, the $\alpha$-complex containing some $\beta$-crystallin subunits plus phosphopeptides has even higher water vapor uptake than $\alpha$-crystallin alone, although again the $\beta$-crystallin alone has low water sorptive capacity. (Fig. 3)

These experiments thus clearly imply that the combination of subunits may occur via hydrophobic interactions. It seems that the hydrophobic regions of the individual subunits can be buried inside a complex combination more successfully than in the simple aggregation of one kind of subunit. Thus, in an aqueous environment such as in the lens fibers it will be thermodynamically more advantageous for the different subunit to combine and form complexes rather than form individual topographic entities.

It would be informative if on the basis of sorption isotherms one could estimate the number of available hydrophylic sites on the surface of the different subunit aggregates and of those of combined subunits. The B.E.T. isotherms have been applied in the past to water vapor sorption of proteins. However, B.E.T. plots (23) usually give straight lines only from 0-0.4. relative vapor pressures. Still monolayer values from B.E.T. plots have been reported in the literature on swelling polymers (25,26) inspite of the fact that the B.E.T. model is much more restrictive.

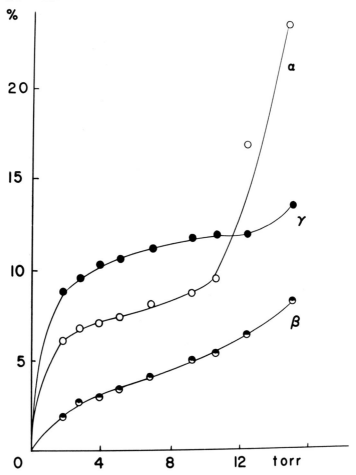

Fig. 3. Water vapor sorption isotherms at 20°C of α Ο——Ο; ●——● β and ●——● γ crystallin complexes.

D'Arcy and Watt (24) have proposed an isotherm in which the total sorption is the result of three simultaneous processes: a) a monolayer type sorption on strongly binding sites b) monolayer type sorption on weakly binding sites and c) the formation of multilayer sorption.

$$W = \sum_{i=0}^{1} \frac{A_i B_i (P/P_o)}{1 + B_i (P/P_o)} + C (P/P_o) + \frac{D E (P/P_o)}{1-E(P/P_o)} \quad (1)$$

In this equation the first term on the right side is the monolayer on strong sites. The second term is the monolayer on weak sites and the third term is the multilayer sorption. The summation is over the number of different kind of sorption sites.

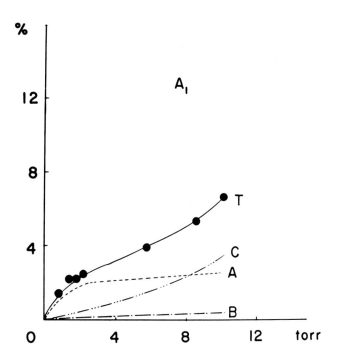

Fig. 4. Total water vapors sorption isotherm ●——●, and the first term ... second term -.-.- and the third term -...- of D'Arcy-Watt equation of $\alpha A_1$ polypeptide.

Since the equation contains five parameters, even if one assumes one kind of sorption site, it is not surprising that the D'Arcy and Watts isotherms fits well the experimental isotherms. For example, we provided in Fig. 4 and 5 the decomposition of the water vapor sorption isotherms of $A_1$ and $B_2$ into the three components mentioned above assuming that i = 1.

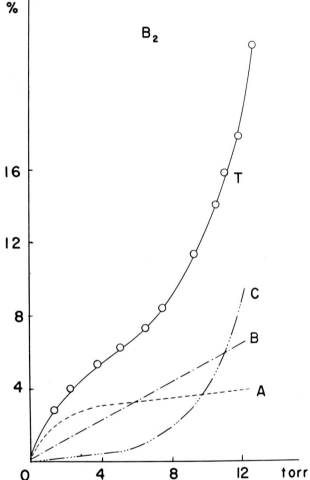

Fig. 5. Total water vapor sorption isotherm O——O, the first term .... second term -.-.- and the third term -...-... of D'Arcy-Watt equation of $\alpha B_2$ polypeptide.

In the Table I the B.E.T. and D'Arcy-Watt parameters are given for all eleven samples.

TABLE I

B.E.T. and D'Arcy-Watt parameters of water vapor sorption isotherms at 20°C on eleven different samples of lens proteins.

| Sample | B.E.T. | | D'Arcy-Watt | | | | |
|---|---|---|---|---|---|---|---|
| | $V_m$ g/100g | C | A g/100g | $Bx10^3$ | $Cx10^3$ | D | $Ex10^3$ |
| $\alpha A_1$ | 3.53 | 12.57 | 2.90 | 129 | 5.9 | 3.6 | 9.1 |
| $\alpha A_2$ | 5.37 | 12.98 | 4.30 | 135 | 42 | 4.2 | 9.7 |
| $\alpha B_1$ | 7.58 | 4.80 | 4.10 | 135 | 85 | 4.0 | 11.5 |
| $\alpha B_2$ | 6.81 | 7.64 | 5.05 | 141 | 103 | 3.1 | 9.0 |
| $\alpha(A_1+A_2)$ | 5.87 | 20.55 | 8.25 | 200 | 87 | 7.0 | 9.6 |
| $\alpha(A_1+A_2+B_2)$ | 7.08 | 8.21 | 7.42 | 202 | 90 | 16.0 | 9.1 |
| $\alpha(A_2+B_2)$ | 7.42 | 12.31 | 8.40 | 205 | 105 | 8.5 | 11.3 |
| $\alpha$ | 8.05 | 32.73 | 10.52 | 205 | 120 | 12.0 | 10.5 |
| $\alpha$ complex | 6.67 | 30.0 | 13.66 | 220 | 84 | 13.0 | 11.8 |
| $\beta$ complex | 3.89 | 13.67 | 4.20 | 120 | 10.5 | 6.5 | 8.2 |
| $\gamma$ complex | 8.65 | 231 | 16.20 | 150 | 60 | 8.2 | 9.5 |

The B.E.T. $V_m$ values and the D'Arcy-Watt A values should be related, representing the amount water
sorbed on strong sites monolayer fashion. Similarly
the B.E.T. C and D'Arcy-Watt B values should be related representing parameters describing the energetics of the sorption on these sites.

One can observe that the D'Arcy-Watt A parameters correspond more closely to the order of the
total sorptive capacity than the B.E.T. $V_m$ values
among our samples. Fig. 4 and 5 also demonstrate
that according to the D'Arcy-Watt equation most of
the strongly sorbing sites are already filled at
about 0.2 relative vapor pressure and swelling increases sorption on weaker sites as well as the
multilayer sorption.

Since equations with five parameters can fit
the experimental data quite well such a fit does not
demonstrate, ipso fact, the uniqueness of the model.
Therefore, no comments are necessary regarding the
other more questionable interpretations of the remaining parameters, especially that of C.

## IV CONCLUSION

General water sorptive capacities and monolayer
parameters have shown that 1. the acidic subunits of
α-crystallin sorb less water than the basic subunits,
2. combination of subunits have greater sorptive
capacity than the individual subunits, 3. the more
complex the combination, especially α and the α-com-
plex of Mörner, the higher the vapor sorptive capacity, 4. the vapor sorptive capacity tend to represent also the amount of water vapor sorbed in strong
polar sites.

The significance of these findings is that in
an aqueous environment the combination of different
polypeptide chains into complexes is more advantageous than their self aggregation and, therefore, it
is likely that the different polypeptides of the
soluble proteins of the lens form specific complexes,
in situ, rather than having either a random distribution or self aggregated domains. Thus, the optical anistropy observed in our laboratory in lens
fibers (27) may represent the special alignment of
such complexes.

## V ACKNOWLEDGEMENT

This research was supported by a grant of the
National Eye Institute EY-00501-08 N.I.H., P.H.S.

REFERENCES

1. Spector, A., Li, L.K., Augusteyn, R.C., Schneider, A., and Freund, T., Biochem J. 124, 337 (1971).
2. Spector, A., Wandel, T., and Li, L.K., Invest. Opthalmol. 7, 179 (1968).
3. Jedziniak, J.A., Kinoshita, J.H., Yates, E.M., Hocker, L.I., and Benedek, G.B., Invest. Opthalmol. 11, 905 (1972).
4. Schoenmakers, J.G.G., Gerding, J.J.T., and Bloemendal, H., Eur. J.Biochem. 11 472 (1969).
5. Stauffer, J., Rothschild, D., Wandel, T., and Spector, A., Invest. Opthalmol. 13, 135 (1974)
6. van der Ouderaa, F.J., DeJong, W.W., and Bloemendal, H., Eur. J. Biochem. 39, 207 (1973).
7. Bloemendal, H., Berns, A.J.M., van der Ouderra, F.J., and DeJong, W.W., Exp. Eye Res. 14, 80 (1972).
8. van der Ouderaa, F.J., De Jong, W.W., Hilderink, A., and Bloemendal, H., Eur J. Biochem. 49, 157 (1974).
9. Benedek, G.B., Appl. Opt. 10, 459 (1971).
10. Li, L.K., and Spector, A., Exp. Eye Res. 15, 179 (1973).
11. Van Kamp, G.J., Van Kleef, F.S.M., and Hoenders, H.J., Biochim. Biophys. Acta, 342, 89 (1974).
12. Bettelheim, F.A., Exp. Eye Res. 14, 251 (1972).
13. Bettelheim, F.A., and Mehrotra, K.N., Exp. Eye Res. 14, 259 (1972).
14. Kramps, H.A., Stols, A.L.H., Hoenders, H. J., and De Groot, K., Eur. J. Biochem. 50, 503 (1975)
15. Bloemendal, H., Zweers, A., Benedetti, E.L., and Walters, H., Exp. Eye Res. 20, 463 (1975).
16. Li, L.K., and Spector, A., Exp. Eye Res., 13, 110 (1972).
17. Stauffer, J., Li, L. K., Rothschild, C., and Spector, A., Exp. Eye Res. 17, 329 (1973).
18. Sluyterman, L.A., and Wijdenes, J., Biochim. Biophys. Acta. 200, 593 (1970).
19. Ornstein, L., Ann. N.Y. Acad. Sci. 121, 321(1964)
20. Mörner, C.T., Hoppe-Seyler's Z. Physiol, Chem. 18, 61 (1894).
21. Burky, E.L., and Woods, A.C., Arch. Opthalmol. N.Y. 57, 41 (1928).
22. Bettelheim, F.A., and Ehrlich, S.H., J. Phys. Chem. 67, 1948 (1963).
23. Brunauer, S., Emmett, P.H., and Teller, E., J. Am. Chem. Soc. 60, 309 (1938).
24. D'Arcy, R.L., and Watt, I.C., Trans. Faraday Soc. 66, 1236 (1970).

25. Bull, H., J. Am. Chem. Soc. 66, 1499 (1944)
26. Bull, H., and Breese, K., Arch. Biochem. Biophys. 137, 299 (1970).
27. Bettelheim, F.A., Exp. Eye Res. 21, 231 (1975).

# INTERACTION BETWEEN LIGNOSULPHONATES AND SIMPLE METAL IONS

Tom Lindström
Christer Söremark
*Swedish Forest Products Research Laboratory*

## I. ABSTRACT

This paper deals with an experimental investigation of the interaction between simple ions ($Li^+$, $Na^+$, $K^+$, $TMA^+$, $Ca^{2+}$) and lignosulphonates (LS) in aqueous solutions. The molecular weight of the polymer and the ionic strength of the simple electrolyte have been varied.

In all the experiments, Donnan equilibrium measurements have been used to detect the interactions. The Donnan equilibrium measurements have been carried out using a gel-filtration technique.

It is shown that the salt exclusion coefficient of the LS decreases with increasing molecular weight. The interaction is also strongly dependent on the valency of the counter-ion and slightly dependent on the specific ion when mono-valent ions are considered. Thus the di-valent ion interacts more strongly with the polymer.

From the results with mono-valent ions it is concluded that the interaction between simple metal ions and the polyelectrolytic lignosulphonate ion is determined by the strong electrostatic field surrounding the poly-ion, rather than by a "site binding" of the counter-ions.

Furthermore, an attempt to determine the size of the polymer in solution has been made from data showing the influence of ionic strength on the salt exclusion coefficient. A radius of approximately 16 Angstrom was found for a molecular weight of 12,600.

## II. INTRODUCTION

The lignin in wood is a macromolecule binding the wood cells together. In the pulping process the covalent linkages between the lignin and the carbohydrates present are split and the lignin is broken down to fragments of varying molecular size. During the pulping process the lignin is also solubilized. In the acidic (sulphite) process this is done by introducing sulphonate groups and the product so obtained is called lignosulphonate.

The current state of knowledge indicates that lignosulphonate (LS) is a branched molecule which may be considered as a spherical macromolecule (1). The LS molecule is also a typical polyelectrolyte and thus shows polyelectrolytic behaviour in viscosity measurements (1).

No serious studies have been carried out concerning the interaction of LS with simple ions, although Stenlund (2) has pointed out that some of the counter-ions may be considered to be site-bound. Rezanowich and Goring (1), using the iso-ionic dilution technique to obtain straight plots in their viscosity measurements, have also shown that only a fraction of the counter-ions in a Na-LS sample is free in solution.

The phenomenon that a considerable proportion of the counter-ions act as though they are "bound" to polyelectrolytes in general has been known for a long time (3,4). Kern (3) first discovered the effect when he observed that the counter-ions showed a lowered osmotic activity. Since both conformational and electrochemical properties of polyelectrolytes are dependent on the distribution of ions in the neighbourhood of the polymeric backbone, a considerable amount of both theoretical and experimental work has been done in this field.

It is usually considered that the lowering of the activity of counter-ions can be due to two types of binding (5). One is due to the strong electrostatic field surrounding the poly-ion and is usually termed "ion-atmosphere binding". This effect may in principle be treated by a modified Debye-Hückel theory. The other type of binding is termed "site-binding" where the desolvated counter-ion is supposed to interact with specific sites on the macromolecule, and this kind of binding is treated by the law of mass action.

Although it involves severe experimental difficulties to distinguish between the two kinds of binding, a considerable amount of experimental evidence has been put forward for the existence of "site-binding" by the extensive studies of Strauss and his co-workers (6-11) on different poly-electrolytes.

One of the most powerful tools in the determination of the interactions between small ions and polyelectrolytes is membrane equilibrium measurements.

The Donnan equilibrium measurements have been carried out using a gel-filtration technique originated by Hummel and Dreyer (12-14) and recently extended by the authors (15) to the measurement of Donnan equilibria. This method has here been used in the study of the interactions between different metal ions and lignosulphonates.


III.  EXPERIMENTAL

The lignosulphonate originated from a normal acidic calcium bisulphite cook based on spruce (picea abies). The cooking time was two hours at $135^{\circ}C$ (16). Three molecular weight fractions were prepared by separating the LS on Sephadex gels. The separation of LS has been reported by Stenlund (2,17-20).

The molecular weights were determined by light scattering (16). The characteristics of the samples are found in Table 1.

TABLE 1
*Characteristic Data of Ca-LS (According to Stenlund (21)).*

| Sample no | $M_w$ | Degree of sulphonation (meq/g) | Phenolic hydroxyls (meq/g) | Methoxy content (%) | $A_{280\ nm}$ $1/g^{-1}cm^{-1}$ |
|-----------|-------|-------------------------------|----------------------------|---------------------|----------------------------------|
| 1 | 6000 | 2.06 | 1.37 | 12.3 | 13.10 |
| 2 | 12600 | 1.76 | 1.41 | 12.7 | 13.40 |
| 3 | 20000 | 1.67 | 1.42 | 12.8 | 13.40 |

In order to convert the Ca-lignosulphonates to mono-valent salts a Dowex 50 WX-8 cation-exchange resin bed was used. In all cases the conversion was better than 99.5 % based on an analysis of calcium and of the corresponding metal ion in the converted sample. After the ion exchange procedure the samples were freeze-dried. The degree of sulphonation agreed with the counter-ion content within $\pm$ 2 %. Stenlund's attempts to detect the presence of carboxylic groups have given a negative result (21).

The gel filtration equipment was the same as previously reported (15), a column packed with G-10 Sephadex gel and a continuously recording differential refractometer for detection of the polymer and salt peaks. In one experiment when $Na_2SO_4$ was used, it was necessary to use the G-25 Sephadex gel to achieve good separation between the polymer and the salt. Although the samples were fractionated by gel filtration the presence of low molecular weight impurities was checked on all samples by eluting the samples with distilled water on a G-10 Sephadex column. In no case could low molecular weight impurities be detected.

The salts (commonly chlorides) that were used were of suprapur grade manufacturered by Merck (Darmstadt, Germany). In all experiments, distilled and ion-exchanged water was used with a specific conductivity less than $2 \times 10^{-6} \, \Omega^{-1}$, $cm^{-1}$. The gels were manufactured by Pharmacia Fine Chemicals.

For the sake of clarity, the gel-filtration method will here be briefly described. In a gel filtration column, the LS is eluted with the desired concentration of the simple electrolyte. If a sample of polyelectrolyte is dissolved in the elution liquid, the activity of the simple electrolyte will be increased by an amount that corresponds to the amount of "free" counter-ions contributed by the polymer. As a result two peaks will be present on the refractive index recording. The first peak belongs to the macromolecule excluded from the gel and the second to the salt.

If the salt concentration in the sample is decreased, an equilibrium state will occur at a certain electrolyte concentration where the activity of the electrolyte in the sample equals that of the elution liquid, which is known. In this case no salt peak will be observed in the recorded elution pattern. If

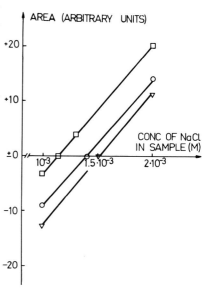

Fig. 1. Area of "salt peak" versus concentration of sodium chloride in the sample for different lignosulphonate concentrations ($M_W$ = 20 000).

□ = 3.0 g/l
o = 2.0 "
▽ = 1.5 "

the salt concentration in the sample is reduced still further, a negative salt peak will be observed.

By an interpolation procedure, the equilibrium salt concentration can be accurately determined. Typical interpolation plots at different polymer concentrations are shown in Fig. 1.

## IV.  THERMODYNAMICS OF MEMBRANE EQUILIBRIUM

Membrane equilibrium data are usually interpreted in terms of the so called salt exclusion coefficient, $\bar{\Gamma}$, (11,22) defined in the following way:

$$\bar{\Gamma} = \lim_{C_2 \to 0} \frac{C_3' - C_3}{C_2 C_p} \tag{1}$$

where $C_3'$ is the concentration of the electrolyte in the "external solution" (polyelectrolyte absent) and $C_3$ is the electrolyte concentration in the "internal solution" (polyelectrolyte present), $C_3'$ and $C_3$ being expressed in moles/litre. $C_2$ is the polyelectrolyte concentration in g/l and $C_p$ is the charge density in equivalent charged groups per gram polymer.

In the following section we denote the water, the polyelectrolyte and the salt as components 1, 2 and 3 respectively; $\mu_j$ and $c_j$ represent the chemical potential and concentration respectively of component j and the primes indicate the external solution where the polyelectrolyte is absent.

At equilibrium the following conditions are fulfilled:

$$\mu_1' = \mu_1 \tag{2}$$

$$\mu_3' = \mu_3 \tag{3}$$

and we can make the following expansion:

$$\mu_3' = \mu_3 + \sigma\mu_3 \tag{4}$$

where:

$$\sigma\mu_3 = \sigma(C_2 C_p) \frac{\partial\mu_3}{\partial(C_2 C_p)} + \sigma C_3 \cdot \frac{\partial\mu_3}{\partial C_3} \tag{5}$$

and the following relationship may thus be derived:

$$\frac{\sigma C_3}{\sigma C_2 C_p} = \frac{\dfrac{\partial \mu_3}{-\partial (C_2 C_p)}}{\dfrac{\partial \mu_3}{\partial C_3}} = \frac{C_3' - C_3}{- C_2 C_p} \tag{6}$$

Suppose the simple salt has the formula $X_A^a Y_B^b$, where $a \cdot A = = b \cdot B$, then its chemical potential is given by

$$\mu_3 = \mu_3^o + RT \ln\left(\frac{C_3}{b}\right)^B + RT \cdot \ln\left(\frac{C_3 + \alpha\, C_2\, C_p}{a}\right)^A + RT \cdot \ln \gamma_3 \tag{7}$$

where $\alpha$ is the degree of dissociation of the polyelectrolyte.

Differentiation of this expression yields:

$$\frac{1}{RT} \cdot \frac{\partial \mu_3}{\partial (C_2 C_p)} = A \cdot \frac{\alpha}{C_3 + \alpha C_2 C_p} + \frac{\partial \ln \gamma_3}{\partial C_2 C_p} \tag{8}$$

$$\frac{1}{RT} \cdot \frac{\partial \mu_3}{\partial C_3} = \frac{B}{C_3} + \frac{A}{C_3 + \alpha C_2 C_p} + \frac{\partial \ln \gamma_3}{\partial C_3} \tag{9}$$

Equations (6), (8), and (9) together with equation (1) then lead to:

$$\bar{\Gamma} = \lim_{C_2 \to 0} \frac{C_3' - C_3}{C_2 C_p} = \frac{A\alpha + C_3 \cdot \dfrac{\partial \ln \gamma_3}{\partial (C_2 C_p)}}{A + B + C_3 \cdot \dfrac{\partial \ln \gamma_3}{\partial C_3}} \tag{10}$$

The activity coefficients are usually omitted and a new parameter called the efficient degree of dissociation ($\beta$) is introduced so that we finally obtain the following working relationship:

$$\bar{\Gamma} = \frac{A \cdot B}{A + B} \qquad\qquad (11)$$

(In this paper the quantity $\dfrac{C_3{}' - C_3}{C_2\, C_p}$ is denoted $\Gamma$.)

## V.   RESULTS AND DISCUSSION

### A.   Molecular Weight

The first factor studied was the molecular weight of the polymer.

Fig. 2 shows that an increased molecular weight of the polymer results in a decreased efficient degree of dissociation ($\beta = 2\bar{\Gamma}$). It can be seen that the efficient degree of dissociation at infinite dilution of the polymer ranges from 58 % to 38 % at the highest molecular weight.

Table 1 shows that another factor, i.e. the charge density, can also influence the efficient degree of dissociation. If the charge density were of decisive importance for the results, it would be expected that an increased charge density would decrease the efficient degree of dissociation. Since the reverse is found to be the case, the effect observed must be mainly due to the influence of the molecular weight.

The increased binding of counter-ions indicates the build-up of the electrostatic field surrounding the macromolecule. This is due to edge effects of the polymer

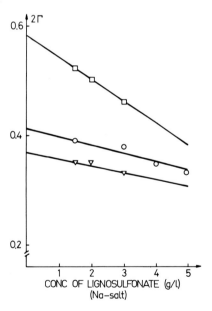

*Fig. 2.  $2\Gamma$ versus concentration of lignosulphonates.*

$M_W$: □ = 6 000

o = 12 000

∇ = 20 000

backbone on the electrostatic field, which are not expected to occur with high molecular weight samples. It has previously been reported (23) that the efficient degree of dissociation is independent of molecular weight at high molecular weights.

The dependence on the molecular weight of the efficient degree of dissociation ($\beta = 3\bar{\Gamma}$, (B = = 2A)) for Ca-LS-samples in calcium chloride is shown in Fig. 3. The efficient degree of dissociation for the di-valent ion also decreases as the molecular weight increases. The di-valent ion is, however, more or less completely trapped in the electrostatic field of the polymer.

In all our experiments it was also observed that in the plots of $2\Gamma$ or $3\Gamma$ versus polymer concentration (Figs 2 and 3) a slightly negative slope was found. In our opinion this must be regarded as being due to a polymer-polymer interaction.

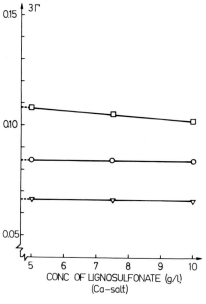

Fig. 3. $3\Gamma$ versus concentration of lignosulphonates.

$M_W$ □ = 6 000

o = 12 600

∇ = 20 000

As we previously pointed out (15), the interpretation of membrane equilibria by the gel filtration method depends to some extent on this slope since a slight dilution of the sample prior to equilibration within the column can hardly be avoided. The relevance of the results are thus maximized if the slope is zero. In our case a considerable slope was observed only in the case of the sodium-lignosulphonate of lowest molecular weight. Studies of Donnan equilibria on other polymers, however, usually show no dependence on polymer concentration of the efficient degree of dissociation (22,23).

B. Specific Ion Interaction

The next factor taken into consideration is the specific nature of the counter-ion. As is evident in Fig. 4, the binding order is $Ca^{2+} \gg Li^+ > Na^+ > K^+ > TMA^+$ for the sample with a molecular weight of 12,600. The difference in inter-action between mono-valent and di-valent counter-ions is due to the stronger electrostatic interaction between the di-valent ion and the polymer.

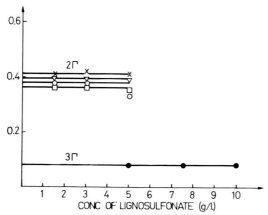

*Fig. 4. 2Γ and 3Γ versus concentration of ligno-sulphonate ($M_W$ = 12 600).*

$x = TMA^+$

$\nabla = K^+$

$o = Na^+$

$\square = Li^+$

$\bullet = Ca^{2+}$

Although the accuracy in the experiments with the mono-valent ions should not be overestimated the regularity cannot be circumvented. The binding order is the same as has been found for polyphosphates (22,24,25) and polycarboxylates (26) but is the reverse of what is found for polysulphonates (24, 27).

The specificity of these metal ions supports evidence for site-binding i.e. the cation leaves its hydration shell and interacts specifically with the macromolecule. As the presence of the sulphonate group itself cannot explain the order found, side-effects due to the presence of phenolic and aliphatic hydroxyls, and methoxy groups may be responsible for the alternation of the binding order.

The presence of a smaller amount of carboxylic groups could have explained the reversed order found, but this idea is not

supported because of the close agreement between the
sulphonate content and the counter-ion content and of the
analysis of Stenlund (21).

In one series of experiments with the sodium lignosulphonate
of $M_W$ = 12,600, the sodium chloride was replaced by sodium
sulphate.  In this case, the "efficient degree of dissociation" remained unaltered within the experimental limits
at a concentration of $2 \cdot 10^{-3}$ mol/l of the simple salts.
It should here be noted that the G-10 gel had to be changed
to G-25 (which separates higher molecular weight species).
Otherwise, the two peaks were more or less superimposed.  The
reason for the change in elution volume with chloride and
sulphate as counter-ions is not understood.  However,  the
influence of salt concentration on its elution volume has
earlier been pointed out (15), indicating the complexity in
the separation of simple salts by GPC.

C.   Influence of Salt Concentration

The influence of ionic strength on the efficient degree of
dissociation at infinite dilution is demonstrated in Fig. 5
for sodium chloride on the
12,600 molecular weight sample.
Fig. 6 illustrates the underlying plots in which the
efficient degree of dis-
sociation is extrapolated to
infinite dilution of the
polymer.

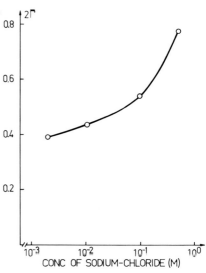

An increase in the efficient
degree of dissociation with
increasing electrolyte concentration (Fig. 5) is
expected from theoretical
considerations (11) and is
also commonly observed (11,
23).  The excluded volume
approach originated by Strauss
et al (22) offers a quantitative explanation of this
phenomenon.  In this model it
is assumed that the simple
electrolyte is excluded by the
polyelectrolyte and "com-
pressed" to an effectively

Fig. 5.  *Efficient degree
of dissociation (2Γ̄) versus
concentration of sodium
chloride (molecular weight of
lignosulphonate 12 600).*

higher concentration $C_3'$. The
simple electrolyte will, however, also shield the electrostatic field surrounding the
polymer. The net result of
these two effects will be an
increase of $2\bar{\Gamma}$ with increasing
electrolyte concentration.

Although not applicable in
their original form these
theories may be extended to
spherical polyelectrolytes in
the following way.

Due to the presence of the
polyelectrolyte the salt is
"compressed" to the higher
salt concentration $C_3'$ which
can be expressed as

$$C_3' = C_3 \left(\frac{V_o}{V_o - V_p}\right) \quad (12)$$

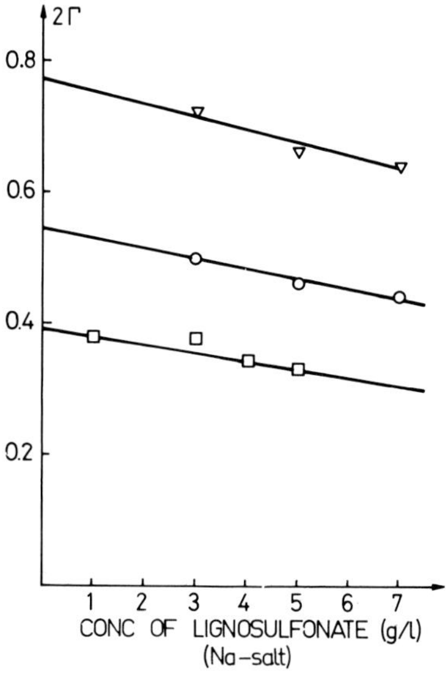

where $V_o$ is the total volume
of the system and $V_p$ is the
volume of the polyions. The
volume of the surrounding
electrostatic field from which the simple ions are excluded
(Donnan exclusion).

Fig. 6. $2\Gamma$ versus concentration of lignosulphonate
($M_W$ 12 600).
$\nabla = 5 \cdot 10^{-1}$ M
$o = 10^{-1}$ "
$\Box = 2 \cdot 10^{-3}$ "

The following rough estimations may be made for the lignosulphonate molecule. Firstly, the molecule is approximated
to a rigid sphere with a radius $r_e$ and assumed to be
relatively monodisperse (16). Secondly, changes in the total
volume of the polyions with changing ionic strength are
attributed to changes in the volume of the electrostatic
atmosphere.

Equation (12) can then be rewritten as

$$C_3' - C_3 = C_3' \cdot \left(\frac{V_p}{V_o}\right) = C_3' \cdot \frac{C_2 \cdot N}{M_W} \cdot \frac{4\pi r^3}{3} \quad (13)$$

where N is the Avogadro number.

227

The radius of a polyion may be written as the sum of its physical radius and the electrostatic radius

$$r = r_e + \frac{s}{\kappa} \tag{14}$$

where s is an arbitrarily chosen form factor with the expected value of unity, and $\kappa$ is the Debye-Hückel's reciprocal shielding length. Equation (13) and (14) can now be rewritten as:

$$\sqrt[3]{\frac{2\bar{\Gamma}}{C_3{}'}} = \sqrt[3]{\frac{N \cdot 8\pi}{M_W \, C_p{}^3}} \, \left(r_e + \frac{s}{\kappa}\right) \tag{15}$$

The Debye-Hückel's reciprocal shielding length $\kappa$ can be calculated at $25°C$ for a salt in which both cation and anion are mono-valent:

$$\kappa = \frac{\sqrt{C_3{}'}}{3.041 \cdot 10^{-9}} \tag{16}$$

If numerical values are inserted, equation (15) can be written:

$$\sqrt[3]{\frac{2\bar{\Gamma}}{C_3{}'}} = 0{,}61 \cdot 10^8 \, \left(r_e + \frac{3{,}041 \cdot 10^{-9}}{\sqrt{C_3{}'}} \cdot s\right) \tag{17}$$

Data from Fig. 5 have been entered into equation (17) and $(2\Gamma/C_3{}')^{1/3}$ is plotted against $(C_3{}')^{-1/2}$ in Fig. 7.

The form factor can be calculated from the slope and was found to be 1.18 compared with the expected value of unity. This indicates that the assumptions regarding the wood-based lignosulphonate polymer are fairly good.

From the intercept, the physical radius ($r_e$) of the ligno-sulphonate molecule ($M_W - 12,600$) can be calculated to 16,3 Angstrom. Viscosity data (28,29) indicated a radius of 21 Angstrom for a Ca-lignosulphonate with a molecular weight of 10,000 (light-scattering, (16)). Our value thus seems to be reasonable for a molecular weight of 12,600.

The experimental results pre-sented support the view that the interaction between the lignosulphonate molecule and the mono-valent salt is due mainly to electrostatic inter-action. Thus, the counter-ions can be regarded as being "trapped" in the electrostatic field ("ion-atmosphere-binding") and not bound to the charged sites on the molecule ("site-binding").

This view is primarily supported by the low specifi-city of the counter-ions and the ionic strength dependency of the efficient degree of dissociation.

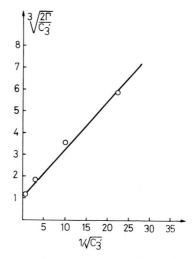

*Fig. 7.  Test plot of equation (17).*

VI.  ACKNOWLEDGEMENTS

Thanks are due to Dr Stenlund of the Central Laboratory of the Finnish Pulp and Paper Industry for preparation and characterization of the calciumlignosulphonate samples. Mrs Gunborg Glad-Nordmark is also due to thanks for her skillful and careful experimental assistance in this study.

VII.  REFERENCES

1.  Rezanowich, A. and Goring, D., J. Colloid Sci. 15, 452-471 (1960).
2.  Stenlund, B., Pap. Puu, No. 10, 671 (1970).
3.  Kern, W., Z. Physik Chem. A181, 240, 283 (1938); A184, 201 (1939).
4.  Huizenga, J.R., Grieger, P.F. and Wall, R.T., J. Am. Chem. Soc. 72, 2636, 4228 (1950).
5.  Morawetz, Fortschr. Hochpol. Forsch. 1.1 (1958).
6.  Strauss, U.P., J. Phys. Chem 66, 2235 (1962).
7.  Strauss, U.P. and Wineman, P.L., J. Am. Chem. Soc. 80, 2366 (1958).
8.  Ross, P.D. and Strauss, U.P., J. Am. Chem. Soc. 82, 1311 (1960).
9.  Strauss, U.P., Woodside, D. and Wineman, P., J. Phys. Chem. 61, 1353 (1957).

10. Strauss, U.P. and Ross, P.D., J. Am. Chem. Soc. 81, 5299 (1959).
11. Gross, L.M. and Strauss, U.P., Chemical Physics of ionic solutions (B.E. Conway and R.G. Barradas, Eds.), 361 (1966). John Wiley and Sons Inc., N.Y.
12. Hummel, J.P. and Dreyer, W.J., Biochim. Biophys. Acta 63, 530 (1962).
13. Fairclough, G.F. and Fruton, J.S., Biochemistry 5 No. 2, 673 (1966).
14. Colman, R.F., Analytical Biochemistry 46, 358 (1972).
15. Lindström, T., de Ruvo, A. and Söremark, Ch., to be published.
16. Forss, K. and Stenlund, B., Pap. Puu, No. 1, 93, 97 (1969).
17. Stenlund, B., Pap. Puu, No. 2, 55 (1970).
18. Stenlund, B., Pap. Puu, No. 3, 121 (1970).
19. Stenlund, B., Pap. Puu, Special issue No. 49, 197 (1970).
20. Stenlund, B., Pap. Puu, No. 5, 333 (1970).
21. Stenlund, B., Personal communication.
22. Strauss, U.P. and Ander, P., J. Am. Chem. Soc. 80, 6494 (1958).
23. Eisenberg, H. and Casassa, E.F., J. Polym. Sci. XLVII, 29 (1960).
24. Strauss, U.P. and Leung, Y., J. Am. Chem. Soc. 87 (7), 1476 (1965).
25. Strauss, U.P., Helfgott, C. and Pink, H., J. Phys. Chem. 71, 2550 (1967).
26. Gregor, H.P., Hamilton, M.J., Oza, R.J. and Bernstein, F., J. Phys. Chem. 60, 263 (1956).
27. Boyd, G.E. and Soldano, B.A., Z. Electrochem. 57, 162 (1953).
28. Le Bell, J., Bergroth, B., Stenius, P. and Stenlund, B., Pap. Puu, No. 56, 463 (1974).
29. Gupta, P.R. and Mc Carthy, J.L., Macromolec. 1 (1968):3, 236-244.

# INTERACTION OF MONOLAYERS OF
## POLYVINYLPYRROLIDONE-POLYVINYL ACETATE GRAFT
## COPOLYMERS WITH P-HYDROXYBENZOIC ACID DERIVATIVES

Joel L. Zatz
*Rutgers College of Pharmacy*

*The penetration of monolayers of polyvinyl-
pyrrolidone-polyvinyl acetate copolymers and of
polyvinyl acetate by p-hydroxybenzoic acid and two
of its esters was investigated by means of surface
pressure measurement. There was correspondence
between the relative effectiveness of the compounds
in increasing the surface pressure of the monolayers and their activity at the air-water surface.
The degree of penetration was also related to the
vinylpyrrolidone content of the monolayers, apparently because of hydrogen bonding with the phenolic
group of the hydroxybenzoates.*

## I.    INTRODUCTION

Polyvinyl acetate is capable of forming stable
monolayers at the air-water interface, while polyvinylpyrrolidone (PVP) is not (1). The two materials
have been combined in the form of copolymers with
vinyl acetate grafted onto a backbone of PVP (2).
These copolymers may be spread to yield stable
monolayers whose properties, such as collapse
pressure and compressibility depend on PVP content
of the copolymers (3).  PVP segments occupy a
portion of the surface, held there by the anchoring
polyvinyl acetate grafts (1,3).  Compression forces
the hydrophilic PVP segments into the subphase.
The p-hydroxybenzoates utilized in this study
are employed as antimicrobial preservatives, and
this use is partly due to the ability of the compounds to interact with the cell membranes of micro-

organisms (4). All of them are bound by PVP in bulk
solution (5). By spreading polyvinyl acetate and
PVP copolymers on aqueous substrates containing
these low molecular weight compounds and measuring
changes in surface pressure of the monolayers, it is
possible to determine whether the affinity for PVP
has any effect on penetration of the p-hydroxybenzoates into the surface layer. Preliminary results
(6) suggested that the anisotropic nature of the
interface may be an overriding factor. Benzoic acid
appears to interact to a greater extent with the
monolayers than does p-hydroxybenzoic acid, although
the latter is more extensively bound by PVP in bulk
solution (5).

## II. MATERIALS AND METHODS

Polyvinyl acetate (Gelva V7, Monsanto Corporation) was purified prior to use (1) by gel filtration. Two PVP/VA copolymers (GAF corporation) were
used in this study. Copolymer 735 (61.2% w/w PVP)
and copolymer 335 (29.2% w/w PVP) were purified
using the procedure previously described (3). Water
was deionized and distilled from an all glass still.
p-Hydroxybenzoic acid, methyl p-hydroxybenzoate and
propyl p-hydroxybenzoate were reagent grade.

The apparatus employed was the same as that
used in previous experiments (1). Subphase temperature was maintained at 25.0 ± 0.1°C.

The p-hydroxybenzoates were dissolved in the
subphase prior to spreading of the monolayer. Polyvinyl acetate was spread from benzene solution.
Isopropanol-benzene was the solvent for the copolymers. The monolayers were compressed manually and
allowed to stand for one minute before each reading
of the surface pressure was made. The readings were
independent of the length of time that the monolayer was permitted to stand prior to compression.
No drifting of surface pressure with time was
observed. When the polymer concentration at the
interface was increased by spreading additional
polymer solution, the surface pressure values agreed
with those obtained by compression to equivalent
polymer surface areas. The surface pressure data
reported in this paper are therefore believed to
represent quasi-equilibrium values.

## III.   RESULTS AND DISCUSSION

Figure 1 shows the surface tension of aqueous solutions of p-hydroxybenzoic acid and of the methyl and propyl esters.  Propyl p-hydroxybenzoate was the most surface active of the compounds studied.  Solutions of p-hydroxybenzoic acid had essentially the same surface tension as pure water at the concentrations studied.

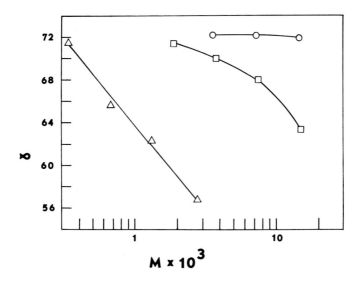

*Fig. 1.  Surface tension, mNm$^{-1}$, as a function of solution concentration at 25°.  (O) p-hydroxybenzoic acid;  (□) methyl ester; (▲) propyl ester.*

The monolayer results shown in Figs. 2 and 3 are represented by plotting $\Delta\pi$, the change in surface pressure at constant area due to penetration of the monolayer, as a function of $\pi_i$, the initial surface pressure of the monolayer on a pure water substrate.  Figure 2 shows the change in surface pressure of copolymer 735 as a result of penetration of propyl p-hydroxybenzoate from aqueous solution. $\Delta\pi$ was a function of subphase concentration, but the three curves exhibited the same general characteristics.  At low values of initial surface pressure, (corresponding to relatively large values of surface area/molecule of the copolymer), the increase in

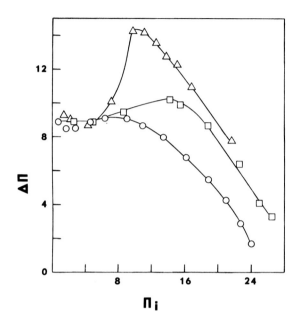

*Fig. 2. Increase in surface pressure, $mNm^{-1}$, of monolayers of copolymer 735 on subsolutions containing dissolved propyl p-hydroxybenzoic acid at various concentrations. (▲) .347 x $10^{-3}$M; (□) .694 x $10^{-3}$M (O) 1.39 x $10^{-3}$M.*

surface pressure was approximately equal to the decrease in surface tension at a clean water surface at the same bulk concentrations.

$\Delta\pi$ rose steeply as the monolayer was compressed reaching a peak at an initial surface pressure of about 11 $mNm^{-1}$. Thereafter, continued compression caused a decrease in $\Delta\pi$, presumably reflecting expulsion of the propyl p-hydroxybenzoate from the monolayer.

In Figure 3, interaction of monolayers of the various polymers studied with the propyl ester at a singly bulk concentration is shown. At low initial surface pressure values, there was not much difference in $\Delta\pi$ among the monolayers. As the initial surface pressure was raised, $\Delta\pi$ increased, though to a different extent for each of the monolayers. Continued compression resulted in a decrease in $\Delta\pi$

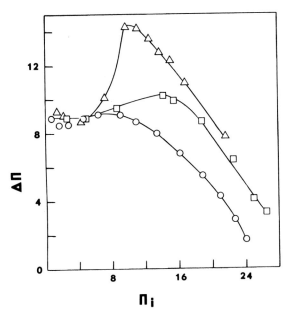

*Fig. 3. Increase in surface pressure, mNm$^{-1}$, of monolayers on subsolutions containing propyl p-hydroxybenzoate, 1.39 x 10$^{-3}$M. (O) polyvinyl acetate; (□) copolymer 335; (▲) copolymer 735.*

in all cases. The polyvinyl acetate monolayer, which is the most compact of the group (1), contained practically no propyl p-hydroxybenzoate at the collapse point. The PVP copolymers, on the other hand, did retain some of the ester, even at collapse. The same was true of other subphase concentrations of propyl p-hydroxybenzoate.

The results shown in Figs. 2 and 3 were typical of those obtained with all of the systems studied. To facilitate comparisons among the polymers, 12 mNm$^{-1}$ was chosen as a standard initial surface pressure. At this value, all of the polymers occupy approximately the same area per vinyl acetate segment (1) and, also, $\Delta\pi$ is nearly maximal in all cases.

The $\Delta\pi$ values due to penetration of copolymer 735 are plotted in Fig. 4. The extent of interaction with the copolymer was directly related to surface activity at the air-water surface, the propyl ester causing the greatest change in surface

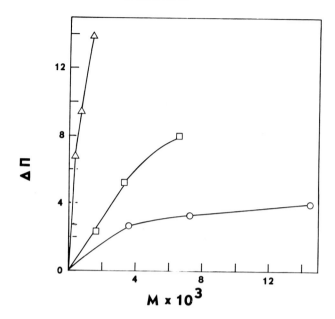

Fig. 4. *Increase in surface pressure, $mNm^{-1}$, of a monolayer of copolymer 735 at an initial surface pressure of 12 $mNm^{-1}$ as a function of subsolution concentration. (O) p-hydroxybenzoic acid; (□) methyl ester; (▲) propyl ester.*

pressure and the acid the smallest change. In bulk solution, the order of binding by PVP is p-hydroxybenzoic acid > propyl ester > methyl ester (5). These findings confirm the importance of amphipathicity in monolayer penetration. Apparently p-hydroxybenzoic acid interacts least with the monolayers because the molecule has polar groups at both ends. Conversion to an ester increases both surface activity and the degree of monolayer penetration, and this effect becomes more pronounced as the ester chain is lengthened.

When $\Delta\pi$ values of the monolayers on the same subsolution are compared (Fig. 5) the influence of the affinity between PVP and the preservatives becomes apparent. With a given compound at a single concentration, copolymer 735, with the highest PVP content, consistently gave the largest change in $\Delta\pi$ while polyvinyl acetate showed the smallest change. Even p-hydroxybenzoic acid caused an increase in

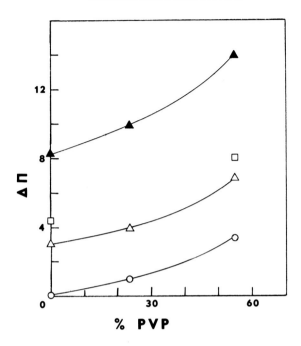

*Fig. 5.   Increase in surface pressure, mNm⁻¹,
of monolayers at an initial surface pressure of
12 mNm⁻¹ as a function of the per cent of PVP
residues in the polymer.   (O) p-hydroxybenzoic acid,
7.25 x 10⁻³M;   (□) methyl ester, 6.57 x 10⁻³M;
(△) propyl ester, .347 x 10⁻³M;   (▲) propyl ester,
1.39 x 10⁻³M.*

$\Delta\pi$ of the PVP copolymers, although there was
negligible influence on polyvinyl acetate.   Some
penetration therefore occurs into the vinylpyrrolidone residues that are constrained to remain at the
surface, most likely because of hydrogen bonding
between the phenolic group of the p-hydroxybenzoates
and the amide function of PVP (7).   The extent of
the interaction with PVP is practically independent
of the surface activity of the subsolution.   We thus
have an interesting "piggy-back" situation, in which
a molecule is held at the surface by PVP segments
that are themselves dependent on the polyvinyl
acetate grafts for their residence at the surface.

IV.  REFERENCES

(1)  Zatz, J.L. and Knowles, B., J. Colloid Interface Sci., 40, 475 (1972).

(2)  PVP/VA Technical Bulletin, GAF Corp., N.Y.

(3)  Zatz, J.L. and Knowles, B., J. Pharm. Sci., 60, 1731 (1971).

(4)  Wedderburn, D.L., "Advances in Pharmaceutical Sciences," Vol. I  p233. Academic Press, New York, 1964.

(5)  Jürgensen Eide, G. and Speiser, P., Acta Pharm. Suecica, 4, 185 (1967).

(6)  Zatz, J.L., J. Pharm. Sci., 61, 977 (1972).

(7)  Higuchi, T. and Kuramoto, R., J. Pharm. Sci., 43, 398 (1954).

V.  ACKNOWLEDGEMENTS

Supported by a Merck Grant for Faculty Development.  The author thanks Ms. Beverly Knowles and Mr. Kam-Bor Ip, for technical assistance, and Ms. Lucille Lear for typing the manuscript.

THE ELASTICITIES OF ALBUMIN MONOLAYERS[1]

Martin Blank and Beatrice B. Lee
Department of Physiology
Columbia University
College of Physicians and Surgeons

ABSTRACT

The elasticities of adsorbed and spread monolayers
of bovine serum albumin and of egg albumin have been
measured by the dynamic method.  At the same surface
pressures, there are distinct differences between the two
albumins, but the differences between the adsorbed and
spread films are due solely to the pH and ionic strength
of the spreading and subphase solutions.  In the case of
rapidly adsorbing films, there is a metastable high elasticity (between 40 and 70 dynes/cm), above a surface
pressure of 4 dynes/cm, that is directly proportional to
the adsorption rate.  The slower surface pressure and
elasticity changes during continued measurements occur
only in films where the surface pressure is greater than
4 dynes/cm.  From these observations we conclude that:
1) the structure and elasticity of adsorbed and spread
albumin films are the same at the same pressures, 2) the
magnitude of the elasticity is determined by the solution
that is the first subphase for the film, and 3) the conformational changes in the formed surface films are relatively rapid up to about $\Pi$ = 4 dynes/cm., but much slower above that pressure.

I.   INTRODUCTION

It has generally been believed that the structure (i.e.,
the relative amounts of adsorbed segments and dissolved loops)
of a protein or polyamino acid molecule at a surface is related to the surface pressure against which it spreads (1).
Thus, the differences between the physical properties of

---

[1]  Supported in part by Research Grant HD 06908 from the
USPHS.

spread and adsorbed protein films are due to the different
film structures that result when molecules are allowed to
reach the surface at about zero surface pressure in the case
of spread films, or under gradually increasing surface pressures. However, it has now been shown that monolayers of
biopolymers reflect the conformations of the molecules in the
solutions prior to their presence at the surface (2,3). The
α-helix structure of the biopolymer that exists in polar solvents is retained in the surface film that it forms, and compression or changes in adsorption do not necessarily lead to
conformational changes in the surface molecules. Both studies, Malcolm's investigation of modified polylysine (2) and
Loeb and Baier's on modified polyglutamate (3) were based on
measurements using transferred monolayers. Since there are
always questions about the relation between the transferred
and original monolayers, we decided to try to demonstrate
similar effects on protein monolayers in situ.

We chose to investigate the structure and mechanical
properties of serum and egg albumins at interfaces, largely
because the monolayers of these substances are already fairly
well characterized. Within the past few years, we have studied the elasticities of spread monolayers by the dynamic
method (4), as well as the factors that influence the adsorption kinetics (5). We have also shown that the interaction
between protein molecules on a surface, especially at higher
surface pressures, can cause the formation of a two dimensional network (6) with a well defined yield (7,8). The
transition to the high surface pressure range is apparently
well defined, and at that point the strength of the intermolecular interactions leads to different physical properties.
In this study, we have found a metastable peak elasticity for
adsorbed films in this range.

Finally, we have been interested in the rheological properties of protein monolayers at interfaces, because these
properties could be characteristic of related proteins at
membrane interfaces. There are many parallels between the behavior of membranes and protein monolayers with regard to
rheology, and there is reason to believe that membrane rheology is really the result of the properties of its protein
subunits (9,10) in monolayer like arrangements.

---

Fig. 1. The elasticity, E, of egg albumin monolayers adsorbed from 0.0002% - 0.0007% solutions in 0.1M NaCl at pH
4.5, vs. the surface pressure, Π. The solid line follows the
data points up to Π = 5 dynes/cm.; the dashed line indicates
the position of data obtained for egg albumin monolayers
spread from the crystal on distilled water (4).

## II. EXPERIMENTAL PROCEDURE

The methods used in this study have been described in earlier papers (4,5). In brief, to study adsorbed monolayers, the solutions (at 25°C.) were allowed to equilibrate in a Teflon trough and the surface was swept to establish t = 0. Spread monolayers were deposited along a glass rod from aqueous solutions that were 0.1M NaCl and had a pH of 4.5. In both cases the movable barrier was started after the surface tension was recorded, and the values of the experimental parameters were generally $\Delta A/A$ = 0.0393 and frequency = 10 per minute. The elasticity was then calculated for each cycle as

$$E = -\frac{\Delta \Pi}{\Delta A/A}$$

The experiments were done with different concentrations of the two proteins: bovine serum albumin (BSA) and egg albumin (EA), both obtained from Sigma Chemical Company. The various pH's were arranged with the following salt mixtures: pH <4 (HCl), 4 <pH <5.5 (Na Acetate-HCl), 5.5 <pH <7 ($KH_2PO_4$-$K_2HPO_4$), 7 <pH <10 ($Na_2CO_3$ - $NaHCO_3$).

## III. RESULTS

During adsorption, the elasticity of a protein monolayer increases as expected. Figures 1 and 2 show the elasticity

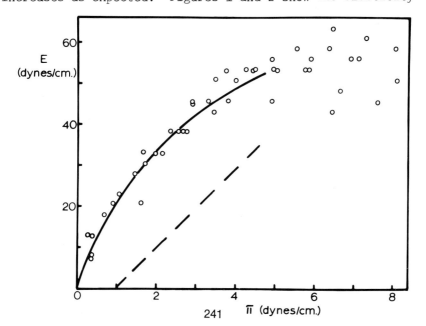

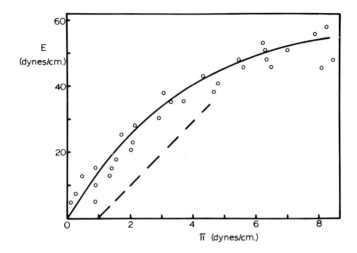

Fig. 2. The elasticity, E, of bovine serum albumin monolayers adsorbed from 0.0003% - 0.0007% solutions in 0.1M NaCl
at pH 4.5, vs. the surface pressure, Π. The solid line summarizes these data points; the dashed line indicates the position of data obtained for serum albumin monolayers spread from
the crystal on distilled water (4).

of the adsorbed protein monolayers as functions of the surface
pressure. Since these data correspond to adsorption rates
that vary by over an order of magnitude (5), the elasticity
appears to be independent of the adsorption rate. At comparable surface pressures, the egg albumin (EA) has a somewhat
(about 20%) higher elasticity than the serum albumin (BSA).
As was noted in an earlier study of spread films (1), EA elasticity also shows an erratic pattern above about Π = 5 dynes/
cm., unlike the BSA. Aside from these relatively small differences, the adsorbed films of the two globular proteins are
quite similar.

For both proteins, the E (Π) curves for the adsorbed
monolayers are quite different from the values obtained earlier for monolayers spread from the crystal on to distilled
water (1), and shown as the dashed lines in figures 1 and 2.
However, these differences are not due to the differences between adsorbed and spread films of the same substances. Comparing the elasticities of adsorbed ($E_a$) and spread ($E_s$) films
on the same aqueous solutions and at the same Π, as in figures 3 and 4, we see that there is no fixed relation between
the two film properties. (Similar results have been obtained
for BSA films under the same conditions.) That is, the elas-

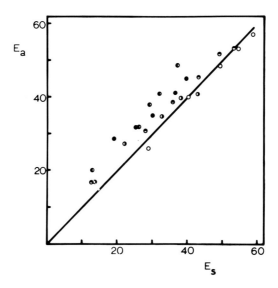

Fig. 3. $E_a$, the elasticity of monolayers of egg albumin, adsorbed from 0.0007% solutions at various pH's vs. $E_s$, the elasticity of the same monolayers when spread from solutions of the same composition as the adsorbed films but on to a subphase at 4.5. The different pH's are as follows: ☻, 1.5; ◑, 2.4; ○ 4.5; ◐, 6.5; ●, 8.0.

ticity for an adsorbed film can be less than, equal to or greater than that for a spread film, depending upon the experimental conditions.

The two sets of points on the correlation lines of figures 3 and 4 represent the experiments where the spreading solutions for the spread films and the subphase solutions for the adsorbed films were at the same pH and ionic strength, i.e. pH = 4.5 and $\mu$ = 0.1. When the spread monolayers were studied on subphases at pH's or $\mu$'s removed from the spreading conditions, they retained the characteristics of films formed under the original conditions. This means that in figure 3, the elasticities of the monolayers are higher when the pH is displaced from the IEP, and in figure 4 we also see the surface "salting in" and "salting out" effects that were noted in earlier studies of the rheology of protein films (6). It appears that the more expanded and "salted in" films give higher elasticities, and the elasticities are characteristic of the subphase upon which the film first forms. From figure 4 we see that the elasticity of a spread film on distilled water gives values that are the same as the earlier re-

243

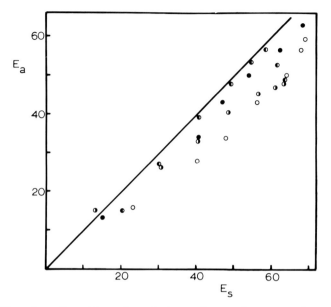

Fig. 4. $E_a$, the elasticity of monolayers of egg albumin, adsorbed from 0.0007% solutions at pH 4.5 and at various salt concentrations vs. $E_s$, the elasticity of the same monolayers when spread from solutions of the same composition as the adsorbed films but on to a subphase of 0.1M NaCl. The different salt compositions are as follows: O, $H_2O$; ◐, 0.01M NaCl; ◑, 0.1M NaCl; ●, 1M NaCl.

sults summarized by the dashed lines in figures 1 and 2.

As a measure of the expected behavior of an adsorbed protein film we can take the elasticity at $\Pi = 5$. This property is shown as a function of pH in figure 5. By and large, the magnitude of the elasticity appears to depend upon the electrical repulsive force in the interface, so that at the IEP and also at high salt concentration, E is at its greatest value.

Spread protein films have a maximum elasticity at a surface pressure of about 5 dynes/cm, the same surface pressures at which the films first show yield phenomena (7,8). These effects occur whether the film is spread from the crystal or from aqueous spreading solutions, and adsorbed films behave in the same way. However, adsorbed films also show an unusual transient response that appears to depend upon the rate of adsorption.

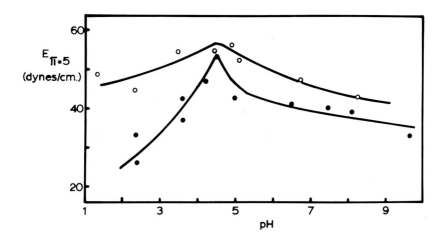

Fig. 5. $E_{\Pi = 5}$, the elasticity at a surface pressure of 5 dynes/cm for spread monolayers of egg albumin ( O ) and serum albumin ( ● ) as a function of pH. The subphase in all cases is at 0.1M ionic strength.

Figure 6 shows the variation of E with the time for films that adsorb at a very fast rate. In these cases, it appears that the elasticity reaches a peak value, Ep. The value of Ep varies between 40 and 70 dynes/cm and occurs above a surface pressure of 4 dynes/cm. All of the different peak values under a variety of experimental conditions appear to depend simply on the rate (see figure 7). Under the conditions of the same adsorption rate, Ep increases with the magnitude of Δ A/A used in the measurement and also with the frequency.

The extrapolated intercepts at zero rate for the transiently elevated elasticity correspond to about Π = 3 dynes/cm, near but below the point of the maximum elasticities under "equilibrium" conditions. It would appear that the further one goes from equilibrium conditions, the greater the magnitude of Ep. There is a higher plateau value as well as a higher peak elasticity, and the plateau values show changes with time (4).

IV. DISCUSSION

A. Determinants of Protein Monolayer Structure

In general, one would expect some structural differences between spread and adsorbed protein monolayers as a result of the variety of spreading solutions and subphases studied.

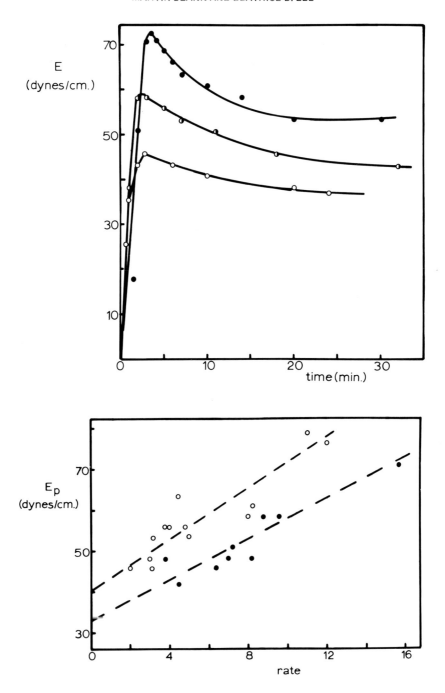

Figure 7

Loeb and Baier (3) showed large structural differences in polymethyl glutamate monolayers that correlated with variations in the polarity of the solvent. In the aqueous systems we have studied, the differences appear to be due solely to the variation of the pH and ionic strength of the solutions. This is true even when comparing films spread from solutions and from the crystal. In all of the systems, the important determinant of film structure was not related to the type of monolayer (i.e., spread or adsorbed), but dependent upon the composition of the solution that forms the first subphase. In particular, the polarity of the solute, i.e., the electrical properties, determined the elasticity and the pressure-area isotherm of the monolayers.

The composition of the first subphase is the major factor in determining albumin monolayer structure, but the time or film age is also important. The time becomes especially important at higher surface pressures, perhaps as a consequence of multilayer formation upon compression (2) or continued adsorption (12). Benjamins et al. (12) have indicated that multilayer formation does not affect the elasticity, so this is an unlikely source of the change. Most probably the cause of the elasticity change is an increase in the intermolecular interactions (6) that occur at higher surface pressures even under constant area conditions.

B. The Transient Peak Elasticity

In an earlier study we noted that the changes in elasticity with time occurred above $\Pi = 4$ dynes/cm, a transition region in many of the surface properties of these protein monolayers (4,8). The transient peak elasticity, shown in figures 6 and 7, extrapolates back to zero rate to give the elasticities that correspond to about $\Pi = 3$ dynes/cm. This is somewhat lower than the point of maximum elasticity and still in the "reversible" range for these proteins. However,

---

Fig. 6. The elasticity, E, of adsorbed monolayers of egg albumin as a function of the time after adsorption starts. The solutions were all at pH 4.5 and contained 0.1M Na Cl and the following concentrations of albumin:  0,  0.001%;  ◐, 0.0015%;  ●, 0.002%.

Fig. 7. Ep, the peak elasticity of an adsorbed monolayer, as seen in figure 6, as a function of the rate of adsorption. The rate is the slope of the $\Pi$ vs $\sqrt{\text{time}}$ curve and has the units dynes/cm. $\sec^{\frac{1}{2}}$. The two dashed lines with the different intercepts relate to ( 0 ) egg albumin and to ( ● ) serum albumin respectively.

we should point out that the peak became smaller as we approached $\Pi = 4$ dynes/cm, and we did not really observe any transient peaks below this surface pressure. The extrapolation may
therefore not be valid, and $\Pi = 4$ dynes/cm probably represents
a true threshold.

The existence of strong interactions in protein monolayers above $\Pi = 4$ dynes/cm has been shown to lead to a two dimensional yield (7,8), as well as large surface pressure gradients (11). These interactions would also contribute to the
occurrence of a metastable supernormal elasticity by slowing
down the respreading process when the barrier is extended.
The surface pressure would drop more than expected, and there
would be a larger $\Delta \Pi$ and a larger value of E. Also, because
the respreading is restricted additional molecules adsorb during the expanded phase, leading to a higher value of $\Pi$. This
picture of the dynamic changes at the surface fits in with the
gradual decrease of the effect with continued measurement as
well as with the effects of the experimental parameters on the
magnitude of Ep. When the adsorption rate is constant, Ep increases with the magnitude of $\Delta A/A$ and with the frequency.

## C. Albumin Monolayers as Models for Membrane Rheology

Natural membranes (13) are composed of lipids and proteins in an approximately 1:1 ratio. The lipids, which can be
phospholipids, neutral lipids, glycolipids, etc., are present
in the form of a bilayer, while the proteins (also present to
some extent in combined form as lipo-proteins or glyco-prot-
eins) are distributed both through the bilayer and on the two
surfaces. There appears to be enough protein for several monolayers in the membrane of the red cell (9), a membrane that
has been studied in great detail both with regard to the physical properties of the intact membrane and the arrangement of
its components. There does appear to be a strong parallel between the rheological properties of albumin films and those of
the membrane, suggesting that a protein layer in the cell (perhaps the "spectrin" that is present on the inner face of the
membrane) has properties that are similar to those of albumin
at an interface (9, 10).

We can summarize some of the similarities as follows:
a. The elasticities of albumin monolayers, when converted to
three dimensional values are equivalent to about $10^7$ dynes/cm$^2$,
and the measured values for erythrocyte membranes are in the
region of $3 \times 10^7$ dynes/cm$^2$ (14).
b. The strong binding of multivalent cations to proteins

causes large changes in the yield stress, a measure of the strength of a surface network that is based on the pressure difference needed to start flow through a two-dimensional channel. For a protein monolayer, the variation of the yield stress due to cations in the aqueous subphase bears a strong resemblance to the variation of erythrocyte rigidity with some of the same ions (8). The recent finding that membrane rigidity decreases when $Zn^{++}$ displaces $Ca^{++}$ parallels the monolayer results.

c. The variation of protein monolayer elasticity with time (4,6) and with changes in surface charge (see Figure 5) shed light on the reported changes of membrane properties with age. The variation of a protein film viscosity with lipid content (15) also fits in with the known membrane changes with age.

All of these observations suggest that the factors which control the rheology of natural membranes are the same as those that cause parallel changes in monolayer properties. It is, therefore, likely that membrane proteins, similar to the albumins with regard to surface properties, are the molecules which affect membrane rheology.

## V.   REFERENCES

1.   Miller, I.R. and Bach, D.  Surface and Colloid Sci. 6, 185 (1973).
2.   Malcolm, B.R.  Biochemical J. 110, 733 (1968).
3.   Loeb, G.I. and Baier, R.E.  J. Colloid Interface Sci. 27, 38 (1968).
4.   Blank, M., Lucassen, J. and Van den Tempel, M.  J. Colloid Interface Sci. 33, 94 (1970).
5.   Blank, M., Lee, B.B. and Britten, J.S.  J. Colloid Interface Sci. 50, 215 (1975).
6.   Blank, M.  J. Colloid Interface Sci. 29, 205 (1969).
7.   Blank, M. and Britten, J.S.  J. Colloid Interface Sci. 32, 62 (1970).
8.   Blank, M., Lee, B.B., and Britten, J.S.  J. Colloid Interface Sci. 43, 539 (1973).
9.   Blank, M. and Britten, J.S.  Biorheology 12, 271 (1975).
10.  Blank M. in "Membranes and Diseases" (L. Bolis, J. F. Hoffman and A. Leaf, Eds.) p. 81.  Raven Press, New York, 1976.
11.  Blank, M. and Lee, B.B.  J. Colloid Interface Sci. 36, 151 (1971).
12.  Benjamins, J., deFeijter, J.A., Evans, M.T.A., Graham, D.E. and Phillips, M.C.  Faraday Discussion 59 (1975).
13.  Bretscher, M.S.  Science 175, 720 (1972).
14.  Skalak, R., Tozeren, A., Zarda, R.P. and Chien, S.  Biophys. J. 13, 245 (1973).

# INTERACTIONS OF ALKALI HALIDES WITH INSOLUBLE FILMS OF FATTY AMINES AND ACIDS

V. A. Arsentiev[1] and J. Leja

*Department of Mineral Engineering*
*University of British Columbia*
*Vancouver, Canada*

## ABSTRACT

*A Langmuir balance was modified, using an additional torsion wire device, in order to measure a vertical force. With this device, the "immersion" forces necessary to submerge solid KCl and NaCl crystals, attached to insoluble alkyl amine or fatty acid films spread on saturated salt solutions were determined and correlated with the surface pressures exerted by the films. From these determinations, employing certain assumptions, some types of interaction energy of surfactants with solid alkali halides were evaluated. In consequence, a mechanism of adsorption for insoluble collectors, leading to the flotation of solid halides from their saturated salt solutions, is suggested.*

## I. INTRODUCTION

Selective separation of solid KCl or NaCl from each other by flotation in their saturated aqueous solutions, and away from various other minerals, is practiced on a very large scale. Nearly 100 million tons of potash ores in the world are being treated by flotation annually for the purpose of producing suitable grades of potash fertiliser. The collectors for floating KCl, established empirically, are straight chain amines. Fatty acids could be used for flotation of NaCl but they are not sufficiently selective with respect to other minerals contained in potash ores.

A number of hypotheses have been suggested to explain the mechanism of soluble salt flotation, which appears to differ from the mechanism of insoluble mineral flotation, such as that of metallic sulphides. Häblich [1] assumed that the

---

[1] On Canada Council Exchange Fellowship, from the Leningrad Mining Institute, Department of Mineral Processing, Leningrad, U.S.S.R.

formation of insoluble compounds is responsible for the adsorption of collector-acting surfactants on halides. Neunhoeffer [2] explained salt floatability in terms of a collector/metal complex formation. Gaudin [3,4] suggested that the similarity of the ionic sizes of $RNH_3^+$ and $K^+$ is responsible for the floatability of KCl with primary amines. Fuerstenau [5] expanded this hypothesis to include other alkali halide salts. Bachmann [6] postulated that the flotation of KCl using primary amines is due to a favourable correlation in the dimensions of the KCl crystal lattice and the quasicrystalline structure of the condensed amine film. Rogers and Schulman [7,8] drew attention to the role of the heat of solvation and the degree of mineral surface hydration. They also recognized the important contribution made by association of the alkyl hydrocarbon chains to the adsorption energy and in the flotation of halides. These ideas were further pursued more recently by Singewald [9], Schubert [10,11], Roman et al [12].

What has not been attempted, as yet, is

1. a determination of a relationship between the structure of the surfactant film and the role of this film in attachment of halide crystals, and in salt flotation, and

2. a measurement of the force required for breaking the contact between a salt crystal in a saturated salt solution and the surfactant film of known structure present at the solution/air interface.

These two aspects were the objectives of the present investigation.

## II. EXPERIMENTAL

The reagents investigated consisted of $C_{12}^-$, $C_{14}^-$, $C_{16}^-$ and $C_{18}^-$ unsubstituted amines, $C_nH_{n+1}NH_2$, and the corresponding carboxylic acids. These reagents are referred to as n-alkyl amines or n-alkyl acids. In addition, oleyl amine $[CH_3(CH_2)_7CH=CH(CH_2)_8NH_2]$, dimethyl-hexadecyl amine $[C_{16}H_{33}N(CH_3)_2]$, tri-methyl octadecylammonium bromide $[C_{18}H_{37}N^+(CH_3)_3Br^-]$, octadecylamine hydrochloride $[C_{18}H_{37}NH_3Cl]$, oleic acid $[CH_3(CH_2)_7CH=CH(CH_2)_7COOH]$, and the sodium salt of octadecanoic acid $[C_{17}H_{35}COONa]$ were used.

All the reagents were of high purity, the main constituent representing 98 to 99% of the total. They were spread on saturated solutions of KCl and/or NaCl. The spreading of surfactants as films was carried out using solutions of surfactants in hexane (for amines and acids) or in a 1:1 mixture of benzene-ethyl alcohol (for salts of amines and of acids). The characteristics of the solutions used are given in Table 1. The temperature was kept at 20±1°C. during all the experiments.

The platelets of NaCl and KCl crystals used were cleaved from an IR window stock material. No additional surface preparation was employed.

The experimental procedure was based on the use of a Langmuir trough with an additional torsion wire device adapted from the du Nouy Ring method to measure the force of submerging the crystal attached to the spread surfactant film.

The procedure was as follows: after a surfactant film was spread and compressed by means of the curved teflon barrier (6), Fig. 1, to attain a definite pressure, $\pi_A$, the support yoke (4) [with the solid crystal of KCl (or NaCl) resting on the support] was moved laterally from the part of the trough free of the film to a position that placed the halide crystal underneath the film in the curved barrier (6). The support yoke with the crystal was then lifted up to establish a contact between the crystal and the surfactant film at the interface. The crystal was subsequently lowered slowly by relaxing the torsion wire of the balance (5), thereby changing the distribution of relevant forces, Fig. 2 (a) and (b). The force required to submerge the solid was measured as the difference between that opposing the gravitational pull on the torsion wire at the start of relaxation and the final measured force at the moment of complete submergence.

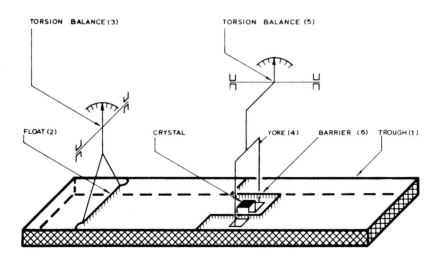

*Fig. 1 Modified Langmuir balance, schematic*

Of the two parameters accessible to measurement, the total force of submergence, T, and the depth of submergence, h, Fig. 2(b), the latter would have required major modifications of the apparatus to achieve sufficient sensitivity.

Since h can be evaluated from the relationships indicated in the Appendix, only the total force of submergence T was measured.

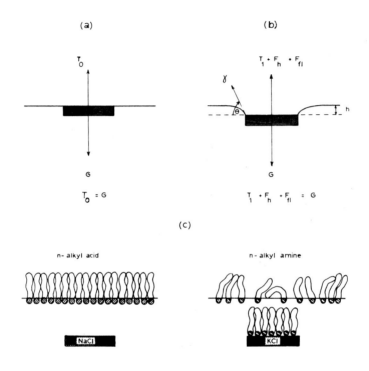

Fig. 2  (a)  *Equilibrium forces at the moment of establishing contact between the halide crystal and the insoluble surfactant film.*

(b)  *Equilibrium forces at the moment of submergence of the crystal.*

(c)  *Two modes of submergence involving severance of adhesion between NaCl crystal and fatty acid film, or fracture at the crystal perimeter of condensed amine film strongly adhering to KCl crystal.*

## III.  RESULTS

### A.  The Pressure/Area Curves for Surfactant Films Spread at Saturated Salt Solution-Air Interfaces

The monolayers of n-alkyl amines become more condensed when spread and compressed on the saturated NaCl solution and on the mixed saturated (KCl+NaCl) solution than they are on the

KCl solution alone, Fig. 3.

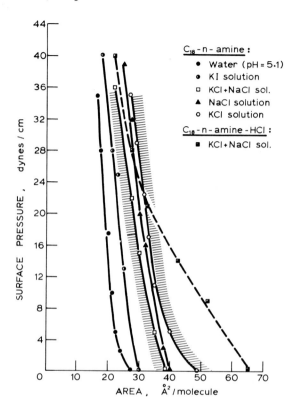

*Fig. 3 Surface pressure versus area per molecule relationship for octadecylamine films spread on different substrate solutions. The two shaded curves represent limits of condensation for saturated chloride solutions of $Na^+$ and $K^+$*

The sequence of compressibility appears to follow that of concentration of the salt solutions given in Table 1.

TABLE 1

THE CHARACTERISTICS OF THE SOLUTIONS USED

| Solution | pH | Concentration NaCl | | Concentration KCl | | Total Salt Conc. | | Density | Surface Tension |
|----------|-----|-----|-------|-----|-------|-----|-------|---------|---------|
| | | % | mol/l | % | mol/l | % | mol/l | gm/ml | dyne/cm |
| NaCl | 5.4 | 26 | 5.32 | - | - | 26 | 5.32 | 1.199 | 82 |
| KCl | 6.2 | - | - | 24 | 3.74 | 24 | 3.74 | 1.164 | 79 |
| KCl+NaCl | 5.6 | 20 | 4.20 | 11 | 1.81 | 31 | 6.01 | 1.224 | 82 |

What is surprising, however, is that the amine films spread on the surface of saturated KCl, (KCl + NaCl) and NaCl solutions *are more expanded than those on the surface of water, in the same pH range and temperature* (Fig. 3). [Only at pH 10.5 is it possible to obtain the same degree of condensation for the amine monolayer on the (KCl + NaCl) solution surface as it is for the same amine on the water surface at pH 5.1].

Long chain n-amines are insoluble in water and for each $\pi_A$ the area per molecule is $A_m = A_{total}/N_{total}$ where $A_{total}$ is the area occupied by the spread monolayer held at the $\pi_A$ pressure, and $N_{total}$ is the number of molecules used in the spreading test. If the amine spread on saturated salt solutions were somewhat soluble, the pressure/area curve would be based on $A^1/N = A_m^1$, [where $A^1 < A$, thus making $A_m^1 < A_m$] instead of the "true" value $A^1/N_{-n} \simeq A_m$. Thus, for a slightly soluble surfactant (for which the surface tension of the uncovered aqueous substrate surface would not be unduly affected, i.e. $\pi_A = \gamma_{film} - \gamma_{substrate} \simeq \gamma_{film} - \gamma_{solution}$), the pressure/area curve would most likely be shifted towards lower $A_m$ values for the same $\pi_A$, i.e. to the left of the curve obtained for the same chainlength amine insoluble in water. Such a finding indicates greater condensation of molecules within the film. A small shift of the pressure/area curve to the right of the amine-on-water reference curve indicates an expanded film that is interacting with the substrate, thus creating lateral resistance to compression forces.

When the chainlength is decreased for surfactants of the same polar group spread on the same substrate, or when the polar group is larger in size, the pressure/area curves are shifted towards larger $A_m$ values, giving more expanded films. This expansion indicates a lower proportion of $CH_2$ interactions in relation to the same degree of polar group interactions (Fig. 4).

A specific interaction between the amino group and the $Cl^-$ ion is the most likely cause of this behaviour. When the same amine films were spread on the surface of a 30% KI solution at pH 5.5, the films were found to be more condensed than those spread on the (KCl + NaCl) solution at the same pH (Fig. 3), and resembled the amine films spread on water at pH 10.5. The above effect could be explained by a different donor-acceptor ability of the $Cl^-$ and $I^-$ ions. These data parallel the counterion effect in the formation of amine micelles. Namely, the cationic surfactant requires a higher surfactant concentration for CMC in the presence of $Cl^-$ ions than in the presence of $Br^-$ ions [14].

The surface pressure/area curves for n-alkyl amines spread on (KCl + NaCl) solution are given in Fig. 4.

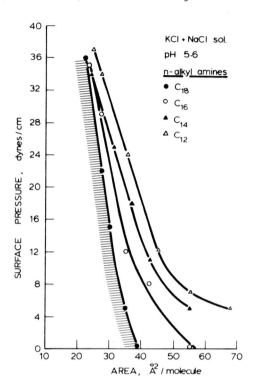

*Fig. 4    Effect of hydrocarbon chainlength on surface pressure/area relationships for amine film on (KCl + NaCl) saturated solution.*

Fatty acids produce slightly more condensed films on the NaCl solution than on the KCl solution. The data for these films on (NaCl + KCl) solution are shown in Fig. 5. The compressibility of acids on salt solutions is practically the same as their behaviour on water. The film of the sodium octadecanoate is, however, much less condensed than that of the octadecanoic acid. Films of fatty acids can be condensed to 20 Å per molecule, whereas films of n-alkyl amines are more expanded on the surfaces of salt solutions than they are on the water surface.

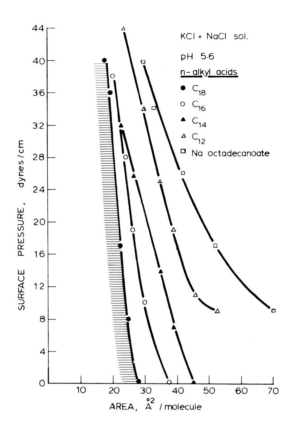

*Fig. 5    Surface pressure/area relationships for fatty acids spread on saturated (KCl + NaCl) solution.*

## B.    Interaction of Surfactant Films with Halides

The results of measurements of the net immersion force T, developed between halide crystals and films of octadecyl amines or octadecanoic acid spread at various surface pressures (surface tensions) on the (KCl + NaCl) solution are given in Table 2, together with other derived data (for derivation see Appendix).

Fig. 6 shows some values of the specific immersion $F_{imm}$ force, per unit length of the perimeter, for the two systems, in relation to the degree of film compression, i.e. area per molecule. The dependence of the derived values of flotation force $F_{fl}$ in relation to the compression of the film (area per molecule is shown in Fig. 7 for several n-alkyl amines and acids. For amines containing branch chains the immersion force was found to be zero, no adhesion between the solid KCl crystal and the spread film (which could not be condensed to below 40 $\text{Å}^2$ per molecule) could be detected.

TABLE 2

INTERRACTION OF THE AMINE FILM

WITH THE KCl CRYSTAL IN SATURATED KCl + NaCl SOLUTION

| Reagent | Area $\text{Å}^2$ | $\gamma_{film}$ dyne/cm | T dyne | h cm | $F_h$ dyne | $F_{imm}$ dyne/cm | $F_{fl}$ dyne/cm | $\theta°$ |
|---|---|---|---|---|---|---|---|---|
| 1 | 2 | 3 | 4 | 5 | 6 | 7 | 8 | 9 |
| Octadecyl-amine | (22) | (42) | (580) | (0.112) | (389) | (76) | (25.1) | (34.8) |
| $(C_{18})$ | 23 | 49 | 815 | 0.153 | 532 | 107 | 37.2 | 44.5 |
| Perimeter KCl crystal P = 7.6 cm | 28 | 62 | 735 | 0.132 | 460 | 96 | 36.2 | 33.8 |
| Surface area | 32 | 72 | 705 | 0.123 | 432 | 93 | 35.9 | 29.2 |
| S = 2.90 cm² | 34 | 77 | 555 | 0.096 | 334 | 73 | 29.1 | 21.8 |

INTERRACTION OF THE ACID FILMS

WITH THE NaCl CRYSTAL IN SATURATED KCl + NaCl SOLUTION

| | | | | | | | | |
|---|---|---|---|---|---|---|---|---|
| Octadecanoic acid | (19) | (46) | (100) | (0.023) | (64) | (16) | (5.8) | (6.8) |
| $(C_{18})$ | 21 | 51 | 150 | 0.035 | 97 | 24 | 8.5 | 9.8 |
| NaCl crystal P = 6.2 cm | 23 | 64 | 200 | 0.044 | 122 | 32 | 12.5 | 11.0 |
| S = 2.36 cm² | 25 | 73 | 280 | 0.060 | 167 | 45 | 18.2 | 14.0 |

Similarly, the immersion force was also found to be zero for KCl crystals being contacted with films of fatty acids spread on (KCl+NaCl) saturated solution. On the other hand, NaCl crystals showed some degree of adhesion to amine films, though this was very much weaker than the adhesion of the same films to KCl. For example, the values of the derived flotation force $F_{fl}$ ranged from 5.1 - 7.5 dynes/cm for the NaCl - octadecylamine system in comparison with 29.1 - 36.2 dynes/cm for the KCl - octadecylamine system, for the same range of areas per molecule.

As shown in Figs. 6 and 7, the two systems, KCl-amine and NaCl-acid, exhibit similar dependence of specific immersion and flotation forces on the area per molecule. With a progressive compression of the film (a decrease in the area per molecule) the immersion and flotation forces tend to reach a maximum, a condition which occurs around 26 $\overset{\circ}{A}{}^2$ per molecule for the NaCl-fatty acid system and 23-24 $\overset{\circ}{A}{}^2$ per molecule for the KCl-amine system.

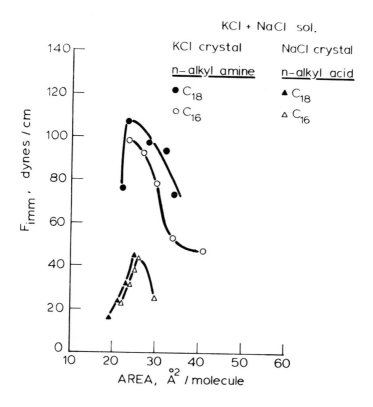

*Fig. 6 Specific immersion forces in relation to area per molecule for the KCl - $C_{18}$ and $C_{16}$ amine system and the NaCl - $C_{18}$ and $C_{16}$ acid system.*

## IV. DISCUSSION

It would appear from Figs. 6 and 7 that the ability of a surfactant to adhere to a given halide depends to a great extent on the degree to which the molecules are condensed, that is, on the structure of the surfactant film at the air/ saturated salt solution interface. Those surfactants which could not be condensed to less than 40 $\overset{\circ}{A}{}^2$ per molecule were found to exhibit no immersion force, that is, showed no adhesion to the halide crystals.

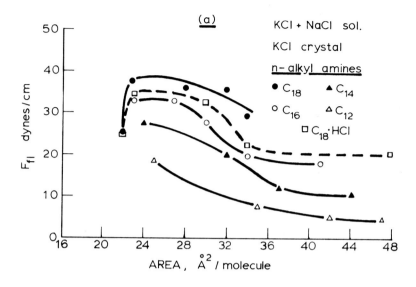

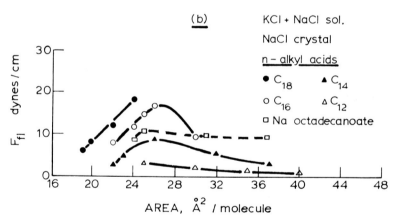

*Fig. 7. Flotation force/area per molecule relation-
ships for*

*(a) KCl crystals with n-alkyl amines*
*(b) NaCl crystals with n-alkyl acids*

The exact nature of the correlation between the crystal
lattice parameters and the structure of the surfactant film
is not easily evaluated. The film is pre-formed at the
solution/air interface and then allowed to interact with the
solid lattice. The compression of the molecules in the film
most likely results in a hexagonal-close-packing (hcp) structure, (Fig. 8), whereas the surfaces of KCl and NaCl crystals
have an ionic face-centre-cubic structure (fcc), whose

lattice parameters are changed on the surface owing to partial hydration of the ions. However, it is not so much the surface coverage of the crystal face by the surfactant molecules which decides the magnitude of the specific immersion and flotation forces, as the packing of these molecules along the linear dimension, the perimeter of the three-phase contact. And the latter is achieved by a two dimensional packing.

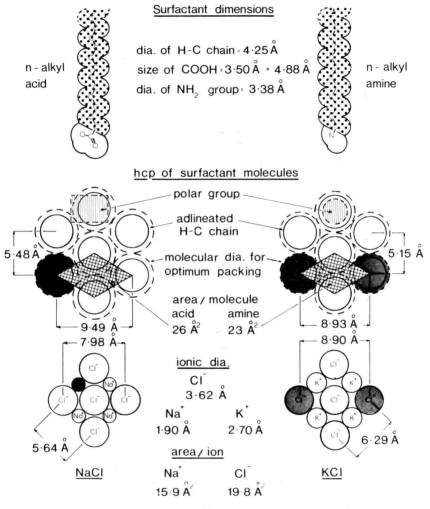

Fig. 8. Comparative dimensions of surfactants and their (probable) hcp arrangements in relation to fcc lattices of NaCl and KCl crystals for optimum specific immersion force.

The total immersion force determined in these experiments is a measure of either the adhesion between the crystal and the film or the cohesion within the film, whichever is the smaller.

a)  If the adhesion is greater than the cohesion bonds within the surfactant film the crystal submerges as a result of a sudden fracture of the cohesion bonds within the film, along the perimeter of the film-crystal contact. A portion of the film thus broken off adheres to the crystal and should make it hydrophobic. Indeed, this is what happens with the KCl n-alkyl amine systems, the face of the submerged KCl crystal responding to tests indicating its hydrophobic character, or

b)  If the adhesion of the surfactant polar group to the crystal surface is smaller than the cohesion within the film, the immersion force is a measure of the adhesion along a strip (of some unknown but very narrow width) around the perimeter of the crystal. The surfactant film peels off the crystal, leaving its surface hydrophilic in character. This is what happens with the NaCl-fatty acid systems, the submerged crystal face being indeed hydrophilic.

The above two cases are schematically depicted in Fig. 2(c).

Fig. 8 gives the lattice parameters (unhydrated), the ionic sizes of the two halides and the dimensions of the surfactants. A comparison of these data makes it clear that

1)  it is impossible for fatty acid molecules to react with every adsorption site ($Na^+$) on the NaCl surface, since this packing would require a mean cross-sectional area of 15.9 $\AA^2$ per site, whereas theoretically, hydrocarbon chains have a cross-section area of 19.6 $\AA^2$. Since amine films spread on salt solutions cannot be condensed below 23 $\AA^2$ per molecule without collapsing, even the available sites ($Cl^-$) on the KCl crystal face, representing 19.8 $\AA^2$ per ion, cannot all be occupied by amine molecules.

2)  the existence of a maximum in the immersion forces for the two systems suggests some degree of coincidence between the appropriate hcp and fcc lattices of the film and the crystal. There is no possibility of a perfect coincidence, only some degree of coincidence, over limited areas of surface, which may occur within a given limit of divergence. These areas are separated by regions of non-coincidence.

The distances between the neighbouring molecules in the two types of lattices are as follows:

for the 26 $\AA^2$ hcp packing - 5.48 $\AA$ side and 9.49 $\AA$ diagonal as compared with the fcc NaCl lattice of 5.64 $\AA$ side and 7.98 $\AA$ diagonal and for the 23 $\overset{\circ}{A}^2$ hcp packing - 5.15 $\overset{\circ}{A}$ side and 8.93 $\overset{\circ}{A}$ diagonal compared with the 6.29 $\overset{\circ}{A}$ side and 8.90 $\overset{\circ}{A}$ diagonal for the fcc KCl lattice.

These theoretical values suggest that for the KCl-amine systems at an area per molecule of 23 $\overset{\circ}{A}^2$ the most likely coincidence occurs between the amine molecules occupying the longer diagonals of parallelograms (Fig. 8) in the hcp structure of the film, and the alternate Cl⁻ lattice sites of the diagonals on the KCl crystal face.

However, the degree of packing (and the consequent cohesion of the film) is not the only factor determining the attachment of the film to the crystal. The energy of interaction, adhesion, between the surfactant polar groups and the crystal lattice appears to play an equally significant role.

A. Estimates of Cohesion or Adhesion Energy

As mentioned above, the specific immersion force may be taken as a measure of the force of cohesion for the KCl-amine system, and a measure of the force of adhesion for the NaCl-acid system. The values range from 73-107 dynes per cm for the KCl-amine and from 24-45 dynes per cm for the NaCl-acid system. These values have to be corrected for the hydrostatic force of buoyancy. The derived values of flotation force $F_{fl}$, as shown in Fig. 7, are also estimates of the cohesion energy obtained by another route. Yet a third way of deriving cohesion energy is to use the incremental $CH_2$ association energies from the surface pressure values for the same area per molecule, making suitable assumptions as regards the number of $CH_2$ groups participating in association for each surfactant. This last method is applicable only to the NaCl-acids system. It is obvious that, firstly, none of these estimates can be considered absolute, and secondly, both the cohesion and the adhesion forces are necessary for attachment of the solid crystal at the air/solution surface. The sum of cohesion and adhesion represents the adsorption energy. A high degree of lateral cohesion between the molecules does not guarantee the attachment of the film to the crystal unless sufficiently strong adhesion forces exist. Equally, a high degree of adhesion present between the isolated polar groups and the scattered sites on the KCl crystal does not lead to attachment either.

The results of estimates (using the incremental $CH_2$ association energies) of the cohesion at the 26 $\overset{\circ}{A}^2$ per molecule packing in the NaCl-fatty acid system show that the maximum adhesion is of the order of 2.6-3.0 Kcal/mole of molecules in the film. This value is close to the hydration energy of Na⁺ ions [15] and suggests that the adsorption of fatty acid

molecules onto NaCl competes with that of the intervening
molecules of water of hydration. The relatively greater
stability of the water of hydration of Na$^+$ in comparison with
that of K$^+$ would confer a higher strength of adhesion between
NaCl and the fatty acids. This selectivity of fatty acid adsorption on NaCl follows from Samoylov's theory (15) and has
been suggested by Schulman and Rogers (7) and later by Schubert (10). In systems containing Na$^+$ ions in solution, the
stronger interaction of fatty acids with Na$^+$ than their interaction with the solid KCl results in no adhesion of fatty
acids to KCl being observed in the presence of Na$^+$, and a
weak adhesion when KCl is in saturated solution of KCl alone.
Singewald (9) found the same behaviour in flotation experiments.

B.  Adsorption Parameters leading to Floatability
    in Saturated Salt Solutions

The tendency to use salts of fatty acids and of alkyl
amines as collectors is mainly due to their ability to be more
easily dispersed in water. Flotation experiments showed that
if a suitable solvent is used for amines (for example, isopropyl alcohol), they could be successfully used as collectors
without their conversion to salts [20]. Experiments in the
flotation of langbeinite, K$_2$Mg$_2$(SO$_4$)$_3$, showed that better flotation results are achieved when fatty acids are used in the
acid form instead of their salts [21].

The data obtained in this study and the results of
other workers emphasize the importance of structural parameters in collector films used in halide flotation. In films
spread on the Langmuir trough the "compression forces" determining the structure of the film are the forces of van der
Waal's interaction between the collector molecules, and the
forces of surface pressure. In flotation systems the *concentration* of the collector and its *hydrocarbon length* determine
the extent of "compression" and the structure of the film at
the liquid/gas interface. The role of nonpolar chain associations in the soluble salt flotation was investigated by
Schubert [11] and was verified by flotation experiments which
indicated that, for example, amines containing less than 14-
16 carbons are very weak collectors for KCl [9-12].

Roman et al [12] found out that an effective salt flotation could be achieved only when the collector concentration was close to the CMC, which is the condition when the
collector film is becoming highly condensed.

The surface pressure depends also on the presence of
other surfactants, like frothers. Frothers change the surface tension of the air/liquid interfaces, influence the
collector film structure, as shown in penetration studies dis-

cussed by Leja [22], and affect the magnitude of the final
surface pressure. The role of nonpolar chain association of
surfactants (collectors and frothers) is of particular importance in those flotation systems where the collector-mineral
interaction is comparatively weak. Hence, the role of surface pressure for these systems is of particular importance.
Two main aspects have been discussed so far, viz.

(1)  the structure of the collector film in salt flotation
     systems in relation to the solid surface parameters,

(2)  the strength of adhesion between the film and the solid.

There is also a third aspect to be considered, namely, the
mechanism by which the attachment of particles to air bubbles
is achieved during halide salts flotation.

     In the case of flotation of sulphides with xanthates
(and of oxidized minerals or silicate minerals with fatty
acids and alkyl amines of shorter hydrocarbon chains than
those used in saturated salt systems) it is accepted that the
collector-acting species dissolve in water, diffuse to the
solid/liquid interface of minerals, adsorb on selected solids
to make them hydrophobic in character. During the actual flotation process the already hydrophobic solid particles attach
to the oscillating air bubble, with the help of frothers.
The frother molecules, partly adsorbed at the air/bubble interface and partly co-existing on the hydrophobic solid and
in the intervening liquid layer, help to thin this intervening liquid layer during the moment of attachment.

     With saturated salt solutions the solubility of long
chain surfactants used as collectors (and that of frothers as
well) is minimal.                         And yet flotation
occurs within kinetically similar time periods of millisec-
onds or centiseconds. The investigations carried out above
with the films of amines spread on saturated salt solutions
and a solid crystal approaching from underneath to establish
a bond, strongly suggest that the details of the mechanism of
flotation in saturated salt solution may be somewhat different from those in solutions of soluble surfactants in lower
electrolyte concentrations. The difference may be due to the
fact that owing to the high insolubility of the surfactant
species it cannot and does not diffuse to the solid to make
it hydrophobic. Instead, the primary mechanism in saturated
salt flotation may be the spreading of the insoluble surfactant particulate at the air/solution interface immediately
the air bubbles are generated. An attachment of the solid in
its hydrophilic state, during collision, becomes fruitful only
when the structure of the spread film and the interaction
energy (between its polar groups and the hydrophilic solid)
are capable of exceeding the hydration energy.

It is not easy to provide unambiguous evidence to support this type of mechanism; evidence that will not be open to different interpretations on account of side effects. However, the possibility of a spectrum of mechanisms in particle-bubble attachment from that operative in sulphide flotation to that suggested above for saturated salt flotation, should be taken into consideration.

## V. GENERAL CONCLUSIONS

Solid KCl and NaCl crystals attach to insoluble n-alkyl amine and fatty acid films with maximum flotation adhesion when the films are compressed to a predetermined area per molecule. The latter appears to be correlated with the lattice parameters of the corresponding halide crystals.

A carboxylic group reacts with NaCl surface with an energy approximately 3 Kcal/mol. suggesting an interaction replacing water of $Na^+$ hydration. The interaction of KCl surface with amine groups appears to be 2 to 3 times stronger. A definite contribution of hydrocarbon chain associations in the overall adsorption energy is necessary for an effective solid-film attachment. This contribution is estimated to be 25% to 50% of the overall adsorption energy. The role of a frother in flotation systems involving weak collector-mineral interaction may be a very significant parameter in the condensation of collector films.

A supplementary mechanism for collector-particle attachment is proposed, based on spreading of the collector as an insoluble (or nearly so) film at the air-bubble interface. This type of mechanism may be operative primarily in the alkali salt flotation systems.

## VI. ACKNOWLEDGEMENTS

The authors would like to thank Canada Council and the Ministry of Higher Education of the U.S.S.R. for the support of exchange fellowship. Also they gratefully acknowledge the assistance from the members of the Department of Mineral Engineering of the University of British Columbia during the course of this work.

## VII. REFERENCES

1. Hälblich, W., _Metall und Erz._, Vol. 30, 431 (1933)
2. Neunhoeffer, D., _Erzmetall_, Vol. 2, 334-38 (1949)
3. Gaudin, A.M., Fuerstenau, D.W., _AIME Trans._, Vol. 202, 66, (1955)
4. Gaudin, A.M., Fuerstenau, D.W., _AIME Trans._, Vol. 202, 258, (1955)
5. Fuerstenau, D.W., Fuerstenau, M.C., _Mining Eng._, No. 8, 302, (1956)

6.  Bachmann, R., Erzmetall, Vol. 4, 316 (1951)
7.  Rogers, J., Schulman, J.H., Proc. II  Int. Congress of Surface Activity, Vol. III, 243-51. Butterworths Sci. Publ.  (1957)
8.  Rogers, J., Trans. Inst. Mining & Met. (London) 66, 439 (1957)
9.  Singewald, A., Quart. Colorado School of Mines, Vol. 56, No. 3, 65-88, (1961)
10. Schubert, H., Eng. and Mining Journal, No. 3, 94-97 (1967)
11. Schubert, H., Freibgerger Forsch. A 514, (1972)
12. Roman, R.J. et al, AIME Trans., No. 3, 56-70 (1968)
13. Padday, J.F., Proc. II,  Int. Congress of Surface Activity Vol. III, 136-42  London, (1957)
14. Jones, M.N., Reed, D.S., "Chimie, Physique et Applications Pratiques des agents de Surface,"Proc. V Intl. Congress Detergency, Vol. II, 1081-89.  Barcelona (1968)
15. Samoylov, O. Y., Struktura vodnych rastvorov electrolitov i gidratacia ionov,  Moscow, (1957)
16. Lin, I.J., Somasundaran, P., Journal of Coll. and Interface Sci., Vol. 37, No. 4, 731-43 (1971)
17. Danielson, I., "Chimie, Physique et Applications Pratiques des agents de Surface", Proc. V, Intl. Congress Detergency, Vol. II, 869-77.  Barcelona (1968)
18. Rosano, H.L. et al,  ibid,  1071-79.
19. Pauling, L., "The Nature of the Chemical Bond", Ithaca, N.Y. (1945).
20. Arsentiev, V.A., Thesis, Leningrad Mining Institute (1973)
21. Zhelnin, A.A., Flotacia Rastovorimych soley.  Leningrad (1972).
22. Leja, J., Proc. II,   Int. Congress of Surface Activity Vol. III, 273-296  London, (1957)
23. Finch, J. A. Smith, G.W., Journal of Coll. and Interface Sci., Vol. 44, No. 2, 387-388 (1973).

VIII.  APPENDIX

The Langmuir balance was equipped with a polyethylene trough and teflon barriers, one of which was ⊓ shaped to accommodate a yoke for a solid crystal.  A schematic diagram is shown in Fig. 1.  The apparatus consists of the trough (1), the float (2), attached to a torsion balance (3), a yoke (4) for the support of a solid halide crystal, and an additional torsion balance (5) to which the yoke (4) was attached, used for measuring the force of immersion.  The yoke with the flat solid crystal of KCl or NaCl was raised sufficiently to establish a contact with the surfactant film spread on the surface of a saturated salt solution.  After the contact had been established, the torsion wire (5) was gradually relaxed to allow the overriding gravitational force to immerse the crystal.  The difference between the initial force $T_0$ balancing the gravitational force of the crystal in contact with the film, Fig. 2(a) and the relaxing force (at the moment of immersion) Fig. 2(b) was expressed as the total force of immersion, $T = T_0 - T_1$. From the equilibrium of relevant forces:

$$T = F_h + F_{fl} \qquad (1)$$

and the specific immersion force

$$F_{imm} = T/P = (F_h + F_{fl})/P \qquad (2)$$

the hydrostatic force

$$F_h = p.g.h.S \qquad (3)$$

the flotation force

$$F_{fl} = P.\gamma.\sin\theta \qquad (4)$$

where

$$\cos = 1 - \frac{\rho.gh^2}{2\gamma} \text{ , as was shown by Padday [14]}$$

$\rho$ = density of the liquid phase, $gm/cm^3$
$g$ = acceleration due to gravity, $cm/sec^2$
$P$ = perimeter of the crystal, cm
$S$ = area of the crystal, subjected to hydrostatic pressure, $cm^2$

From the above relationships:

$$P.\gamma.\sin\theta = T - \rho.g.h.S. \qquad (5)$$

which converts, on utilizing (4), to:

$$\frac{P^2\rho^2g^2}{4}.h^4 + [\rho^2g^2S^2 - P^2\gamma\rho.g]h^2 - 2T.g.\rho.S + T^2 = 0 \qquad (6)$$

Using the appropriate data, the value of $\dfrac{P^2\rho^2g^2.h^4}{4}$ has been found to be an order of magnitude smaller than the other terms in equation (6). Hence, the latter can be simplified to a quadratic equation of the form

$$Zh^2 + Yh + W = 0 \qquad (7)$$

where

$$Z = \rho^2.g^2.S^2 - P^2\rho.g.\gamma$$
$$Y = 2T.\rho.g.S$$

and

$$W = T^2$$

For $\theta < 90°$ the solution of equation (7) for h enables this value of h to be used for estimating $F_h$, $F_{fl}$ and $\theta$. The evaluation of adhesion and association forces, for the NaCl -fatty acid system, is based on the incremental $CH_2$ association energies taken from the surface pressure data in Fig. 5 for the 26 Å$^2$ area per molecule. These values are then substituted into an equation:

$$E_{association} = n.W_{CH_2} + Q + \pi_A \qquad (8)$$

$W_{CH_2}$ = incremental $CH_2$ association energy

n = number of $CH_2$ groups involved in association

Q = lateral interaction between polar groups

$\pi_A$ = surface pressure

At the moment of submergence at $26\overset{\circ}{A}^2$ per molecule, the energy of association within the film is assumed to be equal to the adhesion energy between the film and the NaCl crystal. Q is assumed constant and negligible, since the final degree of condensation appears to be governed by the cross-sectional area of the hydrocarbon chains.

# MONOLAYER AT THE AIR/WATER INTERFACE VS OIL/WATER INTERFACE AS A BILAYER MEMBRANE MODEL

S. Ohki, C.B. Ohki and N. Duzgunes
State University of New York at Buffalo

## Abstract

Surface tension measurements were made for phospholipid monolayers formed at the oil/water interface. It was found that the surface pressure of PS monolayer at the hexadecane/water interface at 70 $\mathring{A}^2$ per molecule was more than 40 dynes/cm. Interaction of water soluble "peripheral" proteins with the oil/water monolayer was studied in terms of the change in surface tension of the monolayer. The surface tension was reduced considerably upon the protein introduction into the subphase solution at a low initial surface pressure (<25 dynes/cm) of the monolayer which was interpreted as protein penetration into the non-polar phase of the monolayer, but was not altered appreciably at a high initial surface pressure (>40 dynes/cm). The phospholipid membrane conductance was increased when the peripheral basic proteins were placed on one side of the PS membrane only, but not altered when placed on both sides.

From the above results, it is deduced that the surface pressure of the lipid bilayer in its liquid crystalline state is quite high (more than 40 dynes/cm), peripheral proteins would interact with lipid bilayers mainly electrostatically at the periphery of the membrane and the oil/water interface monolayer is a proper model representing a physical state of half of a bilayer membrane.

## I.  INTRODUCTION

A number of works concerning the interaction of biologically important ions and macromolecules with model membrane systems have been made in order to elucidate the structure and function of biological membranes (1,2,3,4). These model membrane systems are phospholipid monolayer and bilayer membranes. The monolayer system has been studied as a model membrane representing half of a bilayer or a biological membrane (4-9, (11,13,14). There are two types of monolayer systems; one is

271

the monolayer formed at the air/water interface and the other
formed at the oil/water interface. So far, the former system
has been widely used as a model membrane in order to obtain a
quantitative interaction mechanism of various metal ions and
macromolecules with biological membranes. However, because the
opposing monolayers in a lipid bilayer contact with a hydrophobic environment, it has been pointed out that monolayer
studies at the oil/water interface would better simulate the
bilayer and biological membranes (5,11,12,13,14). The oil/
water monolayer is considered to be a more expanded state than
the air/water monolayer at the same area per molecule, because
of the tail-tail hydrocarbon interaction. It has recently been
reported that the oil/water monolayers are indeed more expanded
than the air/water monolayers for phosphatidyl choline and
other synthetic phospholipids (13,14).

Therefore, there must be some differences in physical
states (such as chain packing, surface pressure and tension)
between these two types of monolayers at the same area per
molecule. So far as one limits his investigation to the interaction mode of ions with the membrane, the use of both types of
monolayer systems may yield similar results since ions would
interact mainly with the polar groups of the lipids. However,
when the interaction between environmental molecules and the
membrane involves a characteristic of the membrane proper (such
as non-polar interaction between protein and membrane hydrocarbon phase), a model study by using one type of the monolayer system will yield a different result from what is obtained by use of the other type of monolayer system.

The previous theoretical study concerning the physical
states of both types of monolayers as well as bilayer membranes
from an energetical point of view, indicates that the monolayer formed at the oil/water interface is a proper model system to represent the physical states of a bilayer in its liquid
crystalline state (15). According to the theoretical analysis
of the physical states of bilayers, the surface pressure of the
bilayer in the liquid-crystalline state is much higher than
those deduced from the results of the air/water monolayer
studies (7,3).

A large amount of work concerning interaction of biologically important macromolecules, such as proteins with membranes,
has been investigated by using the air/water monolayer (3,4,6,
8,9). It has been shown originally by Schulman and co-workers
(16,17) that the increase in surface pressure of phospholipid
monolayers at the air/water interface following the injection
of protein in the subphase depends upon electrostatic interaction. Recently, Quinn and Dawson (8) have demonstrated by

a radio-tracer method that the protein, cytochrome-C, penetrates considerably into the film at a low surface pressure. Also, Kimelberg and Papahadjopoulos (9) have suggested that cytochrome-C and other water soluble basic proteins in the subphase of an acidic phospholipid monolayer both bind electrostatically, and penetrate hydrophobically into the film interior. However, it should be stressed that these experiments have been performed at a relatively low surface pressure, less than 40 dynes/cm, and their data have indicated no surface pressure increase at an initial surface pressure above 40 dynes/cm of the monolayer for all peripheral proteins used. It should be noted therefore, that with the results obtained by using the air/water monolayer, to draw a similar conclusion concerning those protein interactions with the lipid bilayer or biological membranes is not proper, because the lipid bilayer or biological membranes may exert a considerably higher surface pressure than that of the air/water monolayer at the same area per molecule.

In order to provide further experimental support for the idea that the monolayer formed at the oil/water interface is a proper model system to represent the physical state of half of a bilayer in its liquid crystalline state, surface pressure measurements were made for the oil/water monolayer with and without peripheral proteins in the monolayer subphase. Also, the effect of these proteins on electrical conductance of planar lipid bilayers was measured to test if there is a strong interaction between these proteins and the lipid bilayers.

## II.  MATERIALS AND METHODS

### A.  Chemicals

Phospholipids used were phosphatidyl serine (PS) from bovine brain and phosphatidyl choline (PC) from egg, which were both purchased from Applied Science Laboratories (Pa.), or extracted and purified in our laboratory using Papahadjopoulos' method (36). Purity of both samples were confirmed with silica gel thin-layer chromatography.

In the monolayer studies, film spreading solutions consisted of phospholipid dissolved in hexane ($\sim 5 \times 10^{-4}$ M) for the air/water monolayer and in benzene and methanol (5:1 = v:v) ($\sim 5 \times 10^{-4}$ M) for the oil/water monolayer. The exact concentrations of phospholipid in the spreading solutions were determined by the method of phosphate analysis. Hexane and benzene were Baker Instra-Analyzed grade and methanol (Fisher, analytical grade) was redistilled. Hexadecane (purum, >99% GC, Fluka) was used as an upper oil phase, and the subphase

solution was 0.01 M NaCl containing 0.05 mM EDTA (Fisher, reagent grade), and 1.0 mM Tris-base (Ultrapure grade, Mann).
pH of the solution was titrated with HCl (Fisher, reagent
grade) to 7.0. The salt (NaCl, Fisher chemical) used was roast-
ed at 500-600°C. Water was distilled three times, including
the process of alkaline permanganate distilling. The proteins
used were cytochrome-C (horse heart, type VI), protamine sulfate (Salmon, grade I) and albumin(Human,V) which were all purchased from Sigma Chemical Co., and used as received.

The membrane forming solution for the planar bilayer
studies was composed of 10 mg of lipid in 0.5 ml of n-decane
(purum, >99%, Fluka).

B.    Surface Tension Measurements

Surface tension measurements of monolayers at the air/
water interface was done in a similar manner as described in
the previous paper (19). Surface tension was measured by an
electronic microbalance (Beckman) and a chart recorder (omni-
scribe, Houston) using 18 mm x 8 mm x 0.16 mm microscopic cover
slip as a Wilhelmy plate. The trough was a pyrex glass cell
with a constant area ($\simeq$ 60 cm$^2$). For the oil/water interface
studies, two sizes of glass cells were used in order to check
the edge effect of the monolayer. The surface area of one of
them was $\sim$ 60 cm$^2$ and the other was 33 cm$^2$. First, the subphase solution was introduced into the cell and the surface
tension was measured, and then the oil was added on the aqueous phase, keeping the tip of the Wilhelmy plate in the aqueous phase until the plate was immensed completely. After
making correction by a few dynes for buoyant force due to the
immersed Wilhelmy plate, oil/water surface tension was found
to be 51 ± 0.5 dynes/cm. Two methods were undertaken to form
monolayers at the oil/water interface; one was successive
injection of the spreading solution to the oil/water interface
by a microsyringe (Hamilton Co.), and the other was first to
spread the spreading solution (lipid-hexane) on the aqueous
surface and then place the oil gradually on the formed air/
water monolayer. The former method indicated inadequate spreading at the smaller area (<80 $\mathring{A}^2$). However, the latter seemed
to give quite successful results.

The proteins were dissolved in distilled water (1$\sim$10 mg/
ml as stock solutions) for each experiment and a small amount
of the stock solutions or its diluted solution, was injected
into the subphase of the monolayer by a micro-syringe and the
subphase solution was stirred sufficiently. Before the injection of proteins, the subphase solution was also stirred well,
and it was ascertained that the stirring did not result in any

change in surface tension of the monolayer.

## C.   Conductance Measurements

The planar bilayer membranes were prepared on a hole of a delrin cell by the same method as described in the earlier paper (20). The surrounding solutions of the membranes were 0.01 M NaCl containing 0.2 mM Tris and 0.05 mM EDTA of pH 7.0. The membrane conductances were measured by applying dc electrical potential up to 40 mV across the membrane, and measuring the current by a micro-ammeter (Hewlett Packard). The applied electrical potential was monitored with an electrometer (610 C of Keithley Instruments). The applied potential was always more positive in the inner compartment solution against the outer one, while proteins were injected to the outer compartment. Electrodes used were silver-silver chloride electrodes. The concentrated protein solutions were added to either one side compartment or both side compartments of the membrane after the bilayer was formed and the solutions were stirred until a uniform concentration was attained. Experimental temperature was at room temperature of 24°C ± 1°C for both monolayer and bilayer experiments.


## III. RESULTS


A force-area curve for PS monolayers formed at the air/water interface is shown in Fig. 1A. Figs. 1-B and 1-C show the force-area curves for PS monolayers formed at the hexadecane/water interface with successive injection of the spreading solution at the oil/water interface and by placing the oil on the top of the air/water monolayer, respectively. The air/aqueous surface tension was found to be 72.5 ± 0.5 dynes/cm. The value for the interfacial tension of hexadecane/aqueous phases was 51 ± 0.5 dynes/cm after correction due to a buoyant force of the immersed Wilhelmy plate, which was a value close to those reported in the literature (37).

As seen in Figs. 1-B and -C, there was a significant difference in surface tension values between the two methods of forming the oil/water interface monolayers. Successive application of spreading solution directly to the oil/water interface, apparently gave insufficient spreading of the monolayer in the range of the small area (less than 80 Å$^2$) per molecule. This has been recognized by earlier workers (11). It was not certain, however, that the phospholipids did not spread well because of the high film pressure or if the lipid solvents remained at the interface. It seems, therefore, that the

measured surface pressure value is not reliable (Fig. 1-B).
On the other hand, the method of forming the monolayer by
placing oil on the air/water monolayer, seems to give good
results for surface tension measurements. It is seen from Fig.

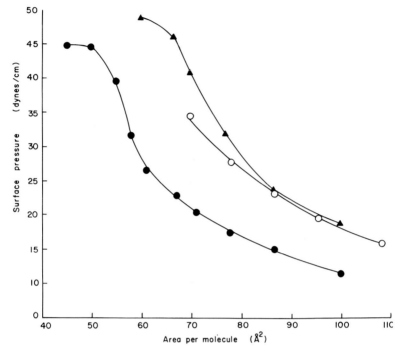

Figure 1.    Force (surface pressure) - area curves of
PS monolayer formed at the air/water and
the oil/water interfaces.  Hexadecane was
used as the oil phase and the subaqueous
phase was 0.01 M NaCl, 1 mM Tris and 0.05
mM EDTA of pH 7.0

● :    air/water interface monolayer
O :    oil/water interface monolayer (spread-
        ing solution was applied successively
        to the interface)
▲ :    oil/water interface monolayer (the oil
        was placed on the air/water monolayer
        formed).

1 that in both methods the PS monolayers at the oil/water interface, are more expanded than those of the air/water interface.
For example, the film pressure corresponding to the area 70 $\overset{\circ}{A}^2$
per molecule of PS monolayer at the air/water interface is
about 21 dynes (see Fig. 1-A).  On the other hand, the film
pressure corresponding to the same area per molecule for the

oil/water interface monolayer is 41 dynes/cm from Fig. 1-C.
The area per molecule of 70 $\text{Å}^2$ approximately corresponds to
those for liquid-crystalline phospholipid bilayers of 67-70 $\text{Å}^2$
estimated from X-ray diffraction studies (21,22).

The effect of peripheral proteins used on the surface tension of the oil/water monolayer was studied at various surface
tension values (Fig. 2). For the monolayers with higher surface tension (lower surface pressure) upon introduction of
$2.5 \times 10^{-6}$ g/ml of proteins, surface tension decreased gradually,

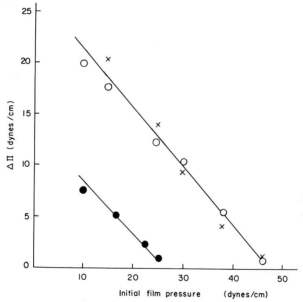

Figure 2. Effects of proteins on surface pressure of PS
monolayer formed at the oil/water interface.
oil: hexadecane, water: 0.01 M NaCl, 1 mM
Tris and 0.05 mM EDTA of pH 7.0. proteins:2.5mg/ml.

O : cytochrome-C
X : protamine
● : albumin

and after a half hour the rate of tension decrease slowed down
and seemed to reach a stationary state after an hour. Monolayers of a low initial surface tension (<5 dynes/cm), however,
did not result in any appreciable decrease or increase in

surface tension with the addition of the same amount of proteins used. Our results correspond qualitatively to those
observed for the monolayer formed at the air/water interface
(9). Surface tension decrease of the monolayer at the same
initial surface pressure was slightly greater ($\sim$3 dynes/cm)
than those reported for the air/water monolayer (9). Cytochrome
C and protamine decreased the surface tension of the monolayer
up to an initial film presure of about 47 dynes/cm. On the
other hand, albumin decreased the surface tension up to an
initial film pressure of about 28 dynes/cm.

Another study was the conductance measurements of the
planar PS bilayer prepared in 0.01 M NaCl with various protein
concentrations on one side (outer solution chamber) and on both
sides of the membrane. The measurements were taken 20 min.
later, after the proteins were introduced into the compartments.
These results are shown in Figs. 3, 4 and 5. It should be
mentioned that there was a slight difference in observed current between the two cases where the voltage was applied in
opposite directions. However, this difference did not affect
the final conclusion for the conductance change with the application of the proteins. Fig. 3 shows the conductance of
the PS membrane with the concentration of cytochrome-C added to
one side compartment of the membrane. The conductance of the
membrane increased with concentrations of cytochrome-C. On
the other hand, the presence of the same concentration of
cytochrome-C on both sides of the membrane did not alter the
membrane conductance, up to a fairly high concentration ($10^{-5}$
M) of the proteins. Similar experiments were done with other
proteins. The results with protamine (Fig. 4) were similar to

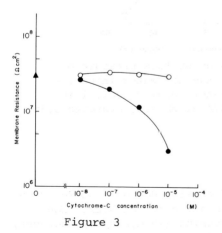

Figure 3

Figures 3,4,5. Membrane
conductances of PS
bilayer in 0.01 M NaCl
(pH:7.0) with the vari-
ation of various protein
concentrations on one
side ($\bullet$) and on both
sides of the membrane (O)
$\blacktriangle$: membrane conductance
without proteins,ordinate
and abscissa: logarithmic
scale.

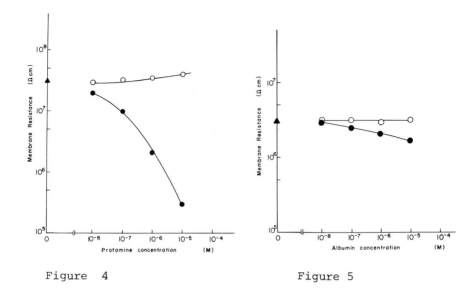

Figure 4                  Figure 5

those of cytochrome-C. When protamines were placed on both side compartments of the membrane, the membrane conductance was decreased even with the increase of protein concentration at least up to $10^{-5}$ M. For the application of albumin, PS membrane conductance was practically unaffected with the protein concentration up to $10^{-5}$ M for both applications on one side and on both side compartments of the membrane (Fig. 5).

Experiments similar to the above were also performed with the PC membranes. However, for both cases where the proteins were placed on one side only and on both sides of the membrane, the membrane conductance was not appreciably altered.

Similar observation has been made for acidic phospholipid membranes with the addition of $Ca^{++}$ to the salt solutions on either side of the membrane (23); the presence of a small amount of $Ca^{++}$ on both sides of the membrane results in decrease of the PS membrane conductance, while the same amount of $Ca^{++}$ on one side of the PS membrane increases the membrane conductance. However, the presence of the same amounts of $Ca^{++}$ does not affect any appreciable change in PC membrane conductance.


IV.  DISCUSSION


The theoretical consideration, that the lipid monolayer formed at the oil/water interface is a proper representation

of the physical state of bilayer membranes in its liquid-
crystalline state, has provided two predictions concerning the
film pressure of the bilayer as well as the number of carbon
atoms involved in an oriented hydrocarbon chain portion of
the bilayer lipid molecules (15). From the knowledge of the
surface tension of the lipid bilayer(0 - several dynes/cm, ref.
38), and that of the monolayer formed at the air/water interface having a corresponding area per molecule to the bilayer,
the pressure of the bilayer has been estimated to be more than
40 dynes/cm and the number of carbon atoms in an oriented
hydrocarbon chain portion of the $C_{18}$ lipid bilayer is calculated to be 7∿8 (15). The latter value is supported by the
results from NMR (24,25) and ESR studies (26,27).

Our experimental results for surface tension (or surface
pressure) on monolayers formed at the air/water and oil/water
interface[s] gave further information concerning the surface
pressure of bilayer membrane. From X-ray studies (21,22), it
is estimated that the area per molecule of the phospholipid
bilayer in its liquid-crystalline state is about 67∿70 Å². 
Suppose that the air/water interface monolayer represents half
of a bilayer, the surface pressure of one side of the bilayer
is deduced to be about 20∿25 dynes/cm, which corresponds to
the surface pressure of the air/water interface monolayer at
area per molecule 70 Å². On the other hand, if the oil/water
interface monolayer, having the corresponding area per molecule to the bilayer state, does represent half of a bilayer,
the surface pressure of a bilayer is estimated to be 40∿45
dynes/cm from Fig. 1-C. This value of surface pressure was
predicted from the previous theoretical analysis (15).

Another evidence to support the idea of the oil/water
monolayer representing a proper physical state for half of a
bilayer state is demonstrated from interaction studies of so-
called "peripheral protein" with both monolayers. It is seen
from Fig. 2 that in the range where the surface pressure of
the monolayer at the oil/water interface is greater than 45
dynes/cm, injection of the peripheral proteins used in the
subphase solution results in no appreciable change in surface
pressure (tension) of the monolayer. This may be an indication that there was no strong interaction (or penetration) of
these proteins with (into) the monolayer hydrocarbon phase. If
half of a bilayer corresponds to the oil/water interface monolayer at the same area per molecule, it is speculated that
these proteins would not appreciably penetrate into the membrane phase since the bilayer surface pressure is considered
to be quite large (more than 40 dynes/cm).

Consistently with our results, there are a number of studies using electron microscope[28,29] x-ray diffraction (30,31,32), optical spectroscopy (33), protein spin labeling (34) and fluorescence spectroscopy (35), all indicating that cytochrome-C is held to the surface layers of the phospholipid bilayers by electrostatic forces, but does not penetrate the hydrophobic interior of the membrane in the higher salt solution ($\geqslant 0.01$ M) at neutral pH.

A similar study on peripheral protein interaction with the PS monolayer at the air/water interface has been done (9) and the results are quite similar to those obtained for the oil/water PS monolayer here, as far as the magnitude of surface tension decrease is concerned at the same initial surface pressure. However, in the case of the air/water monolayer at the area per molecule of 70 $\text{Å}^2$, cytochrome-C produces a considerable decrease in surface tension, which is considered a penetration of cytochrome-C into the monolayer phase, but does not in the case of the oil/water interface at the **same** area per molecule. In the oil/water monolayer, a slight and consistently larger surface pressure than in the air/water monolayer, was observed at the overall initial surface pressures. This is interpreted as the oil/water interface monolayer having a more accessible non-polar part at the interface than the air/water monolayer, because the surface density of polar groups of the oil/water interface monolayer is lower than that of the air/water monolayer corresponding to the state having a same surface pressure.

It should be noted that the surface pressure increase of the monolayer upon the application of proteins may not necessarily indicate protein penetration into the monolayer hydrocarbon phase. In this respect, the bilayer conductance study may provide further evidence whether the protein molecule would interact with the hydrophobic matrix, because a main electrical conductance barrier of the membrane is considered to be the hydrocarbon phase of the membrane. If proteins penetrate into the membrane and disturb the membrane hydrocarbon matrix, it would result in conductance change in the membrane. Experimental results showed that the PS bilayer conductance was increased in the presence of basic proteins in the solution on one side compartment of the membrane only, and the presence of these proteins on both sides of the membrane did not alter membrane conductance appreciably. The presence of protamine on both sides of the membrane even decreased the membrane conductance. These proteins did not affect the PC membrane conductance. From these results, it is suspected that the peripheral basic proteins may not interact strongly with the non-polar membrane proper. Since basic

proteins have positive charges at neutral pH, the electrostatic interaction with the negatively charged phospholipid is
a reasonable assumption. The reason for the change in conductance of the PS membrane, when the basic proteins are placed
only on one side of the membrane, is that the asymmetrical
membrane state created by peripheral basic proteins with their
electrostatic interaction at one membrane surface may result
in instability of the membrane due to asymmetrical binding of
macromolecules on one surface of the membrane rather than non-
polar interaction, as seen in acidic phospholipid membranes
having $Ca^{++}$ on one side of the membrane only (23). The above
situation may correspond to the increase in ionic permeability
of PS vesicle membranes in the presence of basic proteins applied to one side of the vesicle only (18). This possible
electrostatic interaction of peripheral basic proteins with the
negatively charged membrane was strengthened by the experimental results which showed no effect on the conductance of the
PC bilayer membrane, either in the presence of basic proteins
on one side or both sides of the membrane. This also coincides with the results that albumin did not affect PS membrane
conductance as much as the basic protein when it was placed on
one side or both sides of the membrane, because albumin is
considered to be,as a whole, negatively charged at neutral pH.

It should be mentioned here that we are not excluding the
possibility for these peripheral proteins to have non-polar
interaction with the membrane phase. It is quite possible
that, depending on the ionic strength, pH and protein concentration, these proteins could strongly interact not only electrostatically, but also with non-polar attractive force with
the membrane phase (3,4,8). What is to be stressed here is
that at present salt concentration and pH, those peripheral
proteins seem to have mainly electrostatic or polar interaction
with the bilayer membranes, and in order to study the protein
interaction with the bilayer membrane, it is more proper to
use the oil/water interface monolayer as a model for half of a
bilayer. This may also be valid for the study of protein interaction with the biological membrane.

V    REFERENCES


1.    Goldup, A., Ohki, S. and Danielli, J.F., in "Recent Pro-
       gress in Surface Science" (J.F. Danielli, A.C. Riddiford
       and M.D. Rosenberg, Eds.), Vol. 3, p. 193. Academic
       Press, New York, 1970.
2.    Jain, M.K., "The Biomolecular Lipid Membranes", Van
       Nostrand-Reinhold, Princeton, New Jersey, 1972.

3. Papahadjopoulos, D. and Kimelberg, H.K., in "Progress in Surface Science" (S.G. Davison, Ed.), Vol. 4, p. 141, Pergamon Press, Oxford, 1973.

4. Dawson, R.M.C. and Quinn, P.J., in "Membrane-Bound Enzymes" (G. Porcellati and F. Di Jeso, Eds.), p. 1. Plenum Press, New York, 1971.

5. Jackson, C.M., in "Permeability and Function of Biological Membranes (L. Bolis, A. Katchalsky, R.D. Kenyes, W.R. Loewenstein and B.A. Pethica, Eds.), p. 164. North-Holland Pub., Amsterdam, 1970.

6. Colacicco, G., Lipids 5, 636 (1970).

7. Phillips, M.C. and Chapman, D., Biochim. Biophys. Acta 163, 301 (1968).

8. Quinn, P.J. and Dawson, R.M.C., Biochem. J. 113, 791 (1969).

9. Kimelberg, H.K. and Papahadjopoulos, D., Biochim. Biophys. Acta 233, 805 (1971).

10. Hui, S.W., Cowden, M., Papahadjopoulos, D. and Parsons, D.E., Biochim. Biophys. Acta 382, 265 (1975).

11. Brooks, J.H. and Pethica, B.A., Trans. Faraday Soc. 60, 1 (1964).

12. Haydon, D.A. and Taylor, J., J. theoret. Biol. 4, 281 (1963).

13. Taylor, J.A.G., Mingins, J., Pethica, B.A., Tan, B.Y.J. and Jackson, C.M., Biochim. Biophys. Acta 323, 157 (1973).

14. Taylor, J.A.G. and Mingins, J., J. Chemical Soc. Faraday Trans. I, 71, 1161 (1975).

15. Ohki, S. and Ohki, C.B., J. theoret. Biol. in press (1976).

16. Matalon, R. and Schulman, J.H., Discuss. Faraday Soc. 6, 27 (1949).

17. Doty, P. and Schulman, J.H., Discuss. Faraday Soc. 6, 21 (1949).

18. Kimelberg, H.K. and Papahadjopoulos, D., J. Biol. Chem. 246, 1142 (1971).

19. Seimiya, T. and Ohki, S., Biochim. Biophys. Acta 274, 15 (1972).

20. Ohki, S., J. Colloid Interface Sci. 30, 413 (1969).

21. Luzzati, V. and Husson, F., J. Cell. Biol. 12, 207 (1962).

22. Reiss-Husson, F., J. Mol. Biol. 25, 363 (1967).

23. Ohki, S. and Papahadjopoulos, D., in "Surface Chemistry of Biological Systems" (M. Blank, Ed.), p. 155. Plenum Press, New York (1970).

24. Birdsall, J.M., Lee, A.G., Levine, Y.K. and Metcalfe, J.C. Biochim. Biophys. Acta 241, 693 (1971).

25. Darke, A., Finer, E.G., Flook, A.G., and Phillips, M.C., J. Molec. Biol. 63, 265 (1972).

26. McFarland, B.G. and McConnell, H.M., Proc. Nat. Acad. Sci. 68, 1274 (1971).

27. Hubbell, W.L. and McConnell, H.M., J. Am. Chem. Soc. 93, 314 (1971).
28. Stoeckenius, W., Schulman, J.H. and Prince, L.M., Kolloid-Zeitschrift 169, 170 (1960).
29. Kimelberg, H.K., Lee, C.P., Claude, A. and Mrena, E., J. Membrane Biol. 2, 235 (1970).
30. Gulik-Krzywicki, T., Shechter, E., Luzzati, V. and Fauve, M., Nature 223, 1116 (1969).
31. Blaurock, A.E., Biophys. J. 13, 290 (1973).
32. Shipley, G.G., Leslie, R.B. and Chapman, D., Nature, 222, 561 (1969).
33. Steinemann, A. and Läuger, P., J. Membrane Biol. 4, 74 (1971).
34. Van, S.P. and Griffith, O.H., J. Membrane Biol. 20, 155 (1975).
35. Vanderkooi, J., Erecinska, M. and Chance, B., Arch. Biochem. Biophys. 154, 219 (1973).
36. Papahadjopoulos, D. and Miller, N., Biochim. Biophys. Acta 135, 624 (1967).
37. Fowkes, F.M., at Symposium on Properties of Surfaces and Surface Coatings Related to Paper and Wood, State University College of Forestry at Syracuse University, Oct. 1965.
38. Tien, H.Ti, in "Bilayer Lipid Membranes", p. 40. Marcel Dekker, Inc., New York, 1974.

# VERY LOW-ANGLE LIGHT SCATTERING:
## A CHARACTERIZATION METHOD FOR HIGH MOLECULAR WEIGHT DNA

Harold I. Levine*, Robert J. Fiel, and Fred W. Billmeyer, Jr.
*Department of Chemistry, Rensselaer Polytechnic Institute
and Department of Biophysics Research,
Roswell Park Memorial Institute*

A very low-angle light-scattering (VLALS) photometer is described with respect to optical features, scattering cell, correction factors, and absolute calibration in the angular range 2-35°. An improved microfiltration apparatus was used to obtain virtually dust-free aqueous solutions for VLALS. The instrument was calibrated with silicotungstic acid, an absolute molecular weight standard, and the calibration was confirmed with the use of several secondary standards, which were also measured with a conventional wide-angle light-scattering (30-150°) photometer.

VLALS measurements were made to determine the weight-average molecular weight $\overline{M}_w$ and z-average radius of gyration $R_{g,z}$ of a commercial preparation of calf-thymus DNA. Micro-filtration of the solutions allowed measurements down to $\theta=6°$. The value $\overline{M}_w = 20.0 \times 10^6$ obtained by extrapolating 6-9° data to $\theta=0°$ is >3X that from 30-75° data ($6.38 \times 10^6$) but 20% smaller than that from 10-35° data ($23.7 \times 10^6$). The experimental errors in $\overline{M}_w$ and $R_{g,z}$ are estimated to be ±8% and ±14%, respectively. Combined 6-75° data from the two photometers fit well a theoretical scattering curve for a model wormlike coil of the same $\overline{M}_w$ as the DNA sample, and with a persistence length of 60nm.
 (NASA Grant NGL-33-018-003 and USPHS Grant GM 19883.)

# MECHANISMS OF PROTEIN STABILIZATION AND DESTABILIZATION BY SOLVENT COMPONENTS

Serge N. Timasheff, James C. Lee,
and Eugene P. Pittz
*Brandeis University*

## ABSTRACT

A thermodynamic study has been carried out of the interactions of proteins with a variety of structure stabilizing and structure destabilizing solvent systems. The experiments consisted of preferential interaction measurements by densimetry and differential refractometry with the rigorous application of multicomponent thermodynamic theory, paralleled by measurements of protein denaturation and changes in conformation, as monitored by differential uv spectroscopy and circular dichroism. In the case of destabilizing systems, 2-chloroethanol and 2-methyl-2,4-pentane diol (MPD) at pH 3 act by direct interaction with hydrophobic regions on the proteins; guanidine hydrocloride interacts with peptide bonds. In all cases binding precedes unfolding. The structure stabilizing solvents act by being preferentially excluded from contact with the protein surface. At neutral pH, MPD is locally salted out by charged groups on the protein, causing phase separations and, as a result, protein crystallization. (This work was supported in part by NIH grants GM 14603, CA 16707 and NSF grant BMS 72-02572.)

# A STUDY OF LIPID-PROTEIN BINDING BY
## ELECTROPHORETIC LASER DOPPLER SPECTROSCOPY

J. G. Gallagher, J. D. Morrisett and A. M. Gotto, Jr.
Baylor College of Medicine and The Methodist Hospital,
Houston, Texas   77030

## Abstract

Apolipoprotein-alanine (or ApoLP—C-III) from  human very
low density lipoprotein has been previously shown to interact with egg yolk (EYPC) and dimyristoyl phosphatidylcholine (DMPC) bilamellar vesicles.  Furthermore, it has
been demonstrated from hydrodynamic measurements that the
adsorption of this apolipoprotein to these lipid bilayers
is consistent with a model where the apolipoprotein is
bound to hydrophobic sites within the vesicle, with parts
of the protein protruding into the polar head region so
that interactions with hydrophobic sites on the surface
of the bilayer can occur.  Consequently, the interaction
of these lipids with protein also produce changes in the
net charge distribution on the surface of the vesicle
which are readily measured via their electrophoretic
mobilities.  Using a dual beam heterodyne optical mixing
technique, quasi-elastic light scattering has been combined with a free boundary electrophoresis method to
measure the net change in the electrophoretic mobility of
these lipids that results from adsorption of ApoLP—C-III.
In this binding study, a multilamellar structure of the
above lipids (approximately 1000 Å in diameter) was used.
In the absence of ApoLP—C-III, the electrophoretic
mobilities of DMPC and EYPC in $5 \times 10^{-4}$ M NaCl, T = 25° C
were 0.61 and $3.05 \times 10^{-4}$ $cm^2$/v-sec. at pH 7.4 and 6.9,
respectively.  Upon complexing with the apolipoprotein,
the mobility of EYPC was found to increase linearly with
the lipid-protein mass ratio (0.07 and 0.14 g/g) and were
4.02 and $8.87 \times 10^{-4}$ $cm^2$/v-sec, respectively.  In the
binding studies with DMPC, the diameter of the particle
decreased with time in the presence of the apolipoprotein.  The mobility of this complex after a 1 hour incu-
bation period (40% reduction in particle diameter) was
7.06 and $6.38 \times 10^{-4}$ $cm^2$/v-sec for lipid-protein mass
ratios of 0.07 and 0.14 g/g, respectively.  Hence, these
data obtained from multilamellar structures of DMPC and
EYPC are also found to be consistent with the binding
model proposed for single bilayer vesicles.

# THERMODYNAMICS OF THE LIPID BILAYER PHASE TRANSITION

Nan-I Liu and Robert L. Kay
*Carnegie-Mellon University*

The transition temperatures for dipalmitoyllecithin bilayers have been determined at pressures up to 400 bars by following the volume change accompanying the transition. Errors in a previous measurement resulting from temperature gradients in the pressure vessel were avoided by following the volume change with pressure at constant temperature. The transition temperature clearly varies linearly with pressure and the slope dP/dT is in good agreement with the one atmosphere value of $\Delta H/T\Delta V$. Since the volume change accompanying the transition was found invariant with pressure, it is clear that the entropy change is also invariant with pressure. The implications of these results for order-disorder theories will be discussed.

# THE EFFECT OF POLYMERS ON ELECTROLYTE TRANSPORT ACROSS INTERFACES

B. J. R. Scholtens and B. H. Bijsterbosch
*Agricultural University*

## ABSTRACT

*A study has been made of the effect of polymers on the transport of electrolytes across liquid/liquid interfaces in a stirred cell. Mutually saturated water and 1-butanol in the presence of poly(vinylalcohol)-poly(vinylacetate) copolymers were chosen as the model system.*

*It has been shown that the hydrodynamic conditions at the interface play a crucial role in this type of transport. Only when these conditions are well-defined and properly taken into account, can the diffusional and interfacial resistances be calculated. The latter will be discussed in terms of the adsorption properties of the copolymers.*

# PRECIPITATION OF SODIUM CASEINATE WITH DEXTRAN SULFATES, SODIUM-DODECYL SULFATE, SODIUM-DODECYL SULFONATE AND LIGNIN SULFONATE

Kelvin Roberts and Lars Davidsson
*The Swedish Institute for Surface Chemistry*

## ABSTRACT

*The precipitation of sodium caseinate has been studied below the isoelectric point (pH 3) using a series dextran sulfates of molecular weights from 2.000 to 500.000 and with sulfur contents from 2 to 16%, as well as sodium dodecyl sulfate, sodium dodecyl sulfonate and lignin sulfonate. The removal of protein precipitation, and the reversibility of precipitation phenomena were followed by turbidity determination, as well as analysis after micro-flotation or sedimentation of the precipitate. Build up of aggregates was followed using Coulter Counter technique.*

*The results showed that maximum protein removal occurred when the positively charged groups on the protein were neutralised by sulfate/sulfonate groups.*

*The sulfates (dextran sulfate and sodium dodecyl sulfate) showed different domains for protein removal depending on the order of mixing of reactants. The "displacement" between these removal domains for the dextran sulfates containing ca 15% S was relatively independent of molecular weight but decreased with decreased sulfur content.*

*The protein removal domains for both sulfonates were independent of order of mixing of reactants. The results show that denaturation precipitation of sodium caseinate at pH 3 occurs when the charged groups on the protein are neutralised by anionic groups on added sulfates or sulfonates. Reactions with sulfonates are reversible suggesting weaker interactions, whilst reactions with sulfates are irreversible suggesting stronger charged group interactions. The reduction in irreversibility with dextran sulfates as the percentage of charged groups is reduced, suggests that the precipitates formed have hydrated polysaccharide chains and a loose crosslinked structure.*

# THERMODYNAMICS OF THE VOLTA EFFECT FOR SURFACE FILMS

D. G. Hall and B. A. Pethica
*Unilever Research Laboratory*

## ABSTRACT

An explicit thermodynamic analysis of the Volta effect
for adsorbed films and spread monolayers is presented in
terms of surface variables. An appropriate fundamental
equation is obtained and used to derive expressions describing the effect of an applied electric field on some of
the parameters that characterise two-dimensional phase
transitions. It is shown that the approximations inherent
in the Bridgeman, Kelvin and Lorentz assumptions concerning
the relation between Volta and compensation potentials usually lie well within current experimental accuracy, irre-
spective of which set of independent intensive variables to
be regarded as constant is chosen. In the transition region of a first order two-dimensional phase transition these
assumptions can break down when the chemical potentials of
all adsorbed components are among the variables to be held
constant, but are unlikely to do so at constant average
surface densities of all surface active components. Finally
a possible method is forwarded for obtaining the slopes of
adsorption isotherms for spread monolayers which are too
small to measure by conventional means.

# COION INTERACTIONS WITH POLYELECTROLYTES BY SELF-DIFFUSION MEASUREMENTS

Robert A. Sasso, Alexander Kowblansky and Paul Ander
*Seton Hall University*

It has generally been accepted that counterions are affected to a great extent by the presence of polyions in solution, but that coions are little affected by the presence of these polyelectrolytes. Yet several types of measurements have shown that the nature of the coion does affect the magnitude of the solution property determined. A recent investigation of monovalent and multivalent coions interactions with sodium polystyrenesulfonate in water indicated that while the coion diffusion coefficients differed at finite concentrations, the coion self diffusion coefficients extrapolated to infinite dilution were all similar in value to the monovalent coions, independent of the valence of the coion.[1] While the theory of Manning [3,4,5] for a line charge model for the polyion proved successful for the interactions of counterions with the polyion, it was found to greatly overestimate the effect of the valence of the coion. The present study was undertaken to investigate the interactions of coions of different valency types with ionic polysaccharides containing different pendant charge groups in aqueous solution at 25°C. Sodium iota carrageenan and sodium alginate, which contain sulfate and carboxyl groups on their respective repeating units, were used since they are stiffer than the synthetic polyelectrolytes used in previous studies [1,2] and were thought to better represent the rod-like model.

Tracer self-diffusion coefficients of $SO_4^{2-}$ and $Cl^-$ ion were measured in aqueous solutions of sodium iota carrageenan and sodium alginate using the open-end capillary method. By monitoring radioactive activity in the capillary before and after a measured period of diffusion, the tracer diffusion coefficient D was calculated using McKay's approximate solution of Fick's equation

$$D = (\pi/4)(1-C/C_o)^2 L^2/t \qquad (1)$$

where $C/C_o$ is the ratio of final to initial radioactive activity, L is the length of the capillary, and t is the time allowed for diffusion.

The diffusion coefficients of $SO_4^{2-}$ and $Cl^-$ ion were

measured at four different concentrations of simple salt
($Na_2SO_4$ or NaCl), 0.010$\underline{N}$, 0.0050$\underline{N}$, 0.0010$\underline{N}$, 0.00050$\underline{N}$. The
concentration of the polyelectrolyte ranged from 5.0 x $10^{-5}\underline{N}$
to 1.0 x $10^{-1}\underline{N}$ so that $\underline{X}$ ranged from 0.1 to 10, where $\underline{X}$ is
the ratio of the equivalent concentration of polyelectrolyte
to the equivalent concentration of simple salt. The measured
ionic diffusion coefficients $D_i$ were compared to the values
of the diffusion coefficients of $SO_4^{2-}$ and $Cl^-$ ions at infinite
dilution in a polymer-free environment $D_i^\circ$, as calculated using
the Nernst equation giving $D_{SO_4^{2-}}^\circ = 1.06$ x $10^{-5}$ $cm^2/sec$.

The presence of both the sodium iota carrageenan and the
sodium alginate decreased the diffusion coefficient of $SO_4^{2-}$
ion from its value in polymer-free solution at infinite
dilution. The values of $D_{SO_4^{2-}}$ for sodium iota carrageenan
solutions initially decreases with increasing $\underline{X}$ value; at the
higher $\underline{X}$ values, where the ratio of polymer to simple salt
becomes more appreciable, the diffusion coefficient becomes
independent of $\underline{X}$ and the ratio $D_{SO_4^{2-}}/D_{SO_4^{2-}}^\circ$ is essentially
constant. At the most dilute salt concentration, the curve
levels at approximately $D_{SO_4^{2-}}/D_{SO_4^{2-}}^\circ = 0.86$. At a given $\underline{X}$
value, $D_{SO_4^{2-}}/D_{SO_4^{2-}}^\circ$ increases as the simple salt concentration
decreases. This effect is more pronounced at higher $\underline{X}$ values;
much less dependence on salt concentration was found as the
$\underline{X}$ value decreases. The same initial decrease in the diffusion
coefficient of $SO_4^{2-}$ ion with increasing $\underline{X}$ value is also
observed in the presence of sodium alginate. The curve also
levels off as the $\underline{X}$ values increase. At the most dilute
simple salt concentration the curve levels at about $D_{SO_4^{2-}}/$
$D_{SO_4^{2-}}^\circ = 0.83$. At a given $\underline{X}$ value, however, the effect of
increasing value of $D_{SO_4^{2-}}/D_{SO_4^{2-}}^\circ$ with decreasing simple salt
concentration is much less pronounced in the presence of this
polymer. Even at relatively high values of $\underline{X}$, the values of
$D_{SO_4^{2-}}/D_{SO_4^{2-}}^\circ$ cluster within very close proximity at all four
simple salt concentrations.

Sodium iota carrageenan and sodium alginate in solution
also were found to decrease the value of $D_{Cl^-}$ from that
given in polymer-free solution at infinite dilution $D_{Cl^-}^\circ$.
The values of $D_{Cl^-}$ in the presence of sodium alginate showed
the same curve characteristics as seen for $SO_4^{2-}$ ion in the
presence of both polymers; there is an initial decrease in
the value $D_{Cl^-}/D_{Cl^-}^\circ$, with leveling at higher values of $\underline{X}$.
The values of $D_{Cl^-}/D_{Cl^-}^\circ$ in the presence of sodium iota carrageenan seems less dependent on the value of $\underline{X}$ at constant
simple salt concentration than is apparent with the sodium
alginate; that is, the initial decrease of $D_{Cl^-}/D_{Cl^-}^\circ$ with
increasing $\underline{X}$ value is much more suppressed. In the presence
of the alginate, the values of $D_{Cl^-}/D_{Cl^-}^\circ$ level off at about
0.80 at the lowest simple salt concentration. In the presence
of the carrageenan, the values of $D_{Cl^-}/D_{Cl^-}^\circ$ level off at

approximately 0.85. In solutions of sodium alginate, we again note the effect of increasing values of $D_{Cl^-}/D^\circ_{Cl^-}$ with decreasing simple salt concentration at a given $X$ value, with the effect less prevalent at lower $X$ values. Conversely the value of $D_{Cl^-}/D^\circ_{Cl^-}$ in carrageenan solutions is considerably more independent of simple salt concentration at a given value of $X$ over the entire range of $X$.

Finally, the $Cl^-$ ion curves level off at approximately the same values as do the $SO_4^{2-}$ ion curves at their most dilute simple salt concentration for both ionic polysaccharides. Therefore, the diffusion coefficients of both the monovalent and divalent coions seems to be affected to the same extent by the presence of the polyions.

Some interesting correlations of these results can be made to Manning's modern theory of polyelectrolytes. Representing the polyion as an infinitely long line charge, Manning has formulated limiting laws describing the properties of polyelectrolytes in both the presence and absence of simple salt. Central to the Manning Theory is the dimensionless parameter $\xi$ which is related to the charge density of the line charge

$$\xi = e^2/\varepsilon kTb \qquad (2)$$

where $e$ is the charge on the proton, $\varepsilon$ is the dielectric constant, $k$ is the Boltzmann constant, $T$ is the temperature, and $b$ is the average distance between adjacent charged groups on the polyion. At values of greater than a critical value $\xi_c$ given by

$$\xi_c = \left| Z_p Z_1 \right|^{-1} \qquad (3)$$

where $Z_p$ is the charge of a single charge site on the polyion and $Z_1$ is the charge on the counterion, counterions will condense on the polyion to reduce $\xi$ to the critical value. The fraction of charge neutralized on the chain is given by $(1-\xi^{-1})$ and the fraction of uncondensed counterions is $\xi^{-1}$. Debye-Huckel interactions are assumed between all uncondensed ions and the polyion, and for $\xi_c = 1$, the diffusion of all uncondensed simple ions is described by

$$D_i^{(u)}/D_i^\circ = 1 - (Z_1^2/3)A(1;\xi^{-1}x) \qquad (4)$$

where $D_i^{(u)}$ is the diffusion coefficient of uncondensed ion $i$, $D_i^\circ$ is the limiting value at infinite dilution of $i$ in the absence of polyelectrolyte, $Z_i$ is the charge on ion $i$, and $A(1;\xi^{-1}x)$ is a series that is a function of the effective charge density of the polyion and effective $X$ value. Since no condensation will occur with coions, their theoretical Debye-Huckel interactions with the polyion should be expressed by the equation

$$D_2/D_2^o = 1 - (z_2^2/3) A (1; \xi^{-1}X) \qquad (5)$$

where the subscript 2 represents the coions. According to this equation, there is a very great decrease in the diffusion coefficients for higher valent coions. While the values for $D_{Cl^-}$ in both sodium iota carrageenan and sodium alginate solutions agree quite well with the predicted monovalent diffusion curves for these particular polysaccharides, the predicted curves for the diffusion coefficients for $SO_4^{2-}$ ion for the two polyelectrolytes are in considerable disagreement with the experimental data. As previously mentioned, experimental results with the divalent coion closely approximate that of the monovalent coion for both our polysaccharides, and it seems that the effect of the charge of the coion was greatly overestimated in Manning's theoretical considerations.

Manning's theory predicts as well that polyion-simple ion interaction depends solely on X, and not on the concentration of simple salt or polymer. Yet our data shows that at a given X value, $D_i/D_i^o$ does indeed depend on simple salt concentration, and the dependence on simple salt concentration seem more prevalent at higher X values and with the divalent coion than with the monovalent coion.

An assumption made in the Manning theory, since the development is that of limiting laws which hold strictly only at infinite dilution, is that simple ion-simple ion and polyion-polyion interactions are not considered. Thus it should be expected that the experimental results will correlate better with the theoretical predictions at lower concentrations of simple salt and polymer. This is exactly what is observed for Cl⁻ ion with both ionic polysaccharides. A linear relationship exists between the value of $D_i/D_i^o$ at a given X value and the square root of simple salt concentration, thus all values of $D_i/D_i^o$ were extrapolated to infinite dilution, i.e., to zero ionic strength. The extrapolated curves for our data for both Cl⁻ ion and $SO_4^{2-}$ ion in sodium iota carrageenan and sodium alginate closely agree with Manning's predicted values for <u>monovalent</u> coions at infinite dilution. Hence, there is an interaction between the polyion and unbound simple coions in solution which may be interpreted in terms of Debye-Huckel interaction, yet the interaction is much less dependent on the valence of the interacting coion as presently suspected in Manning's theoretical developments.

REFERENCES

1.  M. Kowblansky and P. Ander, J. Phys. Chem., <u>80</u>, 297 (1976).
2.  S. Menezes-Affonso and P. Ander, J. Phys. Chem., <u>78</u>, 1756 (1974).

3. G.S. Manning, J. Chem. Phys., $\underline{51}$ 924, 934 (1964).

4. D.I. Devore and G.S. Manning, J. Phys. Chem., 78, 1242 (1974).

5. D.I. Devore, Ph.D. Dissertation, Rutgers, The State University, N.J. 1973.

# ADSORPTION OF PROTEINS AT SURFACES

H. P. M. Fromageot and J. N. Groves
*General Electric Research and Development Center*

## ABSTRACT

*We have studied the behavior of bovine serum albumin (BSA), human erythrocyte surface glycoprotein (GP) and whole human serum (HS) at solid-liquid interfaces by determining: 1) the adsorption isotherms of tritiated BSA of high specific activity (ca. 8000 dpm/μg) on polystyrene latex spheres (PLS) 0.81 μm in diameter, 2) the composition of octadecylsilylated glass capillaries exposed to tritiated BSA before and after exposure to the other macromolecules, 3) the zeta potential versus pH profiles of capillaries exposed successively to the various macromolecules.*

*Adsorption isotherms of tritiated BSA on PLS at various pH's show that a positive cooperative adsorption occurs between 1 and 100 ng/ml of protein in solution. A monolayer of BSA adsorbed side-on (0.1 to 0.4 μg/cm$^2$) is formed between 0.1 and 10 μg/ml of protein in solution. At higher concentrations, formation of "multilayers" (several monolayer equivalents) occurs. When octadecylsilylated capillaries are exposed to a 10 mg/ml BSA solution, an initial "multilayer" is formed, which is disassembled and reduced to a monolayer upon extensive washing. Capillaries first exposed to either polyvinylpyrrolidone, HS, or GP are still capable of adsorbing "multilayers" of BSA, and these are stable to subsequent extensive washing, yet the electrokinetic measurements indicate that the composition at the plane of shear is predominantly that of the other macromolecule.*

MONOMOLECULAR FILM STUDIES OF ACETYLATED
CARBOHYDRATE POLYMERS AND OLIGOMERS

P. Luner, L. S. Sandell, M. Hittmeier and E. P. Lancaster
*S.U.N.Y. College of Environmental Science and Forestry*

*ABSTRACT*

*F-A and μ-A isotherms of cellulosetriacetate (CTA),
amylose triacetate (ATA), and xylan diacetate (XDA) were
determined on a Langmuir film balance. Comparisons between
CTA and ATA reveal difference between a β,1-4 and α,1-4
linkage while comparisons between CTA and XDA reflects substitution in the C-6 position. Analysis of these results in
conjunction with molecular models shows that in a closely
packed monolayer the CTA and XDA molecules are in an extended chain conformation with the average plane of the rings
approximately normal to the surface and the chain areas
roughly normal to the direction of compression in the plane
of the substrate surface. For the xylan polymer, the absence of the C-6 group permits a much closer packing. The
ATA seems to be in a helical conformation. Supporting evidence for these models was obtained from monolayer studies
of acetylated oligomers of D-glucose and D-xylose.*

# THE SVEDBERG AND THE REALITY OF MOLECULES

Milton Kerker
*Clarkson College of Technology*

ABSTRACT

*The Nobel Prizes in Chemistry for 1925 and 1926 were awarded to Zsigmondy and Svedberg, respectively and the 1926 Nobel Prize in Physics went to Perrin - to a large extent all for work done on Brownian motion. This paper very cursorily considers the importance which was attached to Brownian motion early in this century and more specifically evaluates Svedberg's contribution.*

## I. INTRODUCTION

1926 was a very good year! The ACS Division of Colloid Chemistry, whose 50th anniversary we are now celebrating, was formed; the Nobel Prize in Chemistry was awarded to a colloid physical chemist, The Svedberg, for work on Brownian motion; the Nobel Prize in Physics was awarded to a colloid chemical physicist, Jean Perrin, also for work on Brownian motion. Furthermore, the 1925 Nobel Prize in Chemistry had gone to a colloid chemist, Richard Zsigmondy, whose invention of the ultra-microscope made Brownian movement more vivid than ever. He too devoted much effort to elucidating this phenomenon.

Three Nobel Prizes in two years! What a triumph for colloid science which had only emerged at the turn of the century in association with the new physical chemistry. There are two questions which have stimulated my own interest in this episode: Why was the phenomenon of Brownian motion selected for such singular recognition? What precisely was the role of Svedberg? These questions will be treated ever so briefly here since they have been considered in much greater depth elsewhere (1, 2). The reader is referred to the references for greater detail and for the primary literature.

It should be noted first of all that Svedberg's recognition by the Nobel Prize was based upon studies of Brownian movement rather than of ultracentrifugation. His work on the ultracentrifuge did not start until 1923. J. H. Mathews organized a conference that year which became the first National Colloid Symposium. This present conference constitutes the fiftieth of these Symposia. There were three years during which no meetings were held enabling us now to celebrate simultaneously the golden anniversary of both the Symposium and the Division. Indeed, the Division evolved from the Symposium.

In order to insure the success of his conference and to give it publicity, Mathews invited Svedberg as the foreign guest of honor. He also arranged to have Svedberg spend half a year at Wisconsin to do research and to lecture. The optical method utilized in the ultra-centrifuge and the first instrument was developed by Svedberg at Wisconsin and the work continued intensively after his return to Uppsala, but the first preliminary results with proteins were not obtained until 1926, the year in which the Nobel Prize was presented. It is true that Svedberg chose to speak about those most recent results on the ultracentrifugation of proteins in his Nobel Prize address. Yet, this was not the basis for the award. The results which he presented were still mainly unpublished and were considered by Svedberg to be imprecise and merely illustrative. Yet, it made sense to talk about this new work which was so exciting and of such vast potential, while as we shall see, the Brownian motion work harking back to his earliest student days was erroneous.

## II. BROWNIAN MOVEMENT AND MOLECULAR REALITY

The invention of the microscope brought into view a fascinating new world of motile animalcules and bacteria which differed from dancing dust particles made visible in a sunbeam in that the source of motive power lay in the organism itself rather than in turbulent air currents. And so when in 1827 Robert Brown turned his lens onto cytoplasmic granules dispersed in water there was no reason to anticipate that their chaotic movements might constitute a momentous discovery. His recognition that this was a different kind of motion is a tribute to his genius.

Brown demonstrated that the motion was exhibited by all materials - organic and inorganic - provided that they were appropriately pulverized. He examined coal, atmospheric dust, siliceous fossilized wood, ground glass and a variety of minerals. The vigor of the motion depended only upon particle size.

The subject aroused interest through the middle years of the 19th Century but it was not until 1888 that G. Gouy gave a clear and explicit description of how Brownian movement might arise as a result of molecular impacts, although his work had been anticipated in a little known paper published in 1877 by J. Delsaux. The most important aspect of Gouy's deliberations is his emphasis not so much upon effecting a physical explanation of Brownian movement as upon utilizing this phenomenon as a tool with which to explore molecular dynamics. In his words

"The Brownian movement shows us, therefore, assuredly not the movements of the molecules, but something derived closely from these and furnishes us a direct and visible proof of the exactness of the hypothesis on the nature of heat. If one adopts these views, then the phenomenon, the study of which is far from being terminated, takes on assuredly an importance for molecular physics of the first order."

Molecular physics itself and also atomic chemistry were the issues. Polemics over the question of molecular reality, which began with Dalton's premise of elemental atoms, raged throughout the 19th Century and the opposition to molecularity and atomicity reached a crescendo at the turn of the 20th Century under the leadership of Wilhelm Ostwald, Van't Hoff, Helm, Planck, and Mach. Consider these statements by the "fathers of physical chemistry".

Ostwald wrote: "The atomic theory has in physical science a function which is similar to that of certain mathematical auxiliary representations. It is a mathematic model used for the representation of facts. Although vibration is represented by size curves ... . no one would admit that vibration itself has anything to do with circular or angular functions... . For six years I have searched without success to construct a mechanical theory of chemical affinities, and I am convinced that it is only as one has renounced all mechanical analogy that one can find results of some utility."

Van't Hoff put it more colorfully: "... the kinetic theory barely gives the current 4% interest on capital ... . The representations themselves, atoms, molecules, their dimensions, and perhaps their shapes, are after all something doubtful. Here applies the proverb, Virtue must show itself."

Einstein's 1905 paper on Brownian motion crystallized the issue and set the stage for a crucial experiment. He noted

that previous experiments which attempted to measure the
"velocity" of the particles undergoing Brownian movement were
destined to fail.  The velocity, in the sense of a molecular
velocity, was not observable.  Rather, he focused attention
upon a quantity which might be observed - the root mean square
displacement $x$ during a particular interval of time $t$ and he
showed that as a consequence of the kinetic molecular theory
this was given by

$$x = (2Dt)^{\frac{1}{2}}$$

where $D$ is the diffusion coefficient.  Einstein noted "if the
movement discussed here can actually be observed (together
with the laws relating to it that one would expect to find),
then classical thermodynamics can no longer be looked upon as
applicable with precision to bodies even of dimensions
distinguishable in a microscope; and exact determination of
actual atomic dimensions is then possible.  On the other hand,
had the prediction of this movement proved to be incorrect, a
weighty argument would be provided against the molecular-
kinetic conception of heat."  Certainly, the stakes were high!

## III.  THE SVEDBERG

The (Theodor) Svedberg (1884-1972) went to Uppsala in 1904
and, except for eight months at the University of Wisconsin in
1923, he remained there for all of his career.  Following his
dissertation in 1907, he became a docent, then in 1912,
professor of physical chemistry, and in 1949, upon retirement
from his chair in physical chemistry, Director of the Gustaf
Werner Institute for Nuclear Chemistry, until his complete
retirement in 1967.

At Uppsala he plunged directly into the new colloid
chemistry.  Using the tools of physical chemistry, he tackled
the problem of preparing stable colloids reproducibly in order
to permit quantitative studies of the relation between particle
size and other physical properties.  He refined Bredig's
technique for the preparation of gold and platinum hydrosols
by an electric arc, and he assembled an ultramicroscope
according to the new design of Zsigmondy and Siedentopf.

Brownian movement is certainly the most dramatic aspect
of colloidal behavior made apparent by the ultramicroscope, so
it is no wonder that a study of Brownian movement comprised a
major portion of his dissertation.  He also studied the effect
of particle size upon optical absorption as well as the
coagulation of hydrosols upon the addition of electrolyte,

particularly the influence of temperature.

Upon completion of his dissertation, Svedberg and his students expanded upon these and other topics so that Uppsala became known as a center for quantitative work on inorganic sols.

Svedberg was a precocious student. He completed the undergraduate course in less than two years after arriving at the University; he started doing research in September 1905; and he published his first paper dealing with the preparation of organosols of various metals by a modified Bredig method in December 1905. He succeeded next in building an ultramicroscope, and with it he studied the Brownian movement of the particles in the newly prepared metal sols. His first paper on Brownian movement appeared in 1906.

In this work, Svedberg fell under two illusions regarding Brownian movement that had appeared earlier from time to time. One of these was that it was possible by direct observation to measure the velocity of the particles and to relate this to the velocity that would be assigned by the kinetic molecular theory of gases under the assumption of the equipartition of energy. Actually, because the particles are moving against the viscous drag of the liquid, their velocity is quite different than for a gas-like system. Moreover, the particle traces out a path of such extreme sinuosity and varying velocity that it is quite hopeless to expect to observe this velocity directly, and the best that the investigator can do is to measure the displacement of the particle during a particular interval of time. But this displacement is not a measure of the velocity, for, as Einstein showed, it is not a linear function of the interval of time.

The second illusion, a notion that persisted among some of the previous investigators, was that somehow the movement was of an oscillatory nature rather than a random translatory movement, what we now call a "random walk".

Svedberg devised a clever experiment which was designed to kill two birds with one stone. It would enable him to determine both the period of oscillation and also the velocity.

The experiment consisted in observing the particles moving through the field of view of the ultramicroscope while the sol as a whole flowed through the apparatus at a known linear velocity. Svedberg reasoned that the superposition of the oscillatory Brownian motion and the linear motion of the fluid would result in a sinusoidal curve.

Svedberg estimated the amplitude of the sinusoidal tracks visually using an ocular scale. Experiments were carried out with 40 to 50 millimicron platinum organosols in five liquids having a twelve-fold range of viscosity. He reported that the amplitude decreased with increasing viscosity, and a plot of twice the amplitude versus viscosity fell on a hyperbolic curve with what can only be described as incredible precision. In a second series of experiments with calcium organosols, he found that the amplitude decreased with increasing particle size.

Svedberg also estimated the wavelength of the sinusoidal trajectory visually in the same manner as he determined the amplitude, and then, upon dividing this wavelength by the linear velocity of the liquid through the cell of the ultramicroscope, he obtained what he called the period of the vibration. Thus, he treated these data as if he were observing a simple harmonic vibration in linear translation for which the linear velocity equals the product of the wavelength and the frequency of the vibration. Finally, Svedberg calculated "$h$ the absolute velocity" as

$$\underline{h} = \frac{4\underline{A}}{\tau}$$

where $A$ is amplitude of the sinusoidal path and $\tau$ is the period of the vibration. He notes from the results with the five different platinum organosols that, although $\tau$ and $A$ vary widely in these different liquids, the value of $h$ remains quite constant at about 0.03 centimeters per second.

At this point, Svedberg really piles confusion upon confusion. He now uses this "velocity" which he has calculated and the "velocity" obtained by William Ramsay by direct observation of the displacement of somewhat larger platinum particles in order to extrapolate to the gas kinetic velocity of atomic platinum.

He obtained the value $7.6 \times 10^3$ cm/sec compared to the theoretical value $19.2 \times 10^3$ cm/sec, and concluded that "it is therefore quite probable that the characteristic movement of colloidal particles in fact is to be considered as a manifestation of the general molecular movement of matter."

What can one say about this work? It would be best to dismiss it as an immature and fanciful effort of an imaginative and resourceful young student.  But it cannot be dismissed, for Svedberg linked this work to Einstein's theory of Brownian movement.  It persisted in the colloid chemical literature and it figured centrally in the Nobel Prize award.

## IV.  SVEDBERG AND EINSTEIN

Svedberg learned of Einstein's 1905 theory shortly after publishing his first paper in 1906 and within two months a second paper appeared in which he proclaimed that his own experiments could be directly applied to that theory and that they verified the theory.  Einstein took immediate exception to this in a short paper sent to the same journal within a month.  To be sure, his tone was mild and non-polemic, for as he pointed out in a later letter written to Perrin:  "The mistakes in Svedberg's methods of observation and also in the theoretical methods of treatment had become clear to me at once.  I wrote at that time a small correction which seemed to me to correct the worst errors, because I could not resolve to impair for Herr S the great pleasure in his work."

Einstein's paper itself commences:

"In connection with the researches of Svedberg recently published in the Zeit. f. Elektrochemie, on the motion of small suspended particles, it appears to me desirable to point out some properties of this motion indicated by the molecular theory of heat.  I hope I may be able by the following to facilitate for physicists who handle the subject experimentally the interpretation of their observations as well as the comparison of the latter with theory."

Einstein's criticism is primarily that Svedberg believes that he is measuring velocity.  Einstein points out that the actual trajectory of the particle is such that it changes its direction and velocity in a random fashion on a time scale of about $10^{-7}$ seconds.  This is because the moving colloidal particle would come to a virtual standstill during this time because of the viscosity of the liquid, were it not impelled anew by the continual molecular bombardment.  "It is therefore impossible -- at least for ultramicroscopic particles -- to ascertain the root mean square velocity by observation ... it is clear that the velocity ascertained thus corresponds to no objective property of the motion under investigation -- at least if the theory corresponds to the facts."

## V.   SVEDBERG AND PERRIN

Jean Perrin (1870-1942) received his doctorate at the
Ecole Normale in 1897 with a thesis on cathode rays and x-rays
and then accepted a position at the Sorbonne where he was
asked to develop a course of lectures and laboratory work in
the emergent field of physical chemistry (chimie physique).
This led to the writing of the first physical chemistry text-
book in the French language and a new direction for his
research.

Perrin developed an early interest in colloids,
particularly electrokinetic phenomena, and with the invention
of the ultramicroscope, he too became intrigued with Brownian
movement and how it might provide a clue to the reality of the
molecular hypothesis.

The full story of Perrin's work is the subject of a
recent brilliant monograph (3).  We will only recall that not
only did Perrin establish the vertical distribution of
particles in a colloidal sol but also verified Einstein's
equation for translational Brownian movement as well as that
for rotational Brownian movement.  His studies were meticulous
and thorough.  The particle size, the particle substance, the
fluid viscosity, and the temperature were each varied.  Three
independent experimental techniques, vertical distribution,
translational displacement, and rotational displacement gave
internally consistent values for Avogardro's number.  Perrin's
results were universally acclaimed.  After completion of the
work on the vertical distribution, even that arch critic of
atomism Ostwald was to proclaim that "the agreement of
Brownian movement with the demands of the kinetic hypothesis
... which have been proved ... most completely by J. Perrin,
entitle even the cautious scientist to speak of an experimental
proof for the atomistic constitution of space filled matter."
Henri Poincaré summed matters up in these words:  "The
brilliant determination of the number of atoms made by M.
Perrin  has  completed this triumph of atomism ... .  The atom
of the chemist is now a reality."

Perrin criticized Svedberg much more sharply than Einstein
had done.  He insisted that the difference between Svedberg's
results and Einstein's theory could not be accounted for by
any reasonable experimental error, but: "Fortunately, there
is probably little in common between the magnitude $\xi$ and $\tau$
which enter into the formula, and the ill-defined magnitudes
introduced in their place by Svedberg ... from what is known
of the absolute irregularity of the Brownian movement, it

would appear quite impossible that these trajectories can be, as Svedberg, evidently victim of an illusion, describes them, 'lines regularly undulated, of well-defined amplitude and wave-length.' By utilizing a micrometer eyepiece, Svedberg estimated the magnitude of the quantities (in reality non-existent) which he calls the wavelength and the amplitude of oscillation ... . I do not think it necessary to insist upon the uncertainty, to my mind complete, which results from a method so questionable and from estimations so vague."

Svedberg immediately rose to the defense. He persisted in claiming that the motion is oscillatory. "If one observes a colloidal particle in a still fluid, one gets the general impression of hither and thither movement about a not very differing mean position. If, moreover, the liquid is set in translational motion, we see (because of the persistence of the light impression on the retina) a waveform light line."

Perrin was hardly convinced. In his definitive monograph Les Atomes published in 1913 he stated

"Svedberg believed ... that the Brownian movement becomes oscillatory for ultramicroscopic grains. He measured the wavelength (?) of this motion and compared it with Einstein's displacement. It is obviously impossible to verify a theory on the basis of a phenomenon which, supposing it to be correctly described, would be in contradiction to that theory. I would add that the Brownian movement does not show an oscillatory character on any dimensional scale."

There is no further specific rejoinder by Svedberg who even a full decade later in his Wisconsin Lectures on Colloid Chemistry published in 1924 continued to present his 1906 results as "the first quantitative confirmation of the kinetic theory of Brownian movement."

## VI. RECEPTION OF SVEDBERG'S WORK ON BROWNIAN MOTION

From our vantage point, it seems as if Svedberg viewed his Brownian motion work with a kind of schizophrenia. On the one hand, he persisted in defending the position staked out in his 1906 papers, written as a young graduate student, despite the criticism of Einstein and the scathing denunciation of Perrin. Neither he, nor any of his students, nor anyone else attempted to repeat the experiments, even though Perrin accused him of being the "victim of an illusion." Yet, in his subsequent expository and experimental work, he demonstrated a clear understanding of the physics underlying Einstein's equation.

Svedberg's 1906 work was generally well received, and although it ultimately dropped from view, it persisted for a very long time. Svedberg's papers were published just prior to the appearance of a series of classic textbooks in colloid chemistry by recognized leaders in the field -- Freundlich, Wo. Ostwald, Mecklenberg, Hatschek, Burton, Zsigmondy, Bancroft -- but except for a mild <u>caveat</u> here and there, the main thrust is to treat Svedberg's 1906 experiments as definitive. This persists even into the 1930's. Thomas in his well-know book schematically illustrates the waveform that Svedberg claimed to have seen by a perfect sinusoidal curve. Weiser in his book hardly mentions Perrin's contribution, but discusses instead the so-called "time law in Brownian movement (which) was established by Svedberg, independently of the theoretical considerations of Einstein."

However, in time the record did change. In the later editions of his book, Freundlich rewrote the section on Brownian movement completely in order to discuss Perrin's work in great detail, and omitted almost entirely the earlier discussion of Svedberg's work.

More recent books make no mention of Svedberg's 1906 work. The verification of Einstein's theory is attributed solely to Perrin. Yet, the most horrendous boner of all appears in the 1949 treatise by A. E. Alexander and P. Johnson where, in an introductory chapter on the history of colloid science there occurs, following presentation of Einstein's theory for the translational and rotational Brownian movement displacements, the statement: "Equations (1.3) and (1.4) enabled Perrin and Svedberg, by measuring $\bar{x}$ and $\bar{\theta}$ as a function of time for particles of known size, to determine Avogadro's number, N. The values so obtained were in fair accord with one another and with those from other methods."

Yet, this lapse is understandable if we consider the following citation

"Your Majesty, Your Royal Highnesses, Ladies and Gentlemen. The Academy of Sciences has decided to award the Nobel Prize in Chemistry for 1926 to The Svedberg, Professor of Physical Chemistry at the University of Uppsala, for his work on disperse systems.

Almost a hundred years ago, or more accurately in 1827, the English botanist Robert Brown discovered with the aid of the ordinary microscope that small parts of plants, e.g. pollen seeds, which are slurried in a liquid are in a state of continuous, though fairly slow movement in different.

directions ... . As we have recently heard, Einstein evolved a theory for this so-called Brownian movement which was then developed to a high degree by the now late Smoluchowski ... .

The theory in question has been confirmed convincingly by experimental investigations of several colloid scientists among whom especially two of today's prize winners, Perrin and Svedberg, have occupied and still occupy a leading position."

Professor H. G. Soderbaum,
Secretary of the Royal Swedish
Academy of Sciences

## VII.   REFERENCES

(1)   Kerker, M., J. Chem. Educ. 51, 764 (1974).
(2)   Kerker, M., ISIS 67, ... (1976).
(3)   Nye, M. J., Molecular Reality:  A Perspective on the Scientific Work of Jean Perrin, Elsevier, New York, 1972.

IRREVERSIBILITY IN RARE EARTH ION-EXCHANGE
OF THE SYNTHETIC ZEOLITES X AND Y

By:  Howard S. Sherry
*Mobil Research and Development Corporation*

ABSTRACT

*Irreversible ion-exchange behavior in the LaNaX and LaNaY ion-exchange systems results when $La^{3+}$ cations replace $Na^+$ cations in the sodalite cages of zeolites X and Y. This replacement occurs during high temperature ion-exchange or during calcination of partially exchanged LaNaX or LaNaY. The $La^{3+}$ cations that diffuse into the sodalite cages are so strongly complexed by framework and water oxygen atoms that reexchange cannot occur. Moreover the $Na^+$ ions that are displaced into the supercages are easier to replace. Thus, ion-exchange hysteresis is observed. It is shown that the rate controlling step in $La^{3+}$ ion diffusion into a sodalite cage is the stripping of its hydration shell.*

INTRODUCTION

The synthetic zeolites X and Y have crystal structures that result in unusual ion-exchange properties. These zeolites are isostructural with the natural zeolite faujasite[1], for which the framework structure and known cation positions have already been reviewed in an earlier paper on univalent ion-exchange[2], and again in papers on reversible alkaline earth ion-exchange in these zeolites[3,4].

Olson[5] has shown in an X-ray study of a single crystal of hydrated NaX, that sixteen out of the eighty-five $Na^+$ ions in a unit cell are in the network of sodalite cages and hexagonal prisms. The remainder are in the large supercages. Mortier and Bosmans[6] have shown in an X-ray powder diffraction study of hydrated $K^+$ forms of near faujasites of varying Si/Al atom ratio that this number ranges from fourteen to nineteen. It has been shown in a study of the rate of Na isotope exchange of large crystals of NaX[7] that all of the $Na^+$ ions are mobile but that sixteen out of eighty-five in each unit cell exchange more slowly than the rest. Ion-exchange of these cations is slow with $Ca^{2+}$ ions at 25°C and no ion-exchange takes place with $Ba^{2+}$ ions at 5 and 25°C. There are 85 $Na^+$ ions in a unit cell and thus 16/85 or 18% of the $Na^+$ ions are difficult to replace with divalent cations. It has also been shown that at 25°C $La^{3+}$ ions cannot replace these sixteen cations[8].

Results are similar for zeolite Y. At 25°C $Ca^{2+}$, $Sr^{2+}$, $Ba^{2+}$ and $La^{3+}$ ions[4,8] cannot replace sixteen of the fifty $Na^+$ ions in one unit cell of this zeolite at low temperature. Large univalent ions such as $Cs^+$ and $Rb^+$ also cannot replace sixteen $Na^+$ ions per unit cell. From these data it is concluded that sixteen $Na^+$ ions in a unit cell of NaY are also located in the small cavities. This conclusion is at least consistent with a single crystal structure determination of natural faujasite (predominantly in the Ca form) by X-ray diffraction[1], which placed sixteen $Ca^{2+}$ in the sodalite cages (50% occupancy of a thirty-two fold set of sites) and no cations in the hexagonal prisms, and the study of hydrated KX and KY by Mortier and Bosman cited above[6].

If a cation is to diffuse to sites in the sodalite cages or hexagonal prisms, it must do so through a window 2.4 to 2.5 Å in free diameter. It is obvious that an ion with a large Pauling radius (large bare ion) will be sieved from these positions. Thus, $Rb^+$ (1.48 Å Pauling radius), $Cs^+$ (1.69 Å) and $Ba^{2+}$ (1.35 Å) cannot replace sixteen ions in a unit cell of NaX and NaY at 25°C[2]. It is not so obvious why small bare ions, such as $Ca^{2+}$ (0.99 Å Pauling radius), $Sr^{2+}$ (1.13 Å) and $La^{3+}$ (1.15 Å), should experience difficulty in passing through the small windows leading to the sodalite cages and hexagonal prisms. It is possible that these ions are sieved because they enter the large cavities of zeolites X and Y as hydrated ions and must be partially or completely dehydrated to pass through the small rings of six tetrahedra that separate the large cavities from the small ones (sodalite cages and hexagonal prisms). Simple geometry indicates that hydrated alkaline earth or rare earth ions are too large to pass through these rings. It is, therefore, postulated that the rate of ion dehydration must be the slow step when cations diffuse from the large to the small cavities[8].

In this paper some other features of polyvalent ion-exchange will be demonstrated that result from the presence of ions in different size cavities and that lead to irreversible behavior. An attempt is made to prove that the potential energy barrier to diffusion of ions from the network of large cavities to the network of small cavities is the energy required to dehydrate ions with large hydration energies.

EXPERIMENTAL PROCEDURES

All chemicals were reagent grade, purchased (except for the rare earth compounds) from the J. T. Baker Chemical Company. Mixed rare earth chlorides (commercial grade) and $LaCl_3$ (99.9% pure) were obtained from the Lindsay Division of American Potash and Chemical Corporation, West Chicago, Illinois.

The analyses of the NaX and NaY zeolites used have been given before(2,3,7). The contents of the anhydrous unit cell of NaX is

$$Na_{85}\{(AlO_2)_{85}(SiO_2)_{107}\}$$

and of NaY

$$Na_{50}\{(AlO_2)_{50}(SiO_2)_{142}\}$$

The equilibration procedures have been described(2). After phase separation, wet chemical analysis of both phases was performed. Both cations in both phases were determined, as well as the $SiO_2$ and $Al_2O_3$ content of the zeolite phase. In this way we ascertained that no measurable hydronium ion-exchange took place.

Dehydration studies were made by calcining seven gram samples of $RE_{88}Na_{12}X$ (RE indicates a commercial mixture of rare earths) prepared by exchanging NaX with 0.3 N $RECl_3$ at 82.2°C for various times and at various temperatures and then cooling in desiccators over activated Linde 5-A molecular sieve. The water content of these samples was determined by calcining a portion of them at 800°C for five hours. Another portion was then ion-exchanged with a 1000 fold excess of $(NH_4)_2SO_4$ at room temperature for twenty-four hours using 1 M $(NH_4)_2SO_4$. The metal cations that remained in the zeolite after the $(NH_4)_2SO_4$ treatment were called "fixed".

RESULTS AND DISCUSSION

We have studied the ion-exchange properties of partially dehydrated $La_{82}Na_{18}X$ (82% $La^{3+}$ ion-exchanged NaX) by drying samples at 121 C for twenty-four hours or at 425°C for fifteen minutes and then exchanging them at 25°C for twenty-four hours -- a time sufficient to reach equilibrium in samples that had never been dried. The results obtained in zeolite X are shown in Figure 1, along with the reversible ion-exchange isotherm for materials that had never been dried (taken from reference 8). The triangles and diamonds indicate that heating and partial dehydration of $La_{82}Na_{18}X$ leads to the attainment of lower Na content during subsequent $La^{3+}$ ion-

323

exchange than was obtained by $La^{3+}$ exchange of unheated and undehydrated samples. Similarly, NaCl ion-exchange of the heated samples results in higher La content than was obtained with unheated samples (squares). Thus we see that all points obtained using the calcined $La_{82}Na_{18}X$ lie above the equilibrium ion-exchange isotherm. Similar results were obtained when $La_{66}Na_{34}Y$ and $Ca_{78}Na_{22}X$ were heated at 121°C for twenty-four hours (Figures 2 and 3). The data indicate that drying or calcining the partially exchanged zeolites makes the $La^{3+}$ or $Ca^{2+}$ ions more difficult to replace by $Na^+$ ions and the remaining $Na^+$ ions easier to replace with $La^{3+}$ or $Ca^{2+}$ ions.

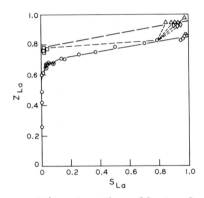

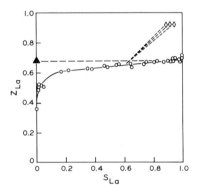

*Fig. 1. The effect of heating on the ion-exchange properties of $La_{82}Na_{18}X$ at 25°C and 0.3 total normality*

○ $La^{3+} + 3$ $NaX$ *(not dried or dried)* ⟶

▽ $3$ $Na^+ + La_{82}Na_{18}X$ *(not dried)* ⟶

△ $La^{3+} + La_{82}Na_{18}X$ *(dried 121°C for 24 hrs)* ⟶

◇ $La^{3+} + La_{82}Na_{18}X$ *(dried 425°C for 15 min)* ⟶

□ $3$ $Na^+$ $La_{82}Na_{18}X$ *(dried 121°C for 24 hrs)* ⟶

*Fig. 2. The effect of heating on the ion-exchange properties of $La_{66}Na_{34}Y$ at 25°C and 0.3 total normality*

○ $La^{3+} + 3$ $NaY$ *(not dried or dried)* ⟵ *(not dried)*

◇ $La^{3+} + \overline{La_{66}Na_{34}Y}$ *(dried 121°C for 24 hrs)* ⟶

△ $3$ $Na^+ + La_{66}Na_{34}Y$ *(dried 121°C for 24 hrs)* ⟶

Whether calcined or not, NaX and NaY can be ion-exchanged with $LaCl_3$ solution at 25°C to give reversible ion-exchange isotherms that terminate at the points $Z = 0.82 \pm 0.02$ and $0.68 \pm 0.01$. The isotherms are believed reversible because uncalcined $La_{82}Na_{18}X$ and $La_{68}Na_{32}Y$ can be reexchanged with 0.3 N NaCl solution, and the same isotherm obtained. That the LaNaX and LaNaY isotherms terminate at points indicating a maximum $La^{3+}$ ion loading of 82 and 68% indicates that at 25°C only the $Na^+$ ions in the large cages have been replaced(8).

That higher polyvalent ion loadings are possible after drying, coupled with the greater difficulty encountered in polyvalent ion removal by $Na^+$ ions, indicates that during the drying process an intercage, or intracrystalline, ion-exchange takes place. Thus, polyvalent ions in the large cages diffuse into the small cages and $Na^+$ ions in the small cages diffuse into the large ones. The $Na^+$ ions in the large cages are then available for exchange with fresh $La^{3+}$ ions, resulting in high $La^{3+}$ ion loadings. If this part of the picture is true then it must also be postulated that the $La^{3+}$ ions or $Ca^{2+}$ ions that diffuse into the small cages are more difficult to remove than those in the large cages -- the ions are "fixed". This picture agrees with the data of Olson, Kokotailo and Charnell[9] who did an X-ray diffraction study of a $Ce^{3+}$ ion-exchanged single crystal of natural faujasite in the hydrated state and after calcining in vacuo for sixteen hours at 350°C (not rehydrated), and of an LaX powder in the hydrated state and after calcining and rehydrating. They found that the sodalite cages contained more rare earth ions after calcining than before calcining.

Irreversibility has also been shown in another way. A sample of $La_{88}Na_{12}X$ was prepared by $La^{3+}$ ion-exchange at 82.2°C for twenty-four hours. At this time and temperature some of the $La^{3+}$ ions replace some of the $Na^+$ ions in the small cavities[8]. This material was then ion-exchanged with NaCl at 82.2°C and at 25°C for twenty-four hours. The points obtained (Figure 4) at 82.2°C lie above the 82.2°C isotherm obtained by exchanging NaX with $LaCl_3$. This hysteresis can be explained in the following way: After the $La_{88}Na_{12}X$ was prepared at 82.2°C it was filtered and washed and stored at room temperature. Ion diffusion into the small cages was quenched. However, when this material was exchanged with NaCl solution at 82.2°C, not only did $Na^+$ ions replace $La^{3+}$ ions in the large cavities, but more $La^{3+}$ ions diffused from the large cavities into the small cavities to replace the residual $Na^+$ ions located there. Since these $La^{3+}$ ions, and those introduced into the small cavities during the initial sample preparation, are "fixed" -- hysteresis results. When another sample of this material was exchanged with NaCl, the expected equilibrium point on the 25°C isotherm was not obtained because the $La^{3+}$ ions that diffused into the sodalite cages at 82.2°C were trapped at 25°C (Figure 4).

Similar results were obtained with $La_{68}Na_{32}Y$ prepared at 82.2°C by ion-exchanging with 0.3 N $LaCl_3$ for twenty-four hours. Reexchange of this material at 25°C for twenty-four hours with 0.3 N NaCl (Figure 5) results in points lying above the equilibrium ion-exchange isotherm at 25°C (taken from

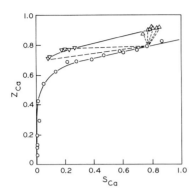

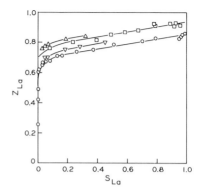

Fig. 3. The effect of drying on the ion-exchange properties of $Ca_{78}Na_{22}X$ at 25°C and 0.1 total normality

◯ $Ca^{2+} + 2\ NaX$ (undried or dried) $\longrightarrow$

△ $Ca^{2+} + Ca_{78}Na_{22}X$ (dried 121°C for 24 hrs) $\longrightarrow$

▽ $2\ Na^+ + Ca_{78}Na_{22}X$ (dried 121°C for 24 hrs) $\longrightarrow$

Fig. 4. Irreversible behavior in the LaNaX system at 0.3 total normality

◯ $La^{3+} + 3\ NaX\ \underset{\overrightarrow{82.2°C}}{\xrightarrow{25°C}}$

☐ $La^{3+} + 3\ NaX\ \xrightarrow{82.2°C}$

△ $3\ Na^+ + La_{88}Na_{12}X\ \xrightarrow{82.2°C}$
$La_{88}Na_{12}X$ prepared at 82.2°C

▽ $3\ Na^+ + La_{88}Na_{12}X\ \xrightarrow{25°C}$
$La_{88}Na_{12}X$ prepared at 82.2°C

reference 8). Again, preparation of the sample at 82.2°C results in some penetration of $La^{3+}$ ions into the small cavities and these will not diffuse out at 25°C. Other samples of this material were reexchanged with 0.3 N NaCl at 82.2°C for twenty-four hours. These points lie on the 82.2°C isotherm obtained when $La^{3+}$ is the ingoing ion. The lack of hysteresis simply indicates that the $La^{3+}$ ion diffuses very slowly into the small cavities even at 82.2°C.

An attempt was made to replace all of the $La^{3+}$ ions from calcined samples of LaX and LaY (425°C for five hours in air) using 1000 fold excess of reagent and it was found that none of the ions were replaced -- all were fixed. This finding is consistent with the X-ray diffraction study of Olson, Kokotailo and Charnell(9). After calcination and rehydration each sodalite cavity was found to contain up to 4 $La^{3+}$ ions and 4 water molecules. They report that each $La^{3+}$ ion is octahedrally coordinated to 3 lattice oxygen atoms and 3 water molecules(9). It would appear that this complex must be extremely stable.

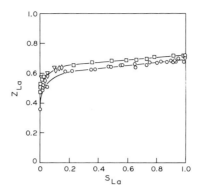

*Fig. 5.  Irreversible behavior in the LaNaY system at 0.3 total normality*

○ $La^{3+} + 3\ NaY \xrightleftharpoons[82°C]{25°C}$

□ $La^{3+} + 3\ NaY \xrightarrow{82°C}$

▽ $3\ Na^+ + La_{68}Na_{32}Y \xrightarrow{82.2°C}$
$La_{68}Na_{32}Y$ prepared at $82.2°C$

◇ $3\ Na^+ + La_{68}Na_{32}Y \xrightarrow{25°C}$
$La_{68}Na_{32}Y$ prepared at $82.2°C$

It is hypothesized that the rate-controlling step in the diffusion of $La^{3+}$ ions, or any other small polyvalent ions, into the small cavities is the stripping of their hydration shells.  A proof of this hypothesis was attempted in the following manner:  A batch of $RE_{88}Na_{12}X$ was prepared by ion-exchanging 100 g of NaX with a 0.3 N solution of mixed rare earth chlorides at 82.2°C.  This material was filtered and washed and stored at room temperature over saturated $NH_4Cl$ solution.  Samples of this material were then placed in an electric furnace at a specified temperature.  At different time intervals a sample was removed and cooled in a desicca-tor, thereby giving a number of samples that were calcined at different combinations of time and temperature.  The $H_2O$ content of these samples was determined by recalcining for five hours in air at 800°C.  In this way curves of $H_2O$ content as a function of calcination time were obtained at constant furnace temperature.  These data are shown in Figure 6. Another set of samples was calcined for various times and at various temperatures; instead of determining their $H_2O$ con-tent, the samples were exchanged with a 1000 fold excess of 1 M $(NH_4)_2SO_4$ solution for twenty-four hours at room tempera-ture.  The rare earth ions that could not be replaced were considered to be fixed.  Figure 7 contains curves that represent the amount of rare earth fixed (as weight percent

$RE_2O_3$) versus calcination time at constant furnace temperature.

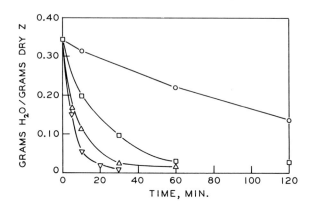

Fig. 6. *The water content of calcined RE₈₈Na₁₂X as a function of time and temperature of calcination*

*○ 150°C,  □ 260°C,  △ 370°C,  ▽ 482°C*

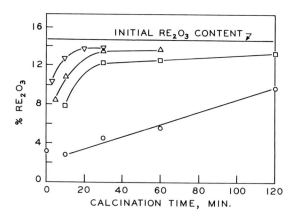

Fig. 7. *Rare earth ion fixation as a function of time and temperature of calcination*

*○ 150°C,  □ 260°C,  △ 370°C,  ▽ 482°C*

A strong correlation between the number of rare earth ions fixed in, and the number of water molecules desorbed from, a unit cell as a result of calcination was expected because it was believed that ion dehydration is the rate-controlling step in $La^{3+}$ ion diffusion from the large to the small cages. In order to find this correlation, the data from Figures 6 and 7 was replotted as the number of rare earth ions fixed in a unit cell versus the $H_2O$ molecules desorbed per unit cell (Figure 8). An excellent linear correlation was obtained, which can be interpreted to mean that the breaking of rare earth ion-water molecule bonds is the rate determining step.

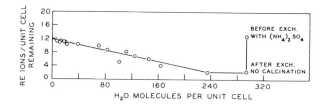

*Fig. 8. Rare earth ion fixation as a function of the water desorption*

There are only four processes that can be visualized as taking place when a sample of $RE_{88}Na_{12}X$ is heated:

1. Water molecules not coordinated to rare earth ions, and therefore weakly sorbed, are desorbed. This desorption gives the flat portion of the curve in Figure 8 at high water contents.

2. Water molecule-rare earth ion bonds in the large cages are broken.

3. These water molecules diffuse out of the crystals.

4. Rare earth ions diffuse through the small windows into the sodalite cavities and $Na^+$ ions diffuse from the small cages into the supercages.

The linear correlation between the number of water molecules desorbed and the number of rare earth ions fixed indicates that as fast as water molecules are removed from the crystals, ions move from large cavities into small cavities. It is highly unlikely that the rates of the third and fourth processes listed above are the same at all temperatures. We must, therefore, conclude that the second process -- the breaking of rare earth ion-water molecule bonds -- is the slowest of the last three processes and controls the rate of ion migration as well as water desorption. As soon as a bond between a cation and a water molecule is broken a water molecule desorbs and the rare earth ion changes position.

It is believed that "ion fixation" simply means that the $La^{3+}$ ions in LaX and LaY and the $Ca^{2+}$ ions in CaX are strongly complexed by lattice oxygen atoms and water molecules in the sodalite cages(9). Our ion-exchange equilibrium studies indicate that the sites in the sodalite cavities are very selective for $La^{3+}$ ions -- probably because of the formation of the stable complex mentioned above. The $La^{3+}$ ions are in a potential energy well that is deeper than the ones in the large cages and the activation energy required for diffusion out of the small sodalite units is greater than that required for diffusion into these cavities. Thus the rate of diffusion out of the small cages is slower than the rate of diffusion into these cages and we observe ion fixation.

Irreversibility when divalent ions are exchanging into, or are present in, zeolites X and Y has been reported by other researchers. Maes and Cremers(10) have shown that, after $Zn^{2+}$ and $Co^{2+}$ ion-exchange of NaX at 45°C, some transition metal ions are fixed in difficulty accessible sites and do not participate in ion-exchange reactions at 5°C. They hypothesize that during ion-exchange at 45°C the transition metal ions selectively enter the small cages and that if the same zeolite sample is subsequently exchanged at 5° the transition metal cations in the small cages are trapped. They have also shown(10b) that after $Co^{2+}$ ion-exchange of NaX and NaY at 45°C the maximum $Ag^+$ ion loading that is attainable is only 90% of the total $Na^+$ ions originally present in these zeolites leading to the conclusion that some $Co^{2+}$ ions are fixed in sites in small cages and reduce the ion-exchange of capacity for $Ag^+$ ion.

Dyer, Gettins and Brown(11) have shown that when zeolites X and Y are fully exchanged with $Ca^{2+}$, $Sr^{2+}$ or $Ba^{2+}$ ions at elevated temperatures a proportion of the divalent cations cannot be reexchanged at room temperature. Dyer and Townsend(12) showed that alternate $Zn^{2+}$ ion-exchange and calcination of NaX and NaY results in almost complete replacement of $Na^+$ ions, which cannot be accomplished solely by ion-exchange at 25°C. Moreover, study of $^{65}Zn^{2+}$ isotope distribution between labeled salt solution and non-radioactive ZnX and ZnY prepared as described above showed that a fraction of the $Zn^{2+}$ in the zeolite did not undergo exchange. These workers also concluded that calcination causes polyvalent ions to enter the small cages from the large cages and that these ions do not exchange at low temperature.

CONCLUSIONS

At 25°C $La^{3+}$ ions only replace the $Na^+$ ions in the large cages. Exchange at higher temperatures results in partial or complete replacement of the $Na^+$ ions in the small cavities by $La^{3+}$ ions depending on the temperature. These $La^{3+}$ ions are in the sodalite cages and form a very stable complex with lattice oxygens and water molecules resulting in irreversibility -- the $La^{3+}$ ions in the sodalite small are very difficult to reexchange and are "fixed" in these sites.

Calcination and concomitant dehydration of a partially $La^{3+}$ ion-exchanged NaX or NaY make the remaining $Na^+$ ions easier, and the $La^{3+}$ ions more difficult, to exchange. During calcination rare earth ions diffuse out of the large and into the small cages and $Na^+$ ions diffuse out of the small and into the large cages. The $Na^+$ ions are then available for subsequent $La^{3+}$ ion-exchange and the $La^{3+}$ ions in the small cavities are fixed.

Thorough calcination and dehydration causes fixation of all the $La^{3+}$ ions because they all diffuse into the sodalite cages of the zeolite. The slow step in the diffusion of $La^{3+}$ ions into the sodalite units -- either during high temperature ion-exchange or calcination -- is the breaking of ion-water molecule bonds.

Even though the companion X-ray studies have not been made in the case of transition metal and alkaline earth ion-exchange, the reasons for ion fixation and irreversibility observed in alkaline earth and transition metal ion-exchange are probably the same.

ACKNOWLEDGMENT

We gratefully acknowledge the support and encouragement of Mobil Research and Development Corporation and the stimulating discussions with many of our colleagues, in particular with Dr. David H. Olson.

REFERENCES

1. Baur, W. H., Am. Mineralogist, 49, 697 (1964).
2. Sherry, H. S., J. Phys. Chem., 70, 1158 (1966).
3. Olson, D. H. and Sherry, H. S., J. Phys. Chem., 72, 4095 (1968).
4. Sherry, H. S., J. Phys. Chem., 72, 4086 (1968).
5. Olson, D. H., J. Phys. Chem., 74, 2758 (1970).
6. Mortier, W. J. and Bosmans, H. J., J. Phys. Chem., 75, 3327 (1971).
7. Brown, L. M. and Sherry, H. S., J. Phys. Chem., 75, 3855 (1971).
8. Sherry, H. S., J. Colloid Interface Sci., 28, 288 (1968).
9. a) Olson, D. H., Kokotailo, G. T. and Charnell, J. F., Nature, 215, 270 (1967).
   b) Olson, D. H., Kokotailo, G. T. and Charnell, J. F., J. Colloid Interface Sci., 28 (1968).
10. a) Maes, A. and Cremers, A., Page 20 in Molecular Sieves, ed. Meier, W. M. and Uytterhoeven, J. B., Adv. in Chem. Series No. 121, Am. Chem. Soc., Washington, D. C., 1973.
    b) Maes, A. and Cremers, A., Recent Progress Report No. 121, Molecular Sieves, ed. Uytterhoeven, J. B., Proc. of Third International Conference on Molecular Sieves, Leuven University Press, Leuven, Belgium, 1973.
11. Dyer, A., Gettins, R. B. and Brown, J. G., J. Inorg. Nucl. Chem., 32, 2389 (1970).
12. Dyer, A., Townsend, R. P., J. Inorg. Nucl. Chem., 35, 2993 (1973).

# DETECTION OF THE PHYSICAL AGING OF TRIGLYCERIDES

James H. Whittam*, Henri L. Rosano, and Anesu Garuba
Department of Chemistry
City College of The City University of New York
New York, NY  10031


* The Gillette Company
Gillette Park  PCD
South Boston, MA.  02106

ABSTRACT

Triglycerides, pure and mixtures, solidify in more than
one crystalline form.  However, not all the polymorphs can be
classified as true thermodynamic states but exist as a
result of the kinetics of formation.  Furthermore, the triglyceride system transform or physically age from one state
to another state even when stored below the observed $\alpha \longrightarrow \beta'$
crystal-crystal transition temperature.  The use of a
microscpectrophotometer which determines phase changes by a
change in scattered light was used to study this physical
aging process.  For pure saturated triglycerides the aging
time is related to the $\Delta T$ below the $\alpha$ transition temperature.
With binary mixtures the relation appears to be more
complex and dependent on the chain length of the triglyceride
used.
When the triglyceride impurity is a high melting
triglyceride there is little change in the physical aging
characteristics.  On the otherhand, low melting triglyceride
impurities intially enhance the alpha stability and then at
higher concentrations prevent the alpha phase of the bulk
triglyceride from forming.  This enhancement of alpha
polymorph stability appears to occur when the binary triglyceride system forms a eutectic like mixture.

## I.  INTRODUCTION

In the past few years we have been studying the physical properties of simple fats and fat emulsions (1).  In particular our efforts have focused on the various polymorphic or crystalline properties of the simple mono, di and triglycerides. While a substantial amount of literature exists on the number or types of polymorphs, little is known about the kinetics of transformation from one structure to another.

Three polymorphs exist for the monoacid saturated triglyceride, that is the $\alpha$, $\beta'$ and $\beta$.  Since only the $\beta$ form is a true thermodynamic state, these systems exhibit what we call Physical Aging. That is, if the triglyceride is initially crystallized in the $\beta'$ state, it may transform to the more stable $\beta$ state even though the temperature is below the phase transition temperature.  Normally one would expect phase transitions to occur with increasing temperature at constant pressure.  A triglyceride rapidly cooled from the melt may be placed in the alpha form at point A (Figure I).

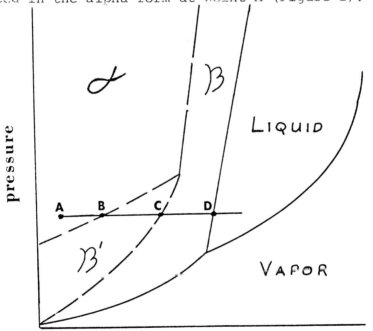

**Figure I** - Typical triglyceride phase diagram exhibiting the three polymorphic states $\alpha$, $\beta'$ and $\beta$.  The $\alpha$ and $\beta'$ are not thermodynamically stable states.

If the triglyceride is heated, a phase transition will occur at Point B, that is $\alpha$ will transform to $\beta'$. Heating to a higher temperature will cause a transformation to $\beta$ at Point C. Finally melting will occur at Point E into the liquid state. This, however, is not the only way a transition will take place for these systems. Even if the triglyceride is left at Point A, a solid transition to $\beta'$ or $\beta$ will take place. The time for this transition (Physical Aging) to occur can be related to the energy of activation of the molecules and the increase in entropy due to chain entangling (1). The higher chain triglycerides having a higher energy of activation and larger entropy in the crystalline state are more stable.

That is

$$\frac{\alpha \longrightarrow B \text{ Transition}}{\text{Time}} = Ae^{S*/R}\, e^{-H*/RT}$$

The magnitude of the energy of activation H* varied from 10kcal/mole for trilaurin to 228 kcal/mole for tribehenin. This is reasonable when one considers that to break the $CH_2$-$CH_2$ interactions requires 250-300 cal/mole. Therefore for trilaurin ($C_{12}$ chain) in the unsymmetric tuning fork orientation, we estimate the HI necessary to break all Van der Waal's attraction is ~20 kcal/mole. As the chains increase in carbon content, it is reasonable to assume that not only should the energy of activation increase due to the additional carbons but also because the chains become more entangled in the crystal structure (therefore, greater Van der Waal's interactions). The absolute entropy in the $\alpha$ or $\beta$ state will be larger as the carbon increases, but the entropy term ($S^I$) in the activated state will be lowered, since it is assumed that the chains are ordered and untangled in this high energy state.

This concept of Physical Aging may have implications in industries concerned with fat and confectionary products as well as biological areas of study as the formation of gall stones and the crystallization and rheology of surface lipids on hair ans skin.

To study this Physical Aging process a number of techniques are open to the researcher. Of these are:

A) X-Ray Diffraction Analysis (2)

B) Infra Red Spectroscopy (3)
C) Differential Thermal Analysis and Scanning Calorimetry (4)
D) Broad Line NMR (5)
E) Dilatometry (6)
F) Microscopy (7,8)

Since the time scale for the Physical Aging process can be extremly short or span over years a problem arises for most of these techniques when determining the aging time. Numerous samples must be prepared then aged at the desired temperature, often in an external source. The samples are then rapidly cooled well below the phase transition temperature to prevent further aging and the crystal structure determined. If the crystal structure is still in its original form a new sample is aged for an extended period and the test repeated. This process consumes a great deal of time and effort before one can pinpoint the transition time exactly. This paper will concentrate on a new time saving technique for studying the Physical Aging process on binary triglyceride mixtures.

## II. EXPERIMENTAL

The monoacid (even) triglycerides were purchased from Appplied Science Laboratories, State College, PA., with the exception of triarachiden which was purchased from the Hormel Institute, Minneapolis, Minn. The purity was estimated by the manufacturers to be 99%. Further purification by crystallization from petroleum ether at $-20°C$. and analysis using GLC confirmed the manufacturers claim.

Mixtures of the triglycerides were prepared by weighing the components within $+$ .01mg. The components were dissolved in petroleum ether to form a homogeneous solution and then the solvent was evaporated under a vacuum. This produced an evenly dispersed solid mixture of the triglycerides.

A useful technique for studying the Physical Aging of triglycerides as a function of temperature and which eliminates the problem discussed above is to observe the light scattering properties of various crystals. This technique has been used by Santoro and Esposito (9) to detect liquid crystal samples in the bulk. Barrall and Guffy (10) have also used a similar technique using a polarized light source.

The instrument used in this study is a quartz

microspectrophotometer. It combines the advantages of the above techniques, in addition to having the ability to measure the absorption of ultraviolet and visible light on thin film samples placed on an isothermal stage. The basic design of the microspectrophotometer, Figure II, was originally by W. Eaton and T. Lewis (11). The main advantage of the instrument is its ability to measure high optical densities on very small crystals. It is built around a Leitz Ortholux Pol polarizing microscope. Monochromatic light is obtained by passing light from an Osram 150x Xenon arc source through a Jarrell Ash grating monochromator (Model 82-410 16. 5A/mm dispersion, 1mm slits). The monochromatic light is plane polarized at the exist slit of the prism monochromator with a Glan prism (Karl Lambrecht Crystal Optics, Chicago, Illinois). Another Glan prism may be inserted in the microscope tube in order to align crystals by extinction on the rotary microscope stage. The light detection system consists of a RCA 1P28 photomultiplier tube, Fluke power supply and Keithley Model 4145 picoameter. The unique microscope stage contains a stannous oxide coating which acts as a heat resistor when a voltage is put across it. This allows the temperature of the thin film to be regulated. A thermocouple is used to determine the temperature of the isothermal stage.

Crystal thickness may be determined from refractive index Measurements using oil immersion methods and from measurements of relative retardation. A Leitz Brace Koehler compensator is used to measure the retardation effect. The advantages are threefold:

1. Thin films rather than bulk quantities of a substance may be studies. This eliminates many of the heat transfer problems associated with bulk systems.
2. The wavelength of the incident light is variable to allow the system to be studied at maximum absorption.
3. Physical Aging processes may take anywhere from a few minutes to days to occur. This instrument is completely automated to plot transmittance as a function of time thus the experiment may be run without the aid of continuous testing or the use of a technician during the aging process.

# MICROSPECTROPHOTOMETER

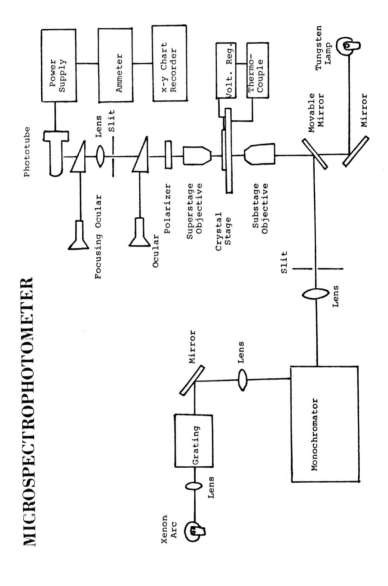

Figure II – Schematic description of the Microspectrophotometer,

338

Triglyceride mixtures of approximately 4 mgs were placed between two No. 1 coverglass slides (Corning, N.Y.) and melted to produce a thin film The samples were kept in an oven at 100°C. for five minutes to provide enough time for all nucleation sites (or crystals) to melt. The triglyceride slide is then rapidly cooled to -70°C. The slide is then placed on the microscope stage at the desired aging temperature and the scattering properties observed.

Differential Thermal Analysis was conducted on a DuPont 900 DTA to compliment the microspectrophotometer data. The heating rate was 15°C./min.

III. RESULTS

Two types of triglyceride mixtures were prepared. In one case, lower melting triglyceride (LTMGI) was used as as the impurity and in the other a higher melting triglyceride impurity (HMTGI) was used (Table I). The concentration ranges studied were from 0 to 30% impurity. More specifically these systems were:

LMTGI

Trilaurin in Tribehenin
Trilaurin in Triarachidin
Trimyristin in Tribehenin
Trimyristin in Triarachidin
HMTGI
Tribehenin in Trilaurin
Tribehenin in Trimyristin
Triarachidin in Trilaurin
Triarachidin in Trilaurin

The aging of various concentrations of HMTGI as a function of temperature below the α transition of the bulk phase was made. These aging curves were almost exactly the same as for the pure bulk components. Thus this type of impurity had little effect on aging. D.T.A. curves of these mixtures indicated slightly lower and broader melting regions for 15% impurity and above.

On the otherhand amd much more significant were the aging curves of the LMTGI. A typical output from the microspectrophotometer is shown in Figure III. In this example, tribehenin was aged at 65°C. or some three degrees below the α phase transition temperature. Information is easily obtained on the stability system as well as how fast or how rapid

## TABLE I

### TRIGLYCERIDE MELTING POINT DATA

| Polymorphic Form | $\alpha$ | $\beta'$ | $\beta$ |
|---|---|---|---|
| Trilaurin | 15 | 34 | 46.4 |
| Trimyristin | 32.8 | 45 | 58.0 |
| Tripalmitan | 45 | 56.6 | 66 |
| Tristearin | 54.7 | 63 | 73.5 |
| Triarachidin | 61.8 | 69 | 78 |
| Tribehenin | 68.2 | 74 | 82.5 |

the transformation is.

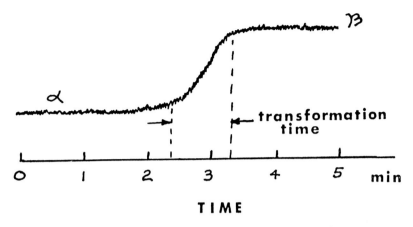

**TIME**

Figure III -- Physical Aging of Tribehenin at 65°C.

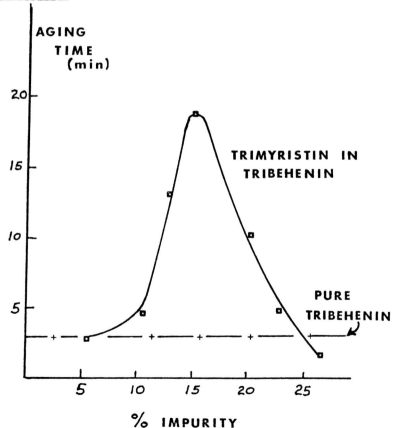

Figure IV - Physical Aging times of Tribehenin at 65°C. as a
function of Trimyristin impurity.

Figure IV plots the Physical Aging time from the metastable alpha state to the stable β state as a function of the trimyristin impurity. Pure tribehenin initially crystallized in the alpha form will transform to the more stable form in approximately eight minutes at 65°C. This transformation is called Physical Aging since the temperature is 3°C. below the transformation temperature. Mixtures of trimyristin (or trilaurin) and tribehenin alter this transformation time. Small quantities of the LMTGI enhance the stability of the tribehenin is prevented from forming. Similar results were also obtained for this system at 64 and 65°C. The maximum in the Physical Aging Curve plot was at 14.5%. When trilaurin was used as the LMTGI in tribehenin and aged at 65°C. a maximum was obtained at 5.5 wt. %. The maximum aging enhancement was 45 minutes or an increase in stability of more than threefold.

Using triarachiden as the bulk component with impurities of trilaurin or trimyristin also produced an aging maximum when the triarachiden was placed in the alpha polymorph and aged at 60°C. These results are indicated in Figure V.

DTA analysis of these mixtures from the alpha form was used to develop a binary phase diagram and also used to reconfirm the aging curves at the observed maximum. A typical series of aging curves shown in Figure VI for trimyristin in tribehenin is given in Figure VII for trimyristen concentrations up to 25%. Notice there is a shallow eutectic region at 15%. Eutectic points were also observed for the other LMTGI systems at maximum aging concentrations. No eutectic was found for the HMTGI systems of up to 25% impurity.

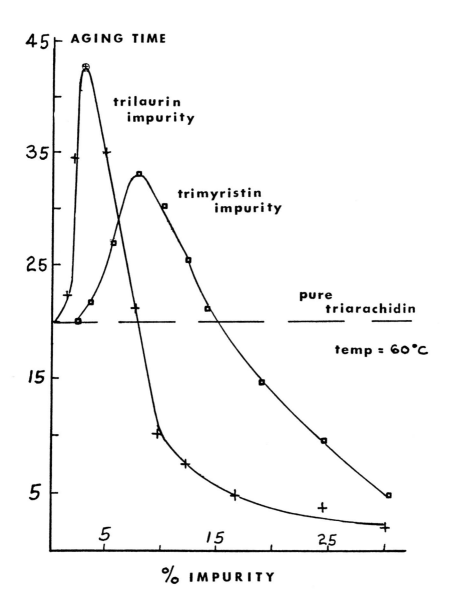

Figure V – Physical Aging times of Triarachidin at 60°C. as a function of Trimyristin and Trilaurin impurity.

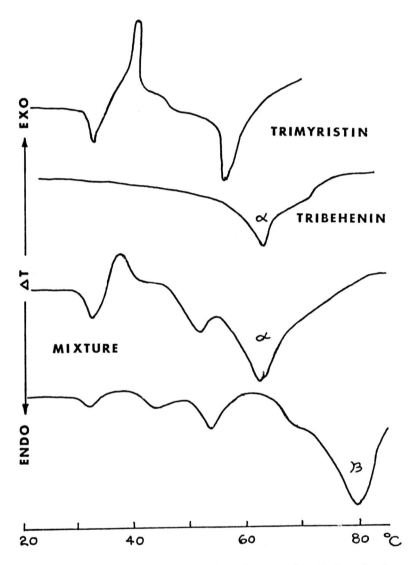

Figure VI – D.T.A. Aging Curves for Trimyristin
in Tribehenin.

Curve 1 – Pure trimyristin in α form.
Curve 2 – Pure tribehenin in α form.
Curve 3 – 25% trimyristin in tribehenin (α).
Curve 4 – 25% trimyristin in tribehenin (β).

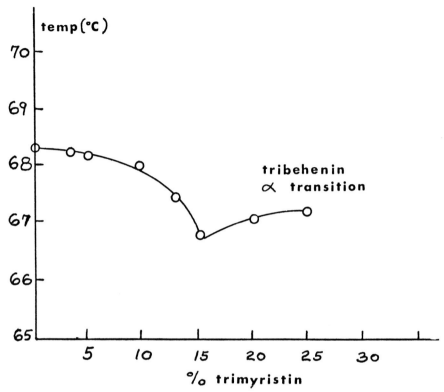

Figure VII – $\alpha \longrightarrow \beta$ Transition Curve for Tribehenin
as a function of Trimyristin impurity.

## IV. DISCUSSION

This study demonstrates that mixtures of triglycerides behave differently than the pure triglycerides. More specifically it is the low melting triglyceride impurities which alter the texture, phase
properties and stability of the triglyceride.

In previous studies on pure triglycerides we
have proved that the stability of the $\alpha$ polymorph
is a logarithmic function of the $\Delta T$ below the
alpha transition point. This correlation also
appears to apply to binary mixtures of triglycerides except in the region of mixed crystal formations or eutectics. When this is the case an
actual increase in $\alpha$ stability is observed. The
DTA results further provide evidence of mixed
crystal formation at the concentrations where the
$\alpha$ stability is longest. This is most likely a

eutectic mixture however due to the many polymorphic forms of the triglyceride it is impossible
to reversily transform from β to α.  Therefore
only melting or α⟶β transformations could be
studied.

Bailey (12) in his monograph series on fats and
oils sites several examples of crystal habit
differences in mixtures.  Generally most of these
differences are observed for the reversible transformations from β ⇌ liquid.  More current work
by Al Mamum (13) on alcohols has indicated that the
chain length difference of the components could
not be more than two carbons for the complex formation to be observed.  In this study of triglycerides, the difference does not appear to be critical
although an optimum difference may exist.  More
importantly, the enhanced aging properties were
observed only when lower melting impurities were
present.  It is also interesting to note that
while there is a concentration for optimum α
stability there is also a crictical concentration
of LMTGI above which the α polymorph of the bulk
phase is prevented from forming.  This occurs
between 26 and 30% of the trilaurin or trimyristin
impurity.

In conclusion the maximum in the alpha polymorph
stability coincides with a minimum melting temperature in the α    β transition temperature indicating
an optimum crystal interaction.  The critical
concentration of LMTGI on the otherhand is based
on an antinucleation phenomenon above which the
unstable α triglycerides are prevented from forming
on rapid cooling from the melt.  These results
provide one with a mechanism for controlling the
stability of the meta stable triglyceride polymorphs.

V.   REFERENCES

1.   Rosano, H. L.; Whittam, J. H.; JAOCS 52, 128
     (1975)

2.   Lutton, E. S.; Fehl, A. J.; Lipids 5:90 (1970)

3.   Chapman, D. J.; Chem. Soc. 2523 (1956)

4.   Hannewijk, J.; Haighton, A. J.; JAOCS 35 457
     (1958)

5.   Chapman, D. J.; Richards, R. E.; Yorke, R. W.;
     Chem. Soc. 436 444 (1960)

6.   Bailey, A. E.; Oliver, G. D.; Oil and Soap 21
     300 (1944)

7.   Quimby, O. T.; J. Am. Chem. Soc. 67 5064 (1950)

8.   Friedman, M.; Rosano, H. L.; Whittam, J. H.;
     J. Coll. Inter. Sci. 40, 123 (1972)

9.   Santoro, A.; Esposito, M.; J. Thermal Analysis,
     6 101 (1974)

10.  Barrall II, E.; Guffy, J. C.; Ordered Fluids
     and Liquid Crystals - ACS Series #63

11.  Lewis, T. P.; Eaton, W. A.; J. Chem. Physics,
     Vol. 53, 6, 2124 (1970)

12.  Bailey, A. E.; Melting and Solidification of
     Fats, Interscience Publication (1950)

13.  Al Mamum, M. A.; JAOCS 51 234 (1974)

# OPTICAL SCATTERING OF SMALL GOLD PARTICLES IN POLYESTER MATRIX. I. EXPERIMENT

C. H. Lin[*+], F. F. Y. Wang[*], H. Herman[*],
P. J. Herley[*] and Y. H. Kao[**]
State University of New York
Stony Brook, N. Y. 11794

## ABSTRACT

Small gold particles (20 Å to 700 Å in diameter) were produced by the chemical reduction of chloroauric acid using a polyester prepolymer as the reducing agent. Polymerization was promoted thermally and catalytically. Subsequent heat treatment at 80°C induced nucleation of gold particles embedded in the rigid polyester medium. Microtoming of the specimen into 1000 Å slices enabled direct measurement of the gold particles by transmission electron microscopy. Optical absorption data were obtained at wavelengths between 3000 Å and 7700 Å.

[*]  Department of Materials Science
[**] Department of Physics
[+]  Present address:  Department of Materials Science
                       National Tsing-Hua University,
                       Hsinchu, Taiwan, Republic of China

## I.  INTRODUCTION

Small gold particles (less than 1000 Å in diameter) have been produced by many methods.  Since gold has a high chemical potential, small gold particles can be easily prepared by chemical precipitation.  Usually water (1) and glass (2) were used as the supporting medium.  In this study, a polyester was used both as the supporting medium as well as the reducing agent.

Previous work on small gold particles exhibit interesting optical scattering and absorption properties.  However, their particle sizes and size distributions have not been routinely measured.  In some studies, the optical scattering results (3) have been used to obtain the particle sizes.  In this study a direct method for particle size measurement was developed by the use of microtoming the specimen and observing it under the transmission electron microscope.  Using the measured particle sizes and their distributions, a comparison can be made with the experimental results obtained from optical absorption measurements and the known theories of optical scattering of gold particles.

## II.  PROCEDURES

### A.  Preparation

A chemical reduction method was employed in  this study to produce small gold particles.  The reducing agent was a polyester prepolymer (trade name "Clear Cast" by American Handicrafts Co., Ft. Worth, Texas).  The polyester prepolymer contained approximately 30% styrene monomer, and the remainder consisted of saturated and unsaturated carboxylic acids in equal portions.  It was found to dissolve up to 10% by weight

of the chloroauric acid with stirring. A catalyst was used
in the form of a solution containing about 11% v/v methyl
ethyl ketone peroxide in dimethyl phthalate.

The resulting polyester solution was poured into a glass
mould, and cast into thin (about 0.4 - 0.6 mm thick) specimens. The polyester was hardened in an oven at 30°C for 24
hours.

Nucleation and growth of the gold particles in the polyester medium was promoted by a heat treatment at 80°C ± 2°C
for various time periods up to 60 hours. At the end of the
required heat treatment period, specimens were kept in a
refrigerator at about 4°C. They were only removed for the
period required for measurements.

Four series of specimens were prepared. The series "0"
represented the blank specimens. The series "4", "5" and "6"
represented specimens with gold concentrations (by weight) of
$0.3 \times 10^{-3}$, $0.6 \times 10^{-3}$, and $1.0 \times 10^{-3}$, respectively. Heat
treatment periods varied in each series from 5 minutes to 60
hours.

B.  Particle size measurement

In this study, specimens were sliced about 1000 Å thick
using an ultra-microtome, model MT2-B, made by Ivan Sorvall
Inc., Connecticut. A Pyrex glass knife was made from a glass
strip (6.35 mm x 25.8 mm x 400 mm) by a knife maker, model
7801A, Lab-Produkter, Sweden. The cutting speed was set at
0.5 mm per second. The thickness of each slice was preset by
the thickness control in the ultra-microtome, and verified by
its own color after slicing.

Direct measurement of the gold particles in the specimens
was made by the use of a Philips EM-300 transmission electron
microscope. Sliced specimens were placed directly on the
copper grids. The electron microscope was operated at 100 kv,
40 ma, with a resolution of about 20 Å at a magnification of
222,000 x. Two size calibration standards were used. One was
the shadow of the base of the central beam interceptor on the
picture. The diameter of the shadow was found to be 150 ±
10 Å. Another was the range of the viewing binocular objective mounted externally on the top of the screen. The distance from the center to the edge view was found to be 500 ±
50 Å. These two standards referenced each particle size
measurement. Deviations were found to be within 15%.

In making the size measurements, careful distinction was made between the gold particles produced through the heat treatment and those produced by electron bombardment during the electron microscopic examination. Only the former were counted in this study.

## C.  Optical measurement

A Cary model 14 Spectrophotometer was used for the optical absorption measurements from 3000 Å to 7700 Å. A zero line basis was obtained by a zero setting of the optical density of pure rigid polyester at 5000 Å. The corresponding blank specimens from the "0" series were placed in the reference chamber for the Cary 14 unit during each run. The difference between the thicknesses of the blank specimen and the test specimen was taken into account in the measured data. The average refractive index of the blank specimens for the "0" series was measured to be 1.576 ± 0.005.

## III.  RESULTS

### A.  Size measurement

Color density indicates that the nucleation process is extremely rapid and occurs virtually instantaneously at 80°C. Consequently the subsequent thermal annealling time is a measure of the growth of the particles. A typical determination of the particle size distribution for specimen 67 (1.0 x 10$^{-3}$ wt. % Au, heat-treated at 80°C for 1/3 hour), as measured by the direct counting and sizing by the transmission electron microscopy, are shown in the form of a histogram in Figure 1. Also shown is a log-normal distribution fitted to the data. The density function for the log-normal distribution can be expressed as

$$P(D) = \frac{1}{\sqrt{2\pi} \ln \sigma} \exp - \frac{(\ln D - \ln D_m)^2}{2 (\ln \sigma)^2} \tag{1}$$

where D is the diameter of the particle, $D_m$ is the average diameter, and $\ln \sigma$ is the standard deviation. In Figure 1, the smooth curve represents the density function P(D) multiplied by the histogram interval (50 Å) and the total number of particles.

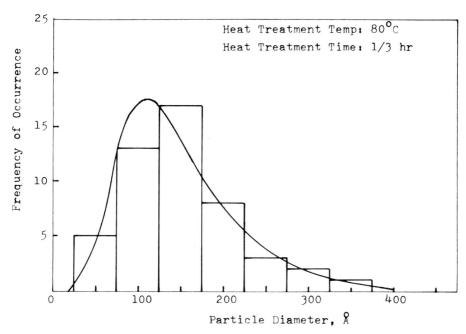

Fig. 1. Histogram and lognormal distribution of particle size in specimen 67.

The measured data of particle sizes of the specimens in the "6" series, fitted to eq. (1), are shown in Figure 2. The average diameters, $D_m$ of the "6" series specimens are found to increase monotonically with the heat treatment time. It changed from a value of 59 Å at 10 min. to 350 Å at 60 hours. Within a standard specimen volume of $5.89 \times 10^{-12}$ cm$^3$, the total number of particles in each specimen was counted. The number of particles in a fixed sample volume decreased monotonically with the heat treatment time. From the number of particles, the average particle volume $V_m$ and the average distance between neighboring particles $S_m$, can be calculated. The results are listed in Table 1.

From the electron micrographs, it was observed that most Au particles were hexagonal in shape in the specimens with a heat treatment time of less than one hour. On increasing the heat treatment time, the corners of the particles became more rounded. However, at the same time the material distribution within each particle became more inhomogeneous. This suggested that the particle size increase is the coalescence of several particles together as distinguished from the growth of a single particle.

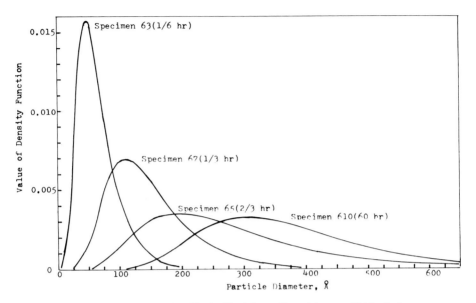

Fig. 2. Lognormal distribution functions fitted to particle size data of particle size for a specimen containing
$1.0 \times 10^{-3}$ wt. % Au subjected to various heat treatment times.

TABLE 1

Average particle diameter, $D_m$, average particle volume, $V_m$,
and average nearest neighbor particle distance $S_m$, of the "6"
series specimen ($1.0 \times 10^{-3}$ wt. % Au).

| Specimen No. | 63 | 67 | 68 | 65 | 66 | 610 |
|---|---|---|---|---|---|---|
| heat treatment time (hr.) | 1/6 | 1/3 | 1/2 | 2/3 | 1 | 60 |
| $D_m$, (Å) | 59 | 136 | 204 | 251 | 261 | 350 |
| No. of particles (specimen volume of $5.89 \times 10^{-12}$ cm$^3$) | 150 | 70 | 45 | 34 | 22 | 15 |
| $V_m \times 10^{12}$, (cm$^3$) | 0.039 | 0.084 | 0.131 | 0.173 | 0.268 | 0.393 |
| $S_m$, (Å) | 3399 | 4382 | 5077 | 5575 | 6445 | 7323 |

B. Optical data

The optical absorption coefficient data of the series "6" specimens are shown in Figure 3. The absorption peak around 2.3 eV grew more pronounced with the increase in the heat treatment time. The blank specimens showed no absorption in this region as did the specimen 60 with zero hour of heat treatment. This peak position decreased first, then increased, and finally reached a steady value with heat treatment, as shown in Figure 4. The peak intensity was found to be a linear function of the weight fraction of gold in the specimens, for the same heat treatment temperature (80°C) and time (60 hrs.), as shown in Figure 5. The area of the absorption peak around 2.3 eV, subtracting the background, was found to increase sharply with heat treatment time and reach a steady value after a heat treatment time of 2/3 hr., as shown in Figure 6.

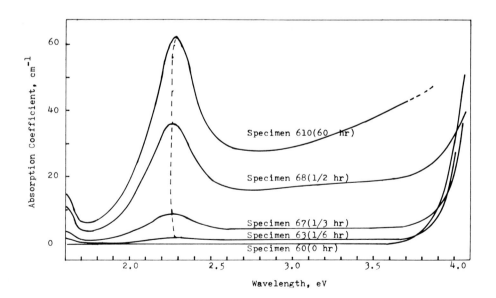

Fig. 3. Special Absorption Coefficient of series "6" specimens (1.0 x 10⁻³ wt. % Au) heat treatment temperature 80°C.

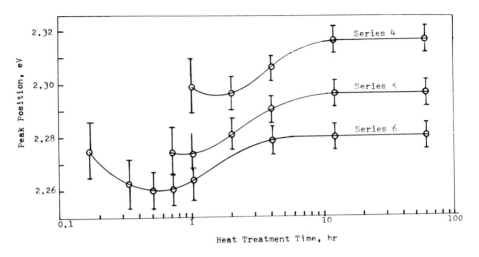

Fig. 4. Peak position versus log heat treatment time for specimens containing various weight % of gold.

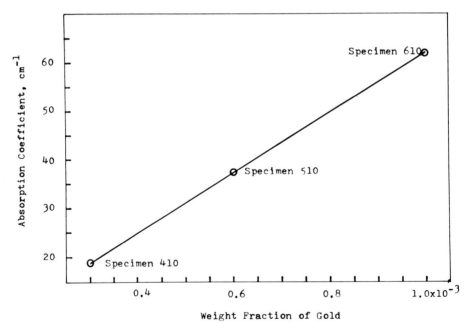

Fig. 5. Peak height versus weight % of gold in the specimens with the same heat treatment time 60 hrs. and temp. 80°C.

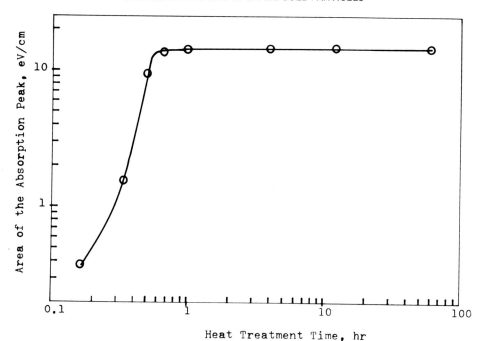

Fig. 6. Log area of the absorption peak versus log heat treatment time.

## IV. DISCUSSION

The results in this study indicate that, in specimens containing $1.0 \times 10^{-3}$ wt. % Au in polyester, (the "6" series), a heat treatment at 80°C in excess of 1 hour caused the complete nucleation of all the available gold atoms. In fact, the gold nuclei decreased with longer heat treatment time. At the same time the particle sizes increased, and apparently coalescence of particles occurred. This phenomenon of particle coalescence remains ambiguous since the average distance between particles was always several tenfolds of the average particle size. At the heat treatment temperature of 80°C, the diffusion coefficient of gold in polyester is insufficient to account for the large distance the particles must travel.

The optical absorption data showed that the absorption peaks around 2.3 eV, at heat treatment times greater than 1/2 hr., shifted to higher energies in agreement with the previous data on waterbased colloidal gold (4). At shorter heat treatment times, the absorption peaks shifted to lower energies. This suggests that the heat treatments at various times may have altered the nature of gold particles, which, in turn, produced the opposite shifts of absorption peaks with time. Longer heat treatment times produced larger gold particle sizes. The absorption characteristics, both the peak position and the peak intensity, are not monotonic functions of the particle size. Theories, such as Kreibig and Fragstein (5), and Doremus (6), were advanced to include the effect of particle size. Their applications have not been totally successful in explaining the experimental results. Results in this study suggest that additional modifications to these theories are required. Results of the theoretical modifications to these data will be published elsewhere.

V.  REFERENCE

1.  For example, Turkevich, J., Hillier, J., and Stevenson, P. C., Disc. Farad. Soc., 11, 55 (1951).

2.  For example, Stovkey, S. D., J. Am. Ceram. Soc., 32, 246 (1949).

3.  Kerker, M., The Scattering of Light, Academic Press (1969), p. 311-396.

4.  Takiyama, K., Bull. Chem. Soc. Japan, 31, 950 (1958).

5.  Kreibig, U. and Fragstein, C. V., Z. Physik, 224, 323 (1969).

6.  Doremus, R. H., J. Chem. Phys. 40, 2389 (1964).

# SCATTERING CROSS SECTION MEASUREMENTS OF RAYLEIGH-GANS SCATTERERS FROM DEPOLARIZED INTENSITY OF DOUBLY SCATTERED LIGHT

J. G. Gallagher, Jr., Dae M. Kim, and C. D. Armeniades
*Rice University, Houston, Texas 77001*

## Abstract

*This work shows that in a weakly turbid dispersion of optically isotropic Rayleigh-Gans scatterers, the scattering cross section can be measured directly from the depolarized component of doubly scattered light. It also shows that the doubly scattered intensity, when corrected for the attenuation of the primary beam, obeys the characteristic $N^2$ dependence associated with incoherent double scattering.*

## I.   INTRODUCTION

The scattering cross section of a particle is usually obtained from measurements of either the Rayleigh ratio or the linear attenuation coefficient in the single scattering region. These measurements require careful instrumental calibrations of the optical system and scattering cells of long path lengths. In addition, the necessity of avoiding multiple scattering places restrictions on the particle number density in the system. Recent theoretical studies (1,2) of the depolarized intensity of doubly scattered light have shown that the scattering cross section of optically isotropic particles can be measured in the double scattering region. In the case of a fluid near the critical point, the scattering cross section has been measured from the ratio of the depolarized (horizontal) and vertical (single scattering) intensities (3).

In this paper we show that the scattering cross section of Rayleigh-Gans scatterers can be obtained from measurements of depolarized, doubly scattered light. In the double scattering region, the incident light beam is attenuated appreciably in a sample cell of short optical path length. This attenuation can be measured directly from the depolarized component of scattered light, provided the cross sectional dimensions of the primary and secondary scattering volumes are much smaller than the photon free mean path. These data, together with the particle number density, can lead to the precise determination of the scattering cross section, $C_{sca}$.

## II. DEPOLARIZED DOUBLY SCATTERED LIGHT

In a weakly turbid medium consisting of $N$ identical iso-tropic scatterers per unit volume, the depolarized intensity that arises from two uncorrelated successive scattering events is found by applying the principle of polarized single scattering to each event. The scattering geometry is shown in Fig. 1.

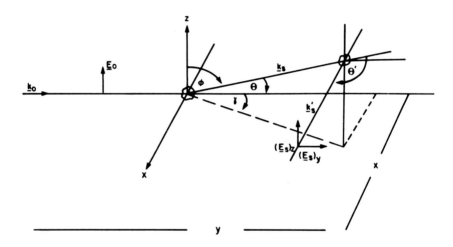

*Fig. 1. Scattering geometry for doubly scattered light.*

The incident field, $\hat{z} E_o \exp i (\omega t - k_o y)$, undergoes the first scattering event within the primary beam path; the second scattering occurs in a volume viewed by the detector, which is placed perpendicular to the incident beam at a distance $y$ along the sample cell. The incident beam is positioned at a distance $x$ from the exit window of the cell adjacent to the detector.

For a spherical scatterer (radius $a$) obeying the Ray-leigh-Gans approximation, the intensity of the scattered light at a distance $r$ is given by (4,5)

$$i = [f(\theta,\varphi)/r^2] \, I_o \tag{1}$$

Here, the scattering indicatrix, $f(\theta,\varphi)$ describes the angular dependence and state of polarization of scattered light and is given by (4,5)

$$f(\theta,\varphi) = (4/9)k_o^4 a^6 |m-1|^2 G^2(u) |\hat{z} \times \hat{k}_s|^2 \tag{2}$$

where $m$ is the relative index of refraction of the scatterer to a solvent, and $G(u)$, a form factor, is expressed as (4,5)

$$G(u) = (3/u^3) \; (\sin u - u \cos u) \tag{3}$$

$$u = 2k_o a \sin (\theta/2)$$

The expression for the depolarized component of doubly scattered light is readily obtained by applying Eq. (1) to the two scatterers involved. Using the far field approximation for the polarization factor (6) one obtains

$$I_{HV}^d = [\frac{4}{9} \frac{k_o^4 a^6 |m-1|^2}{R}]^2 \; N^2 I_o \; \int_{V_1} \int_{V_2} e^{-\tau y} [\frac{\sin \varphi \cos \varphi \cos \gamma}{r_{12}}]^2$$

$$G^2 (2k_o a \sin(\frac{\theta}{2})) \; G^2 (2k_o a \sin(\frac{\theta'}{2})) \; dV_1 \; dV_2 \tag{4}$$

Here, $R, r_{12}, V_1, V_2$, and $\tau$ are the distance to the detector, the distance between two scatterers, the primary and secondary scattering volumes, and the linear attenuation coefficient, respectively. Note that, for vertically polarized incident light, $\tau$ is given by integrating Eq. (2) over the solid angle:

$$\tau = N \, C_{sca}$$

$$= N\pi a^2 \; |m-1|^2 \; [\xi^2 - 0.25 + \frac{1}{2} \operatorname{sinc} 4\xi - \frac{1}{2} (1-\cos 4\xi)/(4\xi)^2] \tag{5}$$

where $\xi \equiv k_o a$.

It can be seen from Eq. (4) that, because of the power law dependence, $r_{12}^{-2}$, and the polarization factor, the major contribution to $I_{HV}^d$ comes from two scatterers within the overlapped region between the primary and secondary scattering volumes. For the case where the cross sectional dimension of the overlapped region, as viewed by the detector, is much less than the photon free mean path, $\tau^{-1}$, Eq. (4) reduces to

$$I_{HV}^d = K \, N^2 e^{-\tau y} \tag{6}$$

$K$ being a constant independent of $N$. It is thus clear from Eq. (6) that the scattering cross section can be obtained in a weakly turbid medium by measuring $I_{HV}^d$ at two different particle number densities.

III. EXPERIMENTAL DETAILS

Latexes of polystyrene microspheres were chosen as the scattering medium because of their known physical properties, uniform particle size, and isotropic polarizability. The density, $\rho$ and relative index of refraction, $m$ of these particles relative to water at room temperature are 1.05 g/cm$^3$ and 1.194, respectively. Three different latexes were used, with respective particle diameters of 0.091, 0.126 and 0.176 $\mu m$. The angular irradiance profiles from these particles are described to within 10% by the Rayleigh-Gans scattering theory (5).

The latexes were obtained from Dow Chemical Co. at a 10% wt. solids concentration. Samples of different concentrations were prepared by adding known volumes of stock latex (using microliter pipets) to 10cc. of filtered distilled water. The particle number density of the scatterers is determined from

$$N = c/\rho \ V_p \qquad (7)$$

where $c$ and $V_p$ are the latex concentration (g/cm$^3$) and the volume of the particle scatterer, respectively.

The intensity measurements were performed at three wavelengths (457.9, 488, and 514.5 $nm$) using an Argon ion laser, with beam power of approximately 100 $mw$. The incident light beam was focused into the center of a 1 cm square × 4.5 cm sample cell using an 8.8 cm focal length lens. This gave a beam diameter of 32 $\mu m$ at the focal plane.

The depolarized component was measured using a Glan-Thompson polarizing prism (extinction 5 × 10$^6$) placed on a rotating mount in front of the optical collection system of the detector. This collection system consists of a pair of small circular apertures 45 cm apart. The distance between the incident beam within the sample cell and the first aperture is approximately 6 cm. In these experiments the primary and secondary scattering volumes, viewed by the detector, are varied by changing the diameter of the first aperture from 0.0025 to 0.06 cm, while the second aperture diameter is fixed at 0.06 cm. The detector is an ITT FW 130 photomultiplier with a dark count at -20°C of 1 count/sec. Standard photon counting techniques were used to measure the depolarized component.

IV. RESULTS AND DISCUSSION

The range of particle diameters in the three polystryene

latexes, used in our study, was determined by photon correlation spectroscopy to be within the manufacturer's specification. We found that double scattering for these latexes occurs in our cell when the ratio of the interparticle distance, $N^{-1/3}$ to the particle radius is less than 100:1. These findings agreed with previous results (7,8). We point out that in the single scattering regime, set by this ratio no depolarized component of scattered light was detected, as expected from the optically isotropic nature of these particles.

In Fig. 2 we present the intensity response of the aperture collection system for the depolarized scattered light as a function of the diameter of the first aperture for four different particle number densities.

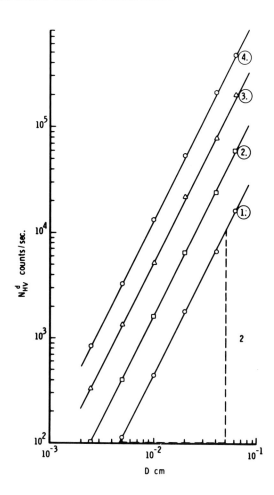

*Fig. 2. Depolarized photoelectron counts as a function of the first aperture diameter for four particle number densities of 0.091 μm diameter spheres ($\lambda_o$ = 457.9 nm). The densities are: (1) 0.243 × $10^{12}$/cm³; (2) 0.486 × $10^{12}$/cm³; (3) 0.972 × $10^{12}$/cm³; (4) 1.823 × $10^{12}$/cm³. The detector is at y = 0.5 cm.*

Note that for a given particle number density, the depolarized scattered intensity increases linearly with the area of the first aperture in our collection system. This implies that the mutual degree of coherence (9) of the doubly scattered light on the first aperture plane is delta function correlated, i.e. no diffraction occurs on the first aperture regardless of its size. This will be analyzed in detail elsewhere.

The scattering cross section of this sphere can be obtained with the use of Eq. (6) from any two data points in Fig. 2, corresponding to two different particle number densities for a fixed first aperture diameter, $D$. In Table 1 we present our measured values of scattering cross section for three different polystyrene spheres using three different wavelengths. As can be seen from Table 1, our measured values of $C_{sca}$ are in excellent agreement with the calculated values (see Eq. (5)).

**TABLE 1**

*Comparison of the calculated and measured single scattering cross sections for three size latexes obtained at three wavelengths from the depolarized (horizontal) component of scattered light. The particle number density of the latexes were: 0.486 and 0.972 x $10^{12}$/cm³ for 0.091μm, 0.915 and 1.83 x $10^{11}$/cm³ for 0.1260μm and 1.68 and 3.36 x $10^{10}$/cm³ for 0.176μm*

| Diameter (μm) | $C_{sca}$ x $10^{12}$ cm⁻² | | | | | |
| --- | --- | --- | --- | --- | --- | --- |
| | $\lambda_o$ = 457.9 nm | | $\lambda_o$ = 488 mn | | $\lambda_o$ = 514.5 nm | |
| | Cal. | Meas. | Cal. | Meas. | Cal. | Meas. |
| 0.091 ± .0058 | 0.81 | 0.82 | 0.65 | 0.68 | 0.54 | 0.53 |
| 0.1260 ± .0043 | 4.50 | 4.32 | 3.70 | 3.68 | 3.12 | 3.23 |
| 0.1760 ± .0023 | 21.91 | 21.42 | 18.72 | 19.03 | 16.25 | 16.95 |

In Fig. 3, we show the intensity of depolarized doubly scattered light, corrected for the attenuation of the primary beam, as a function of the particle number density, $N$. As can be clearly seen, the depolarized intensity is described by the power law, $N^2$, in complete agreement with the fact that the two scatterers involved are uncorrelated.

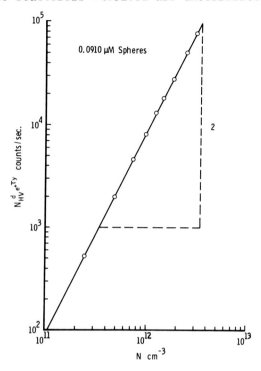

*Fig. 3. Depolarized photoelectron counts, corrected for the attenuation of the primary beam as a function of particle number density; y = 0.5 cm, $\lambda_o$ = 457.9 nm.*

In conclusion, we have shown that the scattering cross section of optically isotropic particles can be measured in the double scattering region directly from the depolarized intensity, when the cross sectional dimensions of the primary and secondary scattering volumes are much less than the photon mean free path. The advantage of our scheme lies in that the measurement can be made in an optical cell of short path length (1 cm). In addition, this technique requires neither extensive calibration of the optical system nor the avoidance of the multiple scattering region. This method should also be applicable to turbid dispersions of Rayleigh and Mie scatterers.

V. REFERENCES

1. Oxtoby, D.W., and Gelbart, W.M., _J. Chem. Phys._ 60, 3359 (1974).

2. Bray, A.J., and Chang, R.F., Phys. Rev. A 12, 2594 (1975).
3. Rieth, L.A., and Swinney, H.L., Phys. Rev. A 12, 1094 (1975).
4. Van De Hulst, H.C., "Light Scattering by Small Particles," p. 85. Wiley, New York, 1957.
5. Kerker, M., "The Scattering of Light and Other Electromagnetic Radiation," p. 414; 428. Academic Press, New York, 1969.
6. Jackson, J.D., "Classical Electrodynamics," p. 271. Wiley, New York, 1962.
7. Woodward, D.H., J. Opt. Soc. Am. 54, 1325 (1964).
8. Napper, D.H., and Ottewill, R.H., J. Colloid Sci. 19, 72 (1964).
9. Beran, M.J., and Parrent, Jr., G.B., "Theory of Partial Coherence," p. 57. Prentice Hall, New Jersey, 1964.

*Support of this work by a research grant from the Robert A. Welch Foundation is gratefully acknowledged.*

# POLARIZATION DEPENDENCE OF EXTINCTION CROSS-SECTIONS FOR ABSORBING AND NON ABSORBING NON-SPHERICAL PARTICLES: A COMPARATIVE STUDY

J.N.Desai,D.B.Vaidya
Physical Research Laboratory

H.S.Shah
S.V.R. College of Engineering and Technology

Changes in the optical transmission, dichroism and birefringence produced in dispersions of bentonite powder when placed in a magnetic field are studied. In order to understand the role of absorption, the dispersions were rendered optically absorbing over a narrow spectral range by adsorption of suitable dyes; viz. Rhodamine and Methylene blue. Comparison of the magnetooptical effects observed for the absorbing and the non-absorbing dispersions reveal several interesting features.

## I    INTRODUCTION

## A    Background

In our earlier work we have reported the changes in the optical transmission and birefringence observed in several Colloids and dispersions when placed in a magnetic field (1-6). Excepting a few cases described below, it is observed that these magnetooptical effects are very well explained if the suspended particles are assumed to scatter and absorb the incident radiation as induced dipoles excited under the influence of the electric field of the incident electromagnetic wave. Validity of this approximation requires that the largest dimension of the particle be less than approximately one-tenth the wavelength of the incident light. We will hereafter refer to this approximation as the "dipolar approximation". Theory of light scattering under this approximation was originally given by Rayleigh and is also known as the Rayleigh approximation (7). We have given earlier the expressions governing the magnetooptical effects and in particular the polarization dependence of the optical transmission in colloids under this approximation (1-4). Interesting cases for which large discrepancies have been observed between the observations and the theory are the following:

1      Freshly prepared Colloids of $Fe_3O_4$ :
The pattern of the magnetooptical transmission changes observed in freshly prepared colloidal solutions of $Fe_3O_4$ show large disagreement with the theory, but after a couple of months of aging the magnetooptical effects change drastically and show a close agreement with the theory (3). Presumably in the freshly prepared colloids primary particles form magnetically linked aggregates which subsequently redisperse into primary particles, the magnetic linking energy being less than the thermal excitation term kT.

2      Dispersion of Bentonite :
The observed polarization dependence of the magnetooptical transmission changes in dispersions of bentonite do not agree with what one expects on the basis of the dipolar scattering (5). Large particle size in bentonite seems responsible for the disagreement.

3      Dispersions of Molybdenum Disulphide :
In the dispersions of molybdenum disulphide (of Acheson Colloid Co.) also, the effects observed are in disagreement with the theory; although the particles are not quite large. This discrepancy has been attributed to the large refractive index of the material and the resulting importance of the contribution from the magnetic dipole scattering term (6).

It may appear somewhat surprising that although the dispersions of aquadag and oildag graphite (of Acheson Colloid Co.) are of fairly larger particles size ($\simeq$ 1 micron diameter) than the particle in dispersions of bentonite ($\simeq$ 0.1 micron diameter), the former dispersions show magnetooptical transmission changes that are in close agreement with the dipolar scattering (2) but the latter show significant disagreement. At the same time, whereas dispersions of bentonite are known to exhibit large birefringence when the particles are oriented, the dispersions of graphite show only very weak birefringence.

The observations indicate a rather important role played by the absorption in determining the pattern of magnetooptical effects observed in dispersions with oriented anisometric particles, particularly when the particles are somewhat larger than the Rayleigh scatterers. In order to understand the role of absorption we carried out a comparative study of the magnetooptical effects in dispersions of ordinary bentonite and the bentonite particles which were rendered optically absorbing over a narrow spectral range by adsorption of a suitable dye.

B       Preparation of the Dispersions :

B.P.grade bentonite marketed by Medical Evans U.K. was used. A few grams of powder was hand milled in an agate mortar for about 30 minutes with water, either with or without an appropriate dye. The paste thus prepared was dispersed in water on a mechanical shaker for about 15 minutes, and then centrifuged at 2000 rpm. For preparing the dispersions of dye adsorbed particles two dyes were selected, viz; rhodamine which gave dispersions having a sharp and strong absorption feature centred at 5800 A and methylene blue which gave dispersions with comparatively weak and broad absorption feature centred at about 6700 A. Each of the three dispersions were further fractionated by centrifugation into four fractions with different particle sizes in an identical manner. These fractions are designated as F-1, F-2, F-3 and F-4 in the text. Excess dye was removed by prolonged dialysis, till the dialysate was completely free of colour. This also confirmed the adsorption of dye on bentonite particles. Average particle size in each fraction was determined by measuring the extinction coefficient at a wavelength outside the absorption band and also the mass concentration for each fraction. The mean particle radii (equivalent spherical) were obtained using extinction coefficients calculated on Mie theory corresponding to the refractive index 1.2, viz; bentonite with respect to water (8). The values obtained for the fractions 1 to 4 were 0.092 micron, 0.08 micron,

0.065 micron and 0.06 micron respectively. Electron microscope study revealed, however, that each fraction also contained some proportion of much coarser particles with diameters as large as 1 micron and that the particles are nonspherical in shape.

Comparative spectral extinction curves of dyed and undyed dispersions are shown in Fig.1 for two of the four fractions.

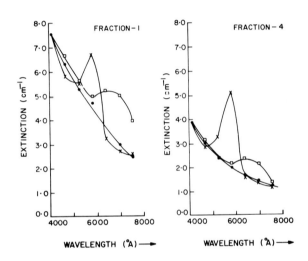

Fig. 1. Spectral extinction curves for dispersions of bentonite. Values for the dye adsorbed dispersions are normalized to that for the corresponding fractions of undyed dispersions at $\lambda$= 4100 A. • -undyed, x-rhodamine dyed, □ -methylene blue dyed.

II    EXPERIMENTAL STUDY:

The dispersions were studied for their transmission changes in a magnetic field and the birefringence at seven wavelengths viz; 4100 A, 4675 A, 5250 A, 5800 A, 6375 A, 6950 A, 7525 A. The transmission changes were further studied for the three principal cases as follows:

(i) Optical transmission through the sample, when the applied magnetic field is perpendicular to the direction of propagation of beam, the incident light being linearly polarized with its electric vector parallel to the field. The extinction coefficient measured in this case is denoted as $C_L$.

(ii) Optical transmission through the sample as in case (i) but incident light being linearly polarized with its electric vector perpendicular to the magnetic field. The extinction coefficient in this case is denoted as $C_R$.

(iii) Optical transmission through the sample when the direction of propagation of light beam is parallel to the direction of applied field. Effects observed in this case are independent of state of polarization. The extinction coefficient measure in this case is denoted as $C_K$.

The magnetooptical transmission changes are described conveniently by introducing dimensionless parameters $Q_L$, $Q_R$ and $Q_K$ which are independent of concentration and are defined as

$$Q_L = \frac{C_L}{C_0}, \quad Q_R = \frac{C_R}{C_0}, \quad Q_K = \frac{C_K}{C_0}$$

where $C_0$ is the extinction coefficient without magnetic field.

Rather strong linear birefringence was observed in all specimens when magnetic was applied perpendicular to the light path. The birefringence can be described conveniently by using the notations common to optics of uniaxial crystal viz. $n_L$ ($= n_e$) to denote refractive index when electric vector of incident light is parallel to the applied magnetic field and $n_R$ ($= n_o$) to denote the refractive index when the electric vector is perpendicular to magnetic field. The phase difference $\delta$, introduced in the birefringent medium between the parallel and perpendicular components was measured by using a method described earlier (3), and the parameter $n_L - n_R$ was calculated as

$$n_L - n_R = \frac{\delta\lambda}{2\pi\ell}$$

where ' $\ell$ ' is path length.

## III    RESULTS

### A    <u>Transmission Changes</u> :

#### 1    Undyed Bentonite

Fig.2 shows the transmission changes observed in dispersions of undyed bentonite in a magnetic field studied at $\lambda$ = 5800 A for the three cases L, R and K. Results obtained at the other wavelengths were similar in both nature and magnitude

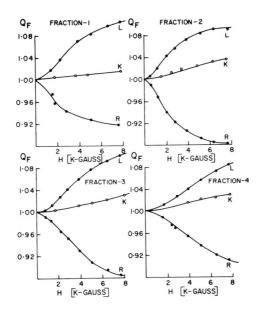

Fig.2. Transmission changes in dispersions of undyed bentonite observed at $\lambda$ = 5800 A.

and are therefore not shown here. It is seen that for light beam propagating perpendicular to the magnetic field the dispersions show strong dichroism and therefore one cannot assume the extinction crosssections of the particles as approximated by the diffraction theory. At the same time the dipolar approximation is also not valid. Firstly, according to the dipolar approximation the ratio $(Q_L-1)/(1-Q_R)$ should equal 2 but the actually observed values are nearly 1. Secondly, under the dipolar approximation the extinction cross-section of a particle depends only on its orientation relative to the direction of the electric field vector of the incident light; direction of propagation being of no consequence (7). Hence the nature of transmission changes for R and K position should be the same, a result which the observations do not confirm. It is interesting to note that since $Q_L-1 \simeq 1-Q_R$ and $Q_K \simeq 1$ (i.e. in K position transmission shows very little change with applied field) it can be inferred that, although large transmission changes are observed when linearly polarized light is used perpendicular to the magnetic field, the dispersions will not show almost any change in the transmission under a magnetic field if unpolarized light is used.

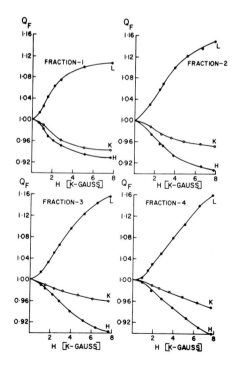

**Fig. 3.** Transmission changes in dispersions of rhodamine dyed bentonite dispersions at $\lambda = 5800$ A.

2    Dispersions of Dye Adsorbed Bentonite :
        Fig.3 shows the transmission changes observed
in dispersions with rhodamine dye adsorbed particles
at the wavelength of peak absorption ( $\lambda$ = 5800 A ).
At the other wavelengths, the transmission changes
observed in rhodamine dyed dispersions do not differ
from those observed in the corresponding fractions
of undyed dispersions; and are similar to those
shown in Fig.2.  Marked difference in the nature of
effects at the wavelength of absorption is, however,
seen very clearly if one compares Fig.3 with Fig.2.
The most interesting points are -

   i) The ratio $(Q_L-1)/(1-Q_R)$ is no longer equal
      to about 1 but approaches a value nearer
      to 2.

   ii) The curve for $Q_K$ also no longer shows a
       slight upward trend but becomes similar to
       that for $Q_R$; i.e. the transmission changes
       for the cases "R" and "K" are now similar.

        In both these respects the results are now in
a closer agreement with the dipolar approximation.

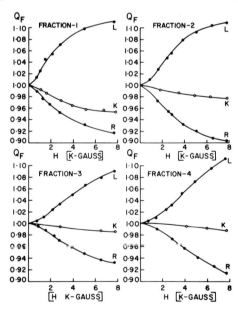

Fig.4. Transmission changes in dispersions of
methylene-blue dyed bentonite at $\lambda$ = 6950 A.

Results for the dispersions of methylene blue dyed particles are shown in Fig.4 for the wavelength 6950 A. Here also it may be mentioned that the results at wavelengths outside the absorption band do not differ from those for undyed dispersions, but are markedly different at 6950 A which is within the absorption band. Absorption band being weaker in this case, the effect is, however, somewhat less marked.

The value of the ratio $(Q_L-1)/(1-Q_R)$ are tabulated below in Tables 1 and 2 for all the seven wavelengths at which the magnetooptical effects were studied for rhodamine dyed and methylene blue dyed dispersions respectively. Corresponding values for undyed dispersions are not tabulated but the values were mostly between 0.75 and 1.2. Feature to be noted is the sharp increase in the value of the ratio at the wavelength of absorption. A comparison of the nature of magnetooptical transmission changes shown by the dispersions within the region of spectral absorption with that shown in the non-absorbing region clearly show that within the absorption region there is a definite closer agreement with the dipolar approximation. This suggests that in this size range, although the scattering cross-section of nonspherical particles are not correctly given by the dipolar approximation, the absorption cross-sections are given fairly accurately.

This inference could be tested on a more quantitative basis by separating the extinction cross-section into its two components, the scattering cross-section and the absorption cross-section for the dye adsorbed dispersions at the wavelengths of absorption. The ratio $(Q_L-1)/(1-Q_R)$ is then expected to be nearer to 2 if calculated for the absorption cross-section alone. This separation was done assuming that for the corresponding fractions of dyed and undyed dispersions the spectral distribution of scattering reamins the same.

TABLE 1

The ratio $(Q_L-1)/(1-Q_R)$ observed at the maximum field strength for the four dispersions of rhodamine dyed bentonite. Values given in brackets are the mean particles radii (microns) in corresponding fractions.

| Wavelength (Å) | Fraction | | | |
|---|---|---|---|---|
| | 1 (0.092) | 2 (0.08) | 3 (0.065) | 4 (0.06) |
| 4100 | 1.02 | 0.93 | 1.03 | 0.91 |
| 4675 | 1.05 | 0.88 | 1.12 | 1.03 |
| 5250 | 1.02 | 0.94 | 1.20 | 0.84 |
| 5800 | 1.55 | 1.61 | 1.65 | 1.70 |
| 6375 | 1.12 | 1.09 | 1.22 | 1.21 |
| 6950 | 0.93 | 1.14 | 1.17 | 1.22 |
| 7525 | 1.01 | 1.15 | 1.22 | 1.29 |

TABLE 2

The ratio $(Q_L-1)/(1-Q_R)$ observed at the maximum field strength for the four dispersions of methylene blue dyed bentonite. Values given in brackets are the mean particle radii (microns) in corresponding fractions.

| Wavelength (Å) | Fraction | | | |
|---|---|---|---|---|
| | 1 (0.092) | 2 (0.08) | 3 (0.065) | 4 (0.06) |
| 4100 | 1.21 | 0.95 | 0.92 | 0.82 |
| 4675 | 1.07 | 0.97 | 0.99 | 0.95 |
| 5250 | 0.95 | 0.97 | 1.22 | 0.98 |
| 5800 | 1.11 | 0.88 | 1.10 | 1.23 |
| 6375 | 1.27 | 1.14 | 1.34 | 1.46 |
| 6950 | 1.36 | 1.17 | 1.29 | 1.35 |
| 7525 | 1.17 | 1.01 | 1.25 | 1.41 |

Table 3 shows the values of $(Q_L-1)/(1-Q_R)$ separately calculated for (i) total extinction and (ii) only for the absorption part, at the respective wavelengths of peak absorption for the dye adsorbed dispersions.

TABLE 3

Comparison of the ratio $(Q_L-1)/(1-Q_R)$ observed for the total extinction with that calculated only for the absorption. $\lambda = 5800$ A for Rh. dyed dispersions. $\lambda = 6950$ A for M.blue dyed dispersions.

| Fraction | Dispersions | | | |
|---|---|---|---|---|
| | Rhodamined Dyed | | M. Blue Dyed | |
| | Ext. | Abs. | Ext. | Abs. |
| 1 | 1.55 | 2.17 | 1.36 | 1.72 |
| 2 | 1.61 | 2.00 | 1.17 | 2.50 |
| 3 | 1.65 | 1.65 | 1.29 | 1.75 |
| 4 | 1.70 | 1.82 | 1.35 | 2.28 |

It is clear that the ratio calculated for the absorption part is now very close to 2. This confirms the inference drawn earlier that the absorption cross-sections for the particles in this size range are given fairly accurately by the dipolar approximation.

B        Birefringence :

Both dye adsorbed and undyed dispersions show fairly strong birefringence in a magnetic field at all the wavelengths. Corresponding fractions with the same concentrations show almost the same value of birefringence even at wavelengths within the absorption band of the dye adsorbed dispersions. Fig.5 shows the comparative results for the two fractions F1 and F2 of undyed and rhodamine dyed dispersions at the wavelengths 5200 A, 5800 A and 6375 A. These results clearly show that unlike the effects observed in transmission changes, the bire-fringence is not affected by the absorption.

377

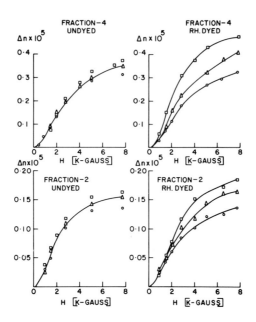

**Fig. 5.** Comparison of birefringence shown by undyed and rhodamine dyed dispersions of bentonite at the three wavelengths; 6375 A ( □ ), 5800 A ( ∆ ) and 5200 A ( ○ ).

The results in Fig.5 indicate, however, a systematic trend for the increased value of birefringence at longer wavelengths, which is possibly due to the decrease of the ratio of particle size to wavelength.  This is further confirmed in Fig.6 which shows average saturation value of birefringence for the four fractions (after normalizing to equal concentrations) against the particle size.  It may be pointed out here, that provided the shape factor of particles remain unchanged, the particle size should not have any effect on the birefringence within the dipolar approximation.  The result obtained here only confirms (as one could have expected) that above the Rayleigh size range, the efficiency of nonspherical particles for causing birefringence decreases.

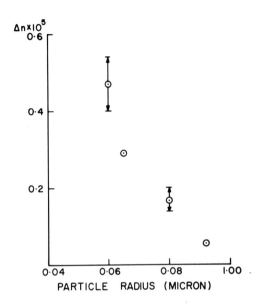

Fig. 6. Mean **saturation value** of birefrin- gence in the four fractions vs. the mean particle size. Vertical bars show the spread observed in the value of birefringence for a given fraction.

## IV  SUMMARY

A comparative study of the magnetooptical effects in dispersions of absorbing and nonabsorb- ing particles that are otherwise similar suggests that, for the particles above the normally accepted Rayleigh range (radius $\leqslant \lambda/10$), the absorption cross-section is still given fairly accurately by the dipolar approximation but the scattering cross- section shows large departure. The results also suggest that in this size range the scattering cross-section becomes almost independent of orient- ation if observed with unpolarised light ( as inferred from $Q_L - 1 \simeq 1 - Q_R$ and $Q_K \simeq 1$), but with polarised light it shows a strong orientation dependence. With the increasing particle size, efficiency for birefringence becomes smaller than what is expected for the particles with the same shape factor under dipolar approximation, and for

particles of dimension $\simeq$ wavelength, the birefringence becomes negligible as observed in case
of dispersions of aquadag graphite (2).

V    REFERENCES

1.    Naik, Y.G., and Desai, J.N., Indian J. Pure
      Appld. Phys., 3, 27 (1965).

2.    Shah, H.S., Desai, J.N., and Naik, Y.G.,
      Indian J. Pure Appld. Phys., 6, 282 (1968).

3.    Dave, M.J., Mehta, R.V., Shah, H.S., Desai,
      J.N., and Naik, Y.G., Indian J. Pure Appld.
      Phys., 6, 364 (1968).

4.    Desai, J.N., Naik, Y.G., Mehta, R.V., and
      Dave, M.J., Indian J. Pure Appld. Phys.,
      7, 534 (1969).

5.    Mehta, R.V., Shah, H.S., and Dave, M.J.,
      J. Colloid Interface Sci., 35, 41 (1971).

6.    Mehta, R.V., and Shah, H.S., J. Phys. D:
      Appld. Phys., 7, 2483 (1974).

7.    Van De Hulst, H.C., "Light Scattering by
      Small Particles", p.63,64, John Wiley &
      Sons, Inc. New York, (1957).

8.    Kerker, M., "The Scattering of Light and
      Other Electromagnetic Radiation", p.325,
      Academic Press, New York, (1969).

# MODEL MICROVOID FILMS

M. S. El-Aasser, S. Iqbal, and J. W. Vanderhoff
*Emulsion Polymers Institute*
*Lehigh University*

## ABSTRACT

*Microvoid coatings derive their opacity from tiny air bubbles dispersed in a continuous polymer matrix instead of the pigment particles used in conventional coatings. Practical microvoid coatings can be made in several ways, but the microvoid size and size distribution is not well-controlled. It is desirable to know the optimum microvoid size for maximum opacity, but the practical systems are not suitable for this purpose and the theoretical calculations of microvoid scattering have not yet given a definitive value. The purpose of this paper is to describe an experimental method to determine the optimum microvoid size.*

*The experimental approach depends upon the fact that monodisperse polystyrene latexes form rhombohedral arrays of particles in contact with one another without significant coalescence. These rhombohedral arrays are model microvoid films: the microvoids are the interstices between the packed particles and the packed particles the continuous polymer matrix. However, these films crumble when removed from the substrate and thus are unsuitable for determination of the optimum microvoid size.*

*The properties of these films were improved by "coating" the polystyrene latex particles with a thin layer of soft, sticky polyethyl acrylate by seeded emulsion polymerization, so that the particles adhere to one another at their points of contact. Uniform-thickness films were formed by ultracentrifugation and their rhombohedral arrangement was demonstrated by scanning electron microscopy. The microvoid size and spacing was varied systematically by variation of the polystyrene seed latex particle size; the microvoid size was varied at constant spacing by variation of the thickness of the polyethyl acrylate layer on the same polystyrene particles. Model microvoid films were prepared using polystyrene latexes of 0.234-1.10μ diameter and polyethyl acrylate layers of 50-250Å.*

*The reflectance-wavelength curves of these model micro-void films were measured over white and black backgrounds, and the specific microvoid scattering coefficients were calculated using the Kubelka-Munk equation. The areas under the specific microvoid scattering coefficient-void size curves showed that the optimum microvoid size was 0.440μ.*

## I. INTRODUCTION

### A. Microvoid Coatings

Microvoid coatings derive their opacity from microscopic or submicroscopic air bubbles instead of the pigment particles used in conventional coatings. In conventional coatings, opacity is obtained by dispersing colloidal pigment particles in a continuous matrix of polymer. Light impinging upon this film is scattered according to the parameters of the <u>pigment</u> system: (i) concentration of particles; (ii) particle size and size distribution (degree of dispersion); (iii) refractive index ratio between the pigment particles and the polymer matrix. In microvoid coatings, opacity is obtained by forming a dispersion of microscopic or submicroscopic air bubbles throughout the continuous matrix of polymer. Light impinging upon this film is scattered according to the parameters of the <u>microvoids</u>: (i) concentration of microvoids; (ii) microvoid size and size distribution; (iii) refractive index ratio between the polymer matrix and air.

The preparation of practical microvoid coatings has been reviewed by Burrell (1) and Clancy (2), and the mechanisms of microvoid formation have been categorized by Seiner and Gerhart (3) as follows:

(i)   prepare microscopic encapsulated air bubbles and disperse them in a continuous polymer matrix (4);

(ii)  form an emulsion of a volatile liquid in a polymer solution, and dry this mixture to form a continuous matrix of polymer containing dispersed droplets of the volatile liquid, then heat the film to evaporate the volatile liquid, leaving a dispersion of air bubbles in the polymer matrix (5, 6);

(iii)    form an emulsion of a volatile liquid (non-solvent for the polymer) in a polymer latex and dry this mixture to form a continuous film of latex polymer containing dispersed droplets of the volatile liquid, then heat the film to evaporate the volatile liquid, leaving a dispersion of air bubbles in the polymer matrix;

(iv)    prepare a film from a mixture of two incompatible polymers (one thermoplastic and the other thermosetting) in the form of two interpenetrating polymer networks, then extract the thermoplastic polymer selectively with a suitable solvent, leaving microscopic voids in the thermosetting polymer matrix (7, 8);

(v)    dissolve the polymer in a solvent-nonsolvent mixture (solvent more volatile than the nonsolvent) of such proportions that it forms a clear solution, then allow the more volatile solvent to evaporate, increasing the proportion of nonsolvent to the point where the polymer precipitates to form a porous, opaque film (9-13).

These methods produce different types of microvoid films, with void sizes ranging from microscopic to submicroscopic, shapes from spherical to irregular and indeterminate, and spacings from random to relatively uniform. Consequently, these films display a wide range of opacities.

B.  Optimum Microvoid Size

The opacity of microvoid films depends upon the size, shape, and spacing of the microvoids. For a given volume fraction of uniformly-distributed microvoids, the opacity of the film should increase with decreasing microvoid size to a maximum and then decrease. This optimum microvoid size for maximum opacity must be known if practical microvoid coating systems are to be developed.

At present, the theoretical derivations of the optical properties of microvoid films (14-18) have not been developed to the point where they can definitively predict the optimum microvoid size for maximum opacity; therefore, these theories cannot serve as a basis for the development of practical microvoid coatings.

It is possible, however, to determine the optimum microvoid size experimentally by developing a model microvoid system. An ideal model microvoid system should comprise a dispersion of uniform-size microvoids of known size and shape distributed uniformly throughout a continuous polymer matrix. Moreover, it should be capable of easy manipulation, i.e., systematic variation of microvoid size, spacing, and concentration. None of the foregoing practical microvoid systems can be adapted to produce uniform-size microvoids and therefore they are not well-suited to the preparation of model microvoid films.

The purpose of this paper is to describe a method developed to prepare model microvoid films and the use of this method to prepare such films of different void size and spacing, followed by the measurement of their opacity, to determine the optimum void size for maximum opacity.

## II.   EXPERIMENTAL RESULTS

## A.   Preparation of Model Microvoid Films

### 1.   *Experimental Approach*
The approach used in this work depends upon the fact that the drying of monodisperse polystyrene latexes under controlled conditions produces rhombohedral arrays of polystyrene spheres in contact with one another without appreciable coalescence (19). These polystyrene latex films are microvoid films in the sense that the interstices between the packed particles form the microvoids. Moreover, since the latex particles are all the same size, the microvoids are uniform in size and spacing. These microvoids are not spherical in shape, but instead comprise a 2:1 ratio of concave tetrahedra and concave cubes between the rhombohedrally-packed spheres (20). The size of the microvoids varies according to the size of the monodisperse polystyrene latex particles.

These polystyrene latex films, however, are too weak to use as model microvoid films. The particles are coalesced only at their points of contact (21), and polystyrene latex films cast on solid substrates cannot be removed without their crumbling; even those cast on mercury often fracture when removed from the mercury surface.

The properties of these polystyrene latex films can be improved by improving the adhesion between the particles at their points of contact, e.g., by polymerizing on the surface of the polystyrene particles a thin layer of a soft, sticky polymer (e.g., polyethyl acrylate) sufficient to give adhesion at the points of contact but insufficient to fill the interstices between the packed particles. The void size and spacing can be varied by using different-size monodisperse polystyrene latexes, and the void size can be varied at constant spacing by using the same latex with increasing thicknesses of the polyethyl acrylate layer.

## 2. *Latex Preparation*

The latexes were prepared by seeded emulsion polymerizations (22) in which a small amount of ethyl acrylate was polymerized in monodisperse polystyrene latexes (23). The polymerizations were carried out in a one-liter four-neck glass flask fitted with a Teflon paddle stirrer, condenser, monomer addition funnel, and nitrogen inlet. Table I gives a typical polymerization recipe.

TABLE I

*Typical Seeded Emulsion Polymerization Recipe*

*15% solids; calculated increase in particle diameter 100Å*

=================================================================

| | |
|---|---|
| seed latex (0.357μ diameter; 31.03% solids) | 120.14 gm |
| deionized water | 146.64 gm |
| ethyl acrylate (inhibitor-free) | 3.22 gm |
| potassium persulfate initiator | 0.20 gm |
| sodium bicarbonate buffer | 0.20 gm |

The polymerization procedure was as follows: (i) the seed latex, water (less 10 ml), and sodium bicarbonate were added to the glass flask partially immersed in a thermostated bath at 60°C, and the ethyl acrylate monomer was placed in the monomer addition funnel; (ii) nitrogen was bubbled through the seed latex for 2 hours under agitation at 60°C; (iii) the potassium persulfate initiator dissolved in 10 ml water was added through the nitrogen inlet, which was then raised above the liquid level so that the atmosphere of the reactor was swept with nitrogen for the remainder of the polymerization; (iv) after 10 minutes, the monomer addition was started at the rate of one drop/minute (monomer addition time of 4-6 hours); (v) at the end of the monomer addition, the agitation rate was decreased, and the polymerization was continued for 16 hours at $60^{\circ}C$; (vi) the latex was then post-stabilized with Aerosol MA (American Cyanamid Company) emulsifier solution and the pH was raised to 9.5 with sodium hydroxide.

These polymerizations used monodisperse polystyrene seed latexes with particle diameters in the range $0.234-1.10\mu$ and calculated thicknesses of the polyethyl acrylate layers in the range $50-250\overset{\circ}{A}$. The uniformity of the "coated" latex particles as well as the absence of a new crop of particles was confirmed by transmission electron microscopy.

## 3. *Preparation of Model Microvoid Films*

There are two requirements that the films of monodisperse polystyrene latex particles must fulfill in order to be used as model microvoid films: (i) the order of packing in the film should be rhombohedral; (ii) the thickness of the film should be uniform in the area used for the optical measurements.

The first requirement can be fulfilled by using monodisperse latexes with a high level of colloidal stability and forming the films in such a way as to insure uniformity of packing. The second requirement can be fulfilled by using the ultracentrifugation method (24) to prepare films of the polyethyl acrylate-"coated" polystyrene latexes.

The ultracentrifugation method was as follows: the latex was diluted in 35% sucrose (pH 9.5) and stabilized with Aerosol MA emulsifier, then centrifuged in 10-ml polycarbonate tubes at 30,000rpm (60,000g) in an IEC B-35 preparative ultracentrifuge equipped with a 6-tube swinging-bucket rotor. Since the particles are lighter than the medium, they rise to the top to form a "cake" at the water-air interface. Only part of this "cake" coalesces to form a continuous film, according to the stability of the latex, the acceleration due to the centrifugation, and the density difference between the particles and the medium. At the end of the run, the films were recovered, washed thoroughly to remove uncoalesced particles and occluded sucrose, then placed on a microscopic slide and dried in a desiccator. This method produced flexible, opaque 1.16-cm diameter films of sufficient strength for handling and measurement of their optical properties. Figure 1 shows typical model microvoid films over black and white backgrounds. The uniformity of packing of the latex particles in these films was confirmed by scanning electron microscopy (Figure 2).

*Fig. 1. Model microvoid films produced by centrif-
ugation of polystyrene latex particles "coated" with a thin
layer of polyethyl acrylate: Films 1 and 2 contain the same
microvoid size (0.795μ polystyrene plus 150Å polyethyl
acrylate) but are of different thickness; Films 3 and 4 are
of the same film thickness (1.6 mils) but contain different
void sizes (0.234μ polystyrene plus 50Å polyethyl acrylate and
1.10μ polystyrene plus 150Å polyethyl acrylate, respectively).*

*4. Characterization of the Model Microvoid Films*
    *a. Film thickness.* The film thickness was measured by
optical microscopy using a 21X objective and the 10X Image-
Splitting Eyepiece.
    *b. Void fraction.* The void fraction $\phi_v$ of the model
microvoid films was calculated from the equation:

$$\phi_v = (d_1 - d_f)/d_1 \tag{1}$$

where $\underline{d_1}$ is the density of the latex particles and $\underline{d_f}$ the bulk
density of the model microvoid film made from these particles.
The density of the polyethyl acrylate-"coated" polystyrene

387

*Fig. 2. Scanning electron micrograph of the surface of a model microvoid film of 0.795μ-diameter polystyrene latex particles "coated" with 120Å-thick layer of polyethyl acrylate, showing the uniformity of packing of the latex particles in the film.*

particles was calculated assuming additive contributions of the polystyrene and polyethyl acrylate densities according to their volume fractions in the particles. The bulk density of the model microvoid films was calculated from their volume and weight. The volume was calculated from the film dimensions; the weight was determined gravimetrically using the Cahn Electrobalance. The average values of the microvoid volume fraction of the various model microvoid films were in the range 34-44%.

    *c. Void size.* The rhombohedral packing arrangement was used as the model for calculation of void size between packed monodisperse latex particles. The average void volume was calculated from the volumes of the tetrahedral and octahedral types of voids weighted according to the ratio 2:1. Allowance was made for the polyethyl acrylate excluded at the points of

contact between the polystyrene particles. Table II summarizes the values of the average void volumes and the
corresponding void diameters of the various model microvoid
films.

TABLE II

*Average Void Volumes and Void Diameters of Model Microvoid
Films*

| Latex Film | Average Void Volume/$10^{-3}\mu^3$ | Average Void Diameter/$\mu$ |
|---|---|---|
| 0.357$\mu$ PS + 100Å PEA | 1.856 | 0.152 |
| 0.795$\mu$ PS + 120Å PEA | 27.85 | 0.376 |
| 0.795$\mu$ PS + 150Å PEA | 26.17 | 0.368 |
| 0.795$\mu$ PS + 200Å PEA | 22.52 | 0.350 |
| 0.96$\mu$ PS + 200Å PEA | 44.38 | 0.439 |
| 1.04$\mu$ PS + 200Å PEA | 58.17 | 0.481 |
| 1.10$\mu$ PS + 150Å PEA | 75.19 | 0.523 |
| 1.10$\mu$ PS + 200Å PEA | 70.20 | 0.512 |
| 1.10$\mu$ PS + 250Å PEA | 63.74 | 0.496 |

## B. Interaction of Visible Light with Model Microvoid Films

*1. Reflectance of the Films in the Visible Region*

The reflectance of each film was measured over the
wavelength range 400-700 nm using the Kollmorgen KCS-40
Spectrophotometer equipped with a mini-computer, teletype unit,
and paper tape-punching unit. This spectrophotometer has
diffuse illumination, with a choice of three different light
sources. The sample and the standard are presented at ports
of an integrating sphere, and the measuring beams are at 8°
from the normal. Monochromatization is achieved by passage
of the reflected beams through a series of interference
filters.

The reflectance-wavelength curve was measured for the
model microvoid films using illumination source Type "A" and
the small port for the small area of view. All measurements
were made with the specular component included. The
measurements were made at 20 nm intervals over the 400-700 nm
range, and the reflectance values were simultaneously printed

out on the teletype and punched on paper tape.

The "working" reflectance standards were two matched squares of non-opaque vitrolite glass, which had been calibrated previously against freshly prepared plaques of Eastman white reflectance standard ($BaSO_4$) assumed to have 100% reflectance. Consequently, the output of the KCS-40 Spectrophotometer is the absolute reflectivity expressed in percent relative to the "working" standards.

Since each model microvoid film was dried on a glass microscope slide, it was possible to position this slide over both the white and black vitrolite glass squares and thus measure the reflectance of the same area over both white and black backgrounds. An optical seal between the glass microscope slide and the vitrolite glass backgrounds was provided by placing a drop of Nujol between the two glass surfaces and pressing them together; the combined sample was then introduced to the sample port. The white and black vitrolite glass squares used as backgrounds were calibrated against the "working" standards over the wavelength range 400-700 nm.

Figure 3 shows typical reflectance-wavelength curves measured over white and black backgrounds for four different model microvoid films of about the same thickness. The shape of the curves varies according to the particle size of the polystyrene seed latex, with those for the 0.795$\mu$ and 1.10$\mu$ sizes showing more maxima and minima than that of the 0.357$\mu$ size.

## 2. *Specific Microvoid Scattering Coefficients of Model Microvoid Films*

The reflectance-wavelength variation (400-700 nm) of the model microvoid films of different thicknesses were analyzed using the general Kubelka-Munk equation (25), which expresses the reflectance $R$ of a film at a given wavelength in terms of the reflectance of the underlying substrate $R_g$, the film thickness $X$, and the thickness-independent scattering $S$ and absorption $K$ coefficients:

$$R = [1 - Rg(a - b \coth b S X)] / [a - Rg + b \coth b S X] \quad (2)$$

where $a = (K + S)/S$ and $b = (a^2 - 1)^{\frac{1}{2}}$.

Since the reflectance-wavelength curves of the model microvoid films were determined over black and white back-

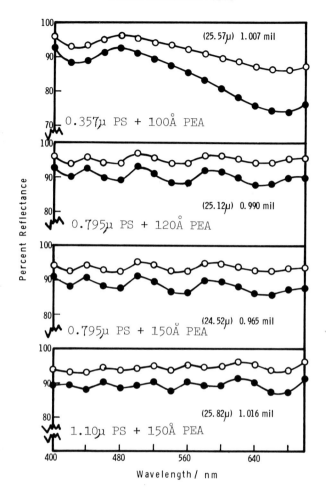

*Fig. 3. Representative reflectance-wavelength curves measured over white (O) and black (●) backgrounds for four different model microvoid films of about the same thickness.*

grounds of known reflectance, Equation 2 gives for each case two simultaneous equations that can be solved by a computer numerical analysis (26). The variation of film scattering coefficient over the wavelength range 400-700 nm was calculated for the model microvoid films of different thickness prepared from nine different latexes. The averages of these film scattering coefficients for films of different thickness were used to calculate the specific microvoid scattering coefficients by dividing the film scattering coefficients by the volume fraction of microvoids. Table III gives the results for films prepared from six of these latexes.

TABLE III

*Specific Microvoid Scattering Coefficients of Model Microvoid Films*

| Wavelength, nm | Specific Microvoid Scattering Coefficient / $mil^{-1}$ $PVC^{-1}$ | | | | | |
|---|---|---|---|---|---|---|
| | 0.357μ PS + 100Å PEA | 0.795μ PS + 120Å PEA | 0.795μ PS + 150Å PEA | 0.795μ PS + 200Å PEA | 1.10μ PS + 150Å PEA | 1.10μ PS + 200Å PEA |
| 400 | 19.028 | 17.264 | 15.461 | 11.702 | 12.057 | 12.809 |
| 420 | 15.538 | 15.574 | 14.024 | 11.165 | 12.130 | 12.811 |
| 440 | 15.154 | 16.773 | 13.962 | 11.061 | 11.204 | 11.962 |
| 460 | 17.036 | 13.793 | 12.682 | 10.158 | 12.198 | 12.532 |
| 480 | 17.736 | 12.922 | 11.832 | 9.663 | 10.819 | 11.682 |
| 500 | 15.965 | 16.062 | 14.405 | 10.260 | 11.056 | 11.420 |
| 520 | 13.933 | 13.715 | 12.608 | 9.884 | 11.299 | 12.015 |
| 540 | 12.134 | 11.331 | 10.523 | 8.743 | 9.706 | 10.587 |
| 560 | 10.687 | 11.127 | 10.202 | 8.381 | 10.833 | 11.389 |
| 580 | 9.404 | 13.205 | 11.810 | 8.774 | 9.819 | 10.602 |
| 600 | 8.280 | 12.535 | 11.533 | 8.987 | 9.498 | 10.341 |
| 620 | 7.288 | 10.959 | 10.196 | 8.478 | 10.461 | 11.064 |
| 640 | 6.517 | 9.828 | 9.175 | 7.844 | 9.643 | 10.452 |
| 660 | 6.049 | 9.617 | 8.879 | 7.503 | 8.225 | 9.116 |
| 680 | 5.954 | 10.461 | 9.524 | 7.503 | 7.984 | 9.407 |
| 700 | 6.362 | 10.359 | 9.510 | 7.593 | 9.591 | 10.248 |

## III.  DISCUSSION

## A.  Model Microvoid Films

The combination of the emulsion polymerization technique for "coating" the monodisperse polystyrene latex particles with a layer of soft, sticky polymer, along with the ultracentrifugation technique for preparing films of uniform thickness, gives model microvoid films suitable for optical measurements.  The soft "shell"-hard "core" combination provides ideal particles for these model films:  the soft "shell" around the particles provides adhesion at their points of contact, and the hard "core" allows coalescence only to the point of contact between the "cores".

In the ultracentrifugation method, the force required for initiation of coalescence of the latex particles is provided by the centrifugal force $\underline{F}$, which is given by Equation 3 (24):

$$F = m \ [1 - (d_1/d_2)] \ a \qquad (3)$$

where $\underline{m}$ is the mass of latex particles, $\underline{d_1}$ and $\underline{d_2}$ the densities of the medium and the polymer, respectively, and $\underline{a}$ the acceleration due to ultracentrifugation.

A latex of given stability must be subjected to a critical minimum force to initiate coalescence of the latex particles (21).  Equation 3 shows that, for given conditions of ultracentrifugation, any particles in excess of the mass corresponding to the critical force will coalesce and form a continuous film.  Thus model microvoid films of different thickness can be prepared by varying the concentration of latex particles charged to the centrifugation tubes.  The absence of lateral movement of the particles during ultra-centrifugation in the swinging bucket rotor ensures the uniform thickness of the latex films formed at the water-air interface close to the center of rotation.

The latexes must be stable in order to attain good packing order before the particles coalesce at their points of contact.  Generally, surfactants were added to these latexes before ultracentrifugation to improve their stability.  The latexes without added surfactant often formed films with poor order of packing.

Examination of the model microvoid films by scanning electron microscopy suggests that the rhombohedral arrangement predominates in the thinner films, but, with increasing film thickness, the order of packing at the underside of the film (i.e., furthest from the water-air interface) is poorer than at the top of the film (i.e., closest to the water-air interface). This poorer order of packing may account for the void fractions of 34-44% observed for these model microvoid films as compared with the theoretical value of 26% for rhombohedrally-packed spheres. Another possible reason is errors in the determination of the void fraction. Actually, however, these values of 34-44% are close to the limiting void fractions of 35-40% observed for random packing of uniform-size spheres (20).

This discrepancy between the measured and theoretical void fractions of the model microvoid films may not affect their optical properties. Careful examination of the reflectance-wavelength variation of model microvoid films prepared from a given latex shows that the positions of the maxima or minima (e.g., as in Figure 3) do not change with increasing film thickness. This suggests that the interaction of the incident light with the rhombohedrally-packed spheres in the top layers of the film is predominant, and that the more irregularly packed particles in the layers underneath do not influence the reflectance. Therefore, the rhombohedral arrangement was used as the model for the analysis of the void distribution in the model microvoid films and for the interpretation of their optical properties.

B. Light Scattering of Model Microvoid Films

The optical properties of the model microvoid films are the result of light scattered by the air interstices between the packed polymer spheres. The absorption coefficients $K$ of the microvoid films calculated using the Kubelka-Munk equation were found to be very small relative to the scattering coefficients $S$, i.e., the values of $K/S$ were in the range 0.2-2%. This result is expected because neither the air interstices nor the polymer spheres should absorb light in the visible range.

These model microvoid films provide a systematic method for investigating the effect of void size and spacing on the light scattering from microvoids. Figures 4, 5, and 6 show the variation of the specific microvoid scattering coefficients with wavelength in the range 400-700 nm for films prepared from monodisperse polystyrene latexes of 0.357, 0.795, and 1.10μ diameter, respectively.

According to the analysis of the void sizes of the models given in Table II, the films prepared from the 0.357μ-diameter

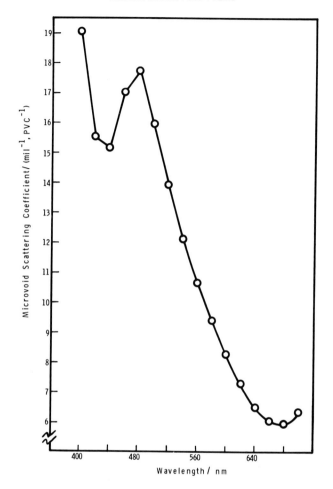

*Fig. 4. Variation of specific microvoid scattering coefficient with wavelength for model microvoid films made from 0.357μ-diameter polystyrene latex particles "coated" with a 100A-thick layer of polyethyl acrylate.*

latexes contained the smallest voids, followed by those prepared from the 0.795μ-and 1.10μ-diameter latexes in that order. Figure 4 shows that the specific microvoid scattering coefficients of the films prepared from the 0.357μ-diameter latexes decreased to a minimum at 440 nm, increased to a maximum at 480 nm, and then decreased rapidly with further increases in wavelength. Figures 5 and 6, however, show a different pattern, i.e., the presence of several maxima and minima, characteristic of the particle size of the seed latex.

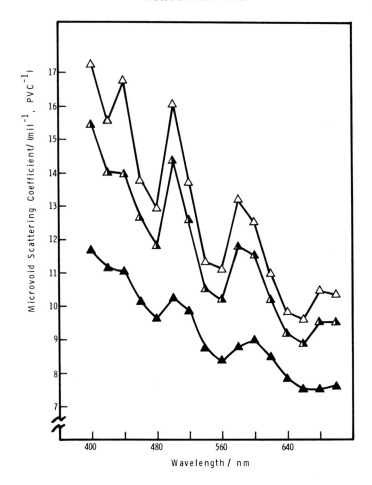

*Fig. 5. Variation of specific microvoid scattering coefficient with wavelength for model microvoid films made from 0.795µ-diameter polystyrene latex particles coated with the following thicknesses of polyethyl acrylate: (△) 120Å; (▲) 150Å; (▲) 200Å.*

Figure 5 shows that the positions of the maxima and minima of the specific microvoid scattering coefficient-wavelength curve do not change with increasing thickness of the polyethyl layer, which indicates that the distance between the centers of the voids is constant, independent of the change in void size. Thus, with this series of films prepared from the 0.795µ-diameter latexes, an increase in the thickness of the polyethyl acrylate layer results in a decrease in the void size at constant void spacing. This decrease in void size results

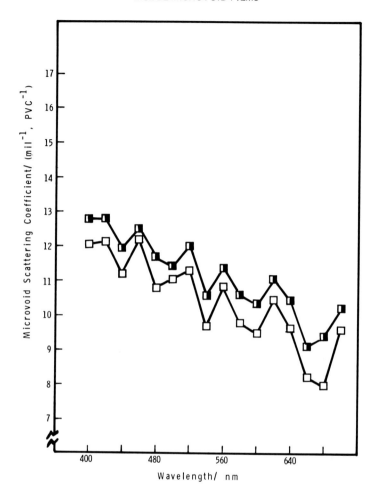

*Fig. 6. Variation of specific microvoid scattering coefficient with wavelength for model microvoid films made from 1.10μ-diameter polystyrene latex particles coated with the following thicknesses of polyethyl acrylate: (□) 150Å; (■) 200Å.*

in a <u>decrease</u> in the specific microvoid scattering coefficient; the highest value of the scattering coefficient was obtained with the thinnest polyethyl acrylate layer (120Å).

Figure 6 also shows that the positions of the maxima and minima of the specific microvoid scattering coefficient-wavelength curve do not change with increasing thickness of the polyethyl acrylate layer, which indicates that in this case as well the distance between the centers of the voids is constant, independent of the change in void size. With this series of films prepared from the 1.10μ-diameter latexes, however, an increase in the thickness of the polyethyl acrylate layer (and the corresponding decrease in the void size) results in an increase in the specific microvoid scattering coefficient; the higher value of the scattering coefficient was obtained with the thicker polyethyl acrylate layer (200Å). Combining the results of the 0.795μ-and 1.10μ-diameter series shows that the optimum microvoid size for maximum opacity is bracketed by the two model systems which give the highest values of the specific microvoid scattering coefficients, i.e., the microvoids from the 0.795μ-diameter particles "coated" with a 120Å-thick layer of polyethyl acrylate as the lower limit and those from the 1.10μ-diameter particles "coated" with a 200Å-thick layer of polyethyl acrylate as the upper limit.

Thus the optimum microvoid size was bracketed between two films with the highest scattering coefficients, one formed from the smaller particle-size seed latex with a relatively thin polyethyl acrylate layer and the other from a larger particle-size seed latex with a relatively thick polyethyl acrylate layer. Therefore, other model microvoid films with voids in the bracketed size range were prepared using mono-disperse polystyrene seed latexes with particle diameters intermediate between 0.795μ and 1.10μ, "coated" with various thicknesses of polyethyl acrylate. Model microvoid films were prepared from these latexes using the ultracentrifugation technique, their reflectance-wavelength curves were measured over the wavelength range 400-700 nm, and their specific microvoid scattering coefficients were calculated.

Since the intensity of light scattered by a coating film is dependent upon the size of the scattering particles relative to the wavelength of light, and since the eye perceives the appearance of the coating film as a result of the scattered light integrated over the visible light range, the areas under the specific microvoid scattering coefficient-wavelength curves of Figures 4, 5, and 6 were determined and correlated with the microvoid size. Figure 7 shows the variation of the area under these curves over the wavelength range 400-700 nm with the average void sizes (from Table II) for the various model microvoid films (e.g., those of Table III). The microvoids of 0.440μ diameter show the highest scattering in the wavelength range 400-700 nm; however, smaller voids of 0.152μ diameter show a relatively high scattering because of their relatively high scattering at the lower end of the wavelength range (See Figure 4).

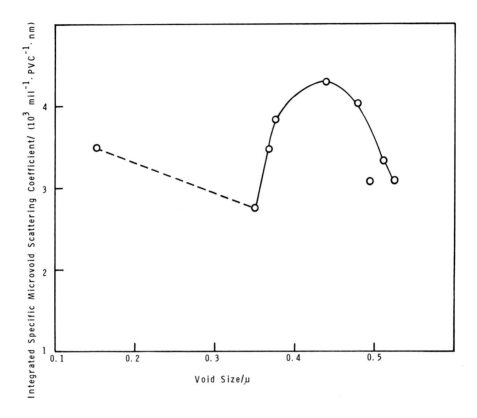

*Fig. 7. Variation of the area under the specific micro-void scattering coefficient-wavelength curve over the wavelength range 400-700 nm with microvoid size.*

This variation of scattering with wavelength is illustrated by taking the areas under the specific microvoid scattering coefficient-wavelength curves for two regions of the visible range: the first over the range 400-560 nm and the second over the range 560-700 nm. Figure 8 shows the variation of the areas under these curves with average void size for the two wavelength ranges. In this case, too, the highest scattering is observed for a void size of about 0.440μ; however, in the wavelength range 400-560 nm, the scattering from the smaller 0.152μ-diameter voids is comparable to that from the optimum 0.440μ-diameter voids, but in the wavelength range 560-700 nm, the scattering from the smaller voids is significantly lower than that from the optimum size.

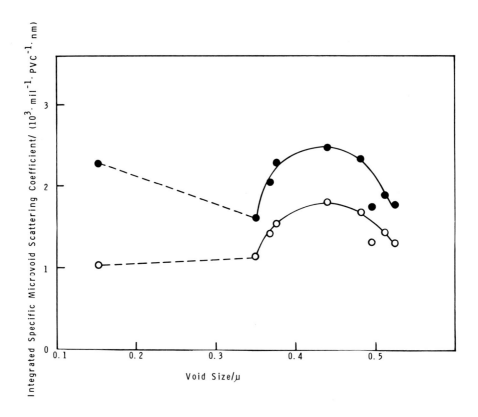

*Fig. 8. Variation of the area under the specific microvoid scattering coefficient-wavelength curve with microvoid size over the wavelength ranges: (●) 400-560 nm; (○) 560-700 nm.*

This difference in behavior can be explained by taking into account the spacing between the microvoids as well as their size. The smaller void sizes in the films prepared from the 0.357u-diameter latexes have a much smaller spacing between the void centers as compared with the larger void sizes and spacings of the films prepared from the 0.960u-diameter latexes, which give the maximum scattering. Thus, in the wavelength range 400-560 nm, the smaller void sizes with smaller distances of separation are as efficient in scattering light as the larger void sizes with larger distances of separation. Clearly, this smaller void size range is worthy of further investigation.

IV.  CONCLUSIONS

A.  Preparation of Model Microvoid Films

1.  Model microvoid films can be prepared by "coating" mono-
    disperse polystyrene latex particles with a thin layer
    of polyethyl acrylate by seeded emulsion polymerization
    and forming uniform-thickness films of these latexes by
    ultracentrifugation.

2.  These model microvoid films comprise monodisperse latex
    particles packed in the rhombohedral arrangement and
    coalesced at their points of contact, with the inter-
    stices between the packed spheres forming the microvoids.


B.  Variation of Microvoid Size and Spacing

1.  Both the size of the microvoids and their distance of
    separation in the model microvoid films can be varied
    systematically by variation of the size of the mono-
    disperse polystyrene seed latex particles.

2.  The size of the microvoids can be varied systematically
    at constant distance of separation by variation of the
    thickness of the polyethyl acrylate layer on a given
    polystyrene seed latex.


C.  Determination of the Optimum Microvoid Size for Maximum
    Opacity

1.  The specific microvoid scattering coefficient for a given
    model microvoid film can be determined by measuring the
    reflectance-wavelength curves over white and black
    backgrounds, calculating the scattering coefficient using
    the Kubelka-Munk equation, and dividing the scattering
    coefficient by the void fraction (determined from the
    film thickness and weight).

2.  The optimum microvoid size for maximum opacity can be
    determined by applying this method to latexes prepared
    from different-size monodisperse polystyrene latexes
    "coated" with different thicknesses of polyethyl
    acrylate.

3.  The optimum microvoid size was found to have an
    equivalent spherical diameter of about $0.440\mu$ using
    monodisperse polystyrene latexes in the size range

0.234-1.10μ and polyethyl acrylate "coating" in the range 50-250Å.

4.  There may be a second, smaller optimum microvoid size, perhaps close to an equivalent spherical diameter of 0.152μ, but this must be confirmed by further experiments.

V.  REFERENCES

1.  Burrell, H., J. Paint Tech. 43, (559), 48 (1971).
2.  Clancy, J. J., Preprints, A.C.S. Div. Org. Coatings & Plastics Chem. 33 (2), 258 (1973).
3.  Seiner, J. A., and Gerhart, H. L., Ind. Eng. Chem., Prod. Res. Dev. 12, 98 (1973).
4.  Kershaw, R. W., Lubbock, F. J., and Polgar, L. (to Balm Painte, Ltd.), Ger. Offen. 1,959,649, June 11, 1970; C.A. 73, 46797a (1970).
5.  Rosenthal, F. (to Nashua Corp.), U.S. 2,739,909, Mar. 27, 1956.
6.  Clancy, J. J., Lovering, D. W., and Wells, R. C. (to Arthur D. Little, Inc.), U.S. 3,108,009, Oct. 22, 1963.
7.  Dowbenko, R., and McBane, B. N. (to PPG Industries, Inc.), Brit. 1,193,865, June 3, 1970.
8.  Dowbenko, R., and McBane, B. N. (to PPG Industries, Inc.), U.S. 3,544,489, Dec. 1, 1970.
9.  Rolle, C. J., and Van Kirk, W. (to Interchemical Corp.), U.S. 2,665,262, Jan. 5, 1954.
10. Larsen, G. H. (to Ludlow Corp.), U.S. 3,031,328, Apr. 24, 1962.
11. Burrell, H., J. Paint Tech. 40 (520), 197 (1968).
12. Murata, K. (to Yuasa Battery Co., Ltd.), U.S. 3,450,650, June 17, 1969.
13. Seiner, J. A. (to PPG Industries, Inc.), Brit. 1,178,612, Jan. 21, 1970.
14. Kershaw, R. W., Aust. Oil Colour Chem. Assoc. Proc., Aug. 1971, p. 4.
15. Ross, W. D., J. Paint Tech. 43 (563), 50 (1971).
16. Lubbock, F. J., Chem. Ind. New Zealand, Dec. 1972, p. 18.
17. Allen, E. M., J. Paint Tech. 45 (584), 65 (1973).
18. Kerker, M., Cooke, D. D., and Ross, W. D., J. Paint Tech. 47 (603), 33 (1975).
19. Bradford, E. B., J. Appl. Phys. 23, 609 (1952).
20. Graton, L. C., and Fraser, H. J., J. Geol. 43, 785 (1935).
21. Vanderhoff, J. W., Tarkowski, H. L., Jenkins, M. C., and Bradford, E. B., J. Macromol. Chem. 1, 361 (1966).
22. Vanderhoff, J. W., Vitkuske, J. F., Bradford, E. B., and Alfrey, T., Jr., J. Polymer Sci. 20, 225, (1956).

23. Prepared by J. W. Vanderhoff, The Dow Chemical Company, 1966-70.
24. El-Aasser, M. S., and Robertson, A. A., *J. Colloid Interface Sci*. 36, 86 (1971).
25. Judd, D. B., and Wyszecki, G., "Color in Business, Science, and Industry," Wiley, New York, 1967, p. 398.
26. Original computer program written by E. M. Allen, Lehigh University, 1972.

# EFFECT OF AN IMPURITY ON THE CRITICAL POINT
## OF A BINARY LIQUID SYSTEM AS A SURFACE PHENOMENON

### O. K. Rice
*University of North Carolina at Chapel Hill*

Many years ago consideration of the effect of small quantities of water as an impurity in the critical mixing point of the cyclohexane-aniline binary-liquid system led me to suggest that this be treated as a surface effect (1-4). At temperatures below the critical such an impurity will be positively or negatively adsorbed at the interface between the two phases which are present at such temperatures, and thus will affect the interfacial tension. The change of interfacial tension can be related to the concentration of impurity and the amount adsorbed at the interface by the Gibbs adsorption equation. Since the critical point always occurs where the interfacial tension vanishes, one will need to change the temperature after the impurity is added, in order to get back to the critical point. Application of these ideas gave the equation

$$dT_c/dc_3 = RT(\Gamma_3/c_3)(\partial\sigma/\partial T)_{c_3}^{-1} , \qquad (1)$$

where $T$ is the temperature, $T_c$ the critical temperature, $c_3$ the concentration of the third component (assumed small), $R$ the gas constant, $\sigma$ the interfacial tension, and $\Gamma_3$ the interfacial adsorption of the impurity, defined with respect to a dividing surface so placed as to make the adsorption of the principal components as small as possible.

Another equation, of purely thermodynamic origin, had been derived much earlier by Wagner (5):

$$dT_c/dc_3 = \frac{(RT/\lambda)d^2\lambda/dx_1^2}{d^2\Delta S/dx_1^2} . \qquad (2)$$

Here $x_1$ is the mole fraction of the third component, $\lambda$ its solubility (e.g., the solubility of the vapor at a standard pressure), and $\Delta S$ is the entropy of mixing of $x_1$ moles of one of the components with $x_2 = 1 - x_1$ moles of the other.

These two equations are equivalent to each other. The interfacial adsorption $\Gamma_3$ depends upon the variation of the solubility across the interface, and the net effect depends upon $d^2\lambda/dx_1^2$. $(\partial\sigma/\partial T)_{c_3}$ is the surface entropy, and it depends on the variation of $\Delta S$ across the interface in much the same way that $\Gamma_3$ depends on the variation of $\lambda$. Since, however, $(\partial\sigma/\partial T)_{c_3}$ can be measured directly this gives Eq. (1) a certain practical advantage over Eq. (2). We still have to know or find out how $\lambda$ varies with concentration of the principal components. We attempt to integrate this quantity across the interface and then find $\Gamma_3$ (by subtracting the amount which would be present if there were a sharp, rather than a gradual change from one phase to another). It turns out that the integral involves $\Delta z$, the thickness of the interface, which is a somewhat fuzzy concept, but which can, nevertheless, be defined reasonably unambiguously. Measurement of $dT_c/dc_3$ then allows us to determine $\Delta z$.

When the procedure outlined above was conceived, our knowledge of critical phenomena was much less precise than it is at present, and $\lambda$ was expanded as a Taylor series in $x_1$. However, it is now known that this is not allowable for most thermodynamic functions, and this is no less true of $\lambda$. In order to determine the behavior of $\lambda$ we investigated the critical line of the binary system. The latter is a line in the three-dimensional space defined by $T$, $x_1$, and $a_3$, the activity of the impurity (which is related to $\lambda$ by $c_3 = a_3\lambda$); it tells how the critical parameters change with $a_3$. Based on the assumption that the critical line is not parallel to any of these three axes it is possible to show that the behavior of $\partial^2\lambda/\partial x_1^2$ should parallel that $\partial^2\Delta S/\partial x_1^2$. Furthermore, we can show that

$$x_2\partial^2\Delta S/\partial x_1^2 = \partial\bar{s}_1/\partial x_1 = -\partial^2\mu_1/\partial T\partial x_1 , \qquad (3)$$

where $\bar{s}_1$ is the partial molal entropy and $\mu_1$ the chemical potential of the designated component. The behavior of $(\partial\mu_1/\partial x_1)_T$ along the coexistence curve [analogous to the pressure coefficient $(\partial P/\partial\rho)_T$ in the liquid-vapor system] is determined by the critical exponent $\gamma'$, it being proportional to $t^{\gamma'}$, where $t = (T_c - T)/T_c$. Thus, along the coexistence curve

$$\partial^2\lambda/\partial x^2 \propto t^{\gamma'-1} , \qquad (4)$$

or

$$\partial^2\lambda/\partial x^2 \propto |x_1-x_{1c}|^{(\gamma'-1)/\beta} , \qquad (5)$$

where $x_{1c}$ is the mole fraction at the critical point, and $\beta$ is another critical exponent

$$\left| x_1 - x_{1c} \right| = t^{\beta} . \tag{6}$$

These exponents have previously been obtained for a similar thermodynamic quantity using a lattice model (6).

As already noted, $\Gamma_3$ is determined by integrating across the surface, the equation being

$$\Gamma_3 = a_3 \rho \left[ \int_{-\infty}^{0} (\lambda - \lambda'') dz + \int_{0}^{\infty} (\lambda - \lambda') dz \right] , \tag{7}$$

where $z$ is the distance from the dividing surface and $\lambda'$ and $\lambda''$ are the respective values in the bulk coexisting phases. The dividing surface is defined by

$$\int_{-\infty}^{0} (x_1 - x_1'') dz + \int_{0}^{\infty} (x_1 - x_1') dz = 0 . \tag{8}$$

We now attempt to evaluate Eq. (7) using Eq. (5). It is not possible to expand $\lambda$ in a Taylor series, since it has a singularity at the critical point, but it is possible to write equations like

$$\lambda - \lambda' = <\partial \lambda / \partial x_1>_{x_1, x_1'} (x_1 - x_1') \tag{9}$$

where the angular brackets denote an average in the range indicated by the subscripts. This can also be done for $\partial \lambda / \partial x_1$. Eq. (7), then, involves a rather complicated set of averages, and can be written symbolically

$$\Gamma_3 = -a_3 \rho <\partial^2 \lambda / \partial x_1^2><x_1 - x_{1c}>^2 \Delta z \tag{10}$$

where the averages are defined by the original integral expression. The thickness, $\Delta z$, is introduced by approximating the profile of the surface (i.e., $x_1$ as a function of $z$) by a line of constant slope, taking $x_1$ from its constant value $x_1''$ in one phase to $x_1'$ in the other $(x_1' > x_1'')$. So within the surface

$$dz = [\Delta z / 2 (x_1' - x_{1c})] d(x_1 - x_{1c}) \tag{11}$$

where of course $2(x_1' - x_{1c}) \approx x_1' - x_1''$.

We now need the experimental data for use in our equations. The solubility data for water in cyclohexane-aniline is not too certain (4), and we have represented it by

$$\lambda_c^{-1} \partial^2 \lambda / \partial x_1^2 = 39.6 (x_1 - x_{1c})^{0.75} . \tag{12}$$

The exponent 0.75 comes from evaluation of the exponents in Eq. (5), which are about the same for binary systems as for the liquid-vapor systems (7). Although Eq. (5) is, strictly, for evaluation along the coexistence curve, we have assumed that we can use it to get relative solubilities within the interface.

The interfacial tension in this system has been measured (2), and we have, approximately,

$$\partial \sigma / \partial T = -T_c^{-1} \partial \sigma / \partial T = -0.15 t^{0.3} \text{dyne cm}^{-1} \text{deg}^{-1} \tag{13}$$

We also have, near the critical point of the cyclohexane-aniline system,

$$\rho = 9.8 \times 10^{-3} \text{ mole cm}^{-3} , \tag{14}$$

and (4)

$$dT_c / dc_3 = 285° . \tag{15}$$

When Eqs. (11), (12) and (14) are used in Eq. (10), or, rather, in the equation which it symbolizes, and the result is used with Eqs. (13) and (15) in Eq. (1) we obtain an expression for the thickness of the interface in the cyclo-hexane-aniline system,

$$\Delta z = 2.1 t^{-0.62} \text{Å} . \tag{16}$$

For orientation, at $t = 0.1$, Eq. (16) gives $\Delta z = 8.7$Å.

Huang and Webb (8) have made optical measurements of the interface thickness in a related binary system, cyclohexane-methanol. Their result is

$$\Delta z = 9.4 t^{-0.67} \text{Å} , \tag{17}$$

so their estimate of the thickness of the surface is about four times as great as ours. I do not believe that all the discrepancy arises from the difference in the systems. Part of it may be due to the roughness of the solubility measurements in the cyclohexane-aniline system. We hope to be able to improve these measurements.

A more detailed account has just been published (9). This also includes consideration of some scaling relations and the renormalization of exponents under conditions of fixed composition.

The work was supported by the National Science Foundation.

REFERENCES

1. Rice, O. K., Chem. Revs. 44, 69 (1949); J. Phys. Colloid Chem. 54, 1293 (1950).
2. Atack, D., and Rice, O. K., Disc. Faraday Soc. 15, 210 (1953).
3. MacQueen, J. T., Meeks, F. R., and Rice, O. K., J. Phys. Chem. 65, 1925 (1961).
4. MacQueen, J. T., and Rice, O. K., J. Phys. Chem. 66, 625 (1962).
5. Wagner, C., Z. physik. Chem. 132, 273 (1928).
6. Wheeler, J. C., Ber. Bunsenges. 76, 308 (1972).
7. Stanley, H. E., "Introduction to Phase Transitions and Critical Phenomena," Chap. 3. Oxford University Press, New York and Oxford, 1971.
8. Huang, J. S., and Webb, W. W., J. Chem. Phys. 50, 3677 (1969).
9. Rice, O. K., J. Chem. Phys. 64, 4362 (1976).

# COLLOIDAL PYRITE GROWTH IN COAL

Raymond T. Greer
Iowa State University

Pyrite ($FeS_2$) constitutes the major source of sulfur in coal. As an outgrowth of experiments to assess sulfur removal techniques, this work has led to the recognition of colloidal size pyrite as being a primary constituent of the various sized pyrite concretions found in coal samples. This significant characteristic relates to choice and improvement of processing and purification procedures, as well as to the importance of taking into account the chemical reactivity of these colloidal particles in pyrite refuse disposal schemes.

Microscale characterization of sulfide phases has been performed for complete channel samples on a micron size scale utilizing combined scanning electron microscope - energy dispersive X-ray analysis techniques for obtaining the microstructural and the microchemical information, respectively. Remarkable observations have been made on electrochemically etched surfaces of pyrite nodules or concretions. A nodule is commonly comprized of an assembly of various size framboids which have been joined ("cemented") by an infilling of additional pyrite (established by wavelength dispersive X-ray analysis). Within each spherical assembly of crystallites (framboid), the pyrite crystallite size is exceptionally uniform (usually 1 μm or less in diameter), and yet when comparing different framboids within even a small region of a specimen, the pyrite crystallite size varies somewhat from framboid to framboid. Independent nucleation and crystallization processes occurred for each framboid. The pyrite material condensed from a gel. Numerous nucleation sites were present. A crystallite grew from each nucleus consuming material. There is evidence that larger aggregations are stable and very small aggregations of atoms and embryos have reverted back to the gel phase, consequently a critical size seems to develop for these conditions of deposition ultimately resulting in a representative primary pyrite crystallite constituent size of about a micrometer in diameter. The crystallites join and form boundaries, predominantly contacting at points rather than at faces with the crystallite octahedra in apposition.

## I.  INTRODUCTION

One aspect of microstructural characterization work for coal is described:  microscale distribution of iron pyrite (FeS$_2$) crystallites as components of large (centimeter size) pyrite nodules and vein fillings, as components related to macerals and as significant constituents contributing problems of mechanical removal and refuse disposal.  Inorganic sulfur in coal usually refers to pyrite, as other phases such as gypsum are relatively low in abundance or rapidly appear as weathering products from the pyrite.

## II.  EXPERIMENTAL PROCEDURE

Microscale characterization of sulfide phases is performed utilizing combined scanning electron microscope - energy dispersive X-ray analysis techniques for obtaining the microstructural and the microchemical information, respectively.  This provides accurate and precise fundamental information, which is obtainable by no other technique, concerning size, shape, orientation, and distribution of phases within the coal.

Subsamples for scanning electron microscope studies were obtained from complete channel samples.  The specimens were marked so that orientation in the face and vertical position in the bed were maintained.  A traverse, for fresh fracture surfaces normal to the stratification, was made across a specimen and for the sequential specimens.  All 1133 channel specimens received a vacuum deposited coating of approximately 200 Å of gold (in some cases, a 20 Å pre-coating of graphite was required, as well) to minimize charging problems.

Additional subsamples were chosen for comminution experiments, etching studies, etc., where other types of analyses (such as wavelength dispersive analyses using an electron microprobe for carbon, sulfur, iron, calcium, and so on) were performed as well.  Electrochemical etching experiments were performed using 70 volts D.C. for 180 seconds as detailed later in this report.  In certain cases it was useful to clean the coal surfaces ultrasonically prior to the microstructural studies.  Placing the coal in distilled water in an ultrasonic cleaning unit was found to be satisfactory.  It should be emphasized that great care must be exercized in the preparation of specimens.  Fresh fracture surfaces must be used.  Coal will oxidize in a relatively short time while being stored in a laboratory.  Also, the effect of various solvents (cutting fluids, etc.) must be closely monitored.  New phases can easily be formed on surfaces.

## III. RESULTS

One of the earliest studies of the occurrence of sulfur in coals was done by Thiessen,[1,2] whose comments are appropriate with minor modification to the findings for Iowa coal and for most coals. Work in sections to follow will detail new specific information concerning occurrence, distribution factors, and interrelations with coal microstructure which are of interest in planning sulfur removal schemes.

### A. Distribution of Pyrite Crystallites

Iron sulfides occur in coal as nodules and narrow seams and in finely disseminated form. Work to date indicates that a significant portion of pyrite in coal occurs either as individual micron size crystallites, or as assemblies of these crystallites forming framboids (a word coined by Rust in 1935 meaning berry-like)[3] generally 20 to 30 microns in width, or as assemblies of these framboids interconnected by additional pyrite. These framboidal masses are seen to comprise even the largest nodules.

Although mean pyrite particle sizes have been reported for Iowa coal (156 micrometers)[4] for example, for over half the currently operating Iowa mines, most of the pyrite is less than 50 micrometers in diameter for channel samples studied. This raises special characterization problems for routine conventional optical microscope studies. What is clear is that there are great inherent difficulties in making these estimates.

### B. Microstructure

The micrographs to follow represent the typical forms of pyrite found in Iowa coal (and for that of other states, as well). The primary constituent for the various sizes of pyrite particles, nodules, concretions, etc., is shown to be colloid sized pyrite. In the Lovilia mine, for example, 72% of the individual pyrite crystallites are approximately 1 $\mu$m in diameter, and 96.6% are less than 2 $\mu$m in diameter for the section of the mine studied. Consequently, this has important implications in the choice of comminution treatment or sulfur removal scheme, in the relatively high reactivity of the pyrite which is related to the very small particle size, as well as in the reliability of "organic" sulfur determinations as associated with the inability to remove all of the pyrite.

For the crystallites of Figure 1, energy dispersive X-

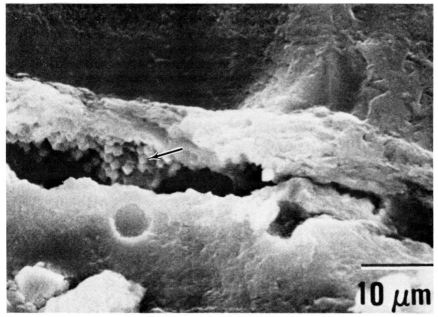

Fig. 1. Secondary electron image of coal with a horizontal crack containing individual pyrite crystallites with diameters of the order of a micron in size (indicated by arrow). ICO mine.

(a)                                    (b)

Fig. 2a. Individual pyrite crystallites and a globular assembly of relatively smaller crystallites (framboid). Scale bar is three microns.

Fig. 2b. Opening in coal filled with framboids of the order of tens of microns in diameter. Scale bar is 20 microns. Otley mine.

ray analyses establish the presence of pyrite by monitoring the $Fe_{K\alpha}$ and $S_{K\alpha}$ X-rays and their signal ratios. In this view of a fresh fracture surface, pyrite crystallites are seen attached to the upper surface of a horizontal opening in the coal. In Fig. 2, both individual crystals and spherical assemblies of crystallites (framboids) are shown. Notice that there is no infilling of material between the crystals and framboid in Fig. 2a, or between framboids in Fig. 2b. Individual pyrite euhedra occur predominantly in a size range of about one to 40 microns in diameter, with most of the crystallites being one to two microns or less in diameter.

The polished smooth surface of fusinite in Fig. 3 indicates coalified plant cell features (similar to a honeycomb appearance) with the roughly circular interiors filled with pyrite (or occasionally, calcite). Also, throughout this field of view numerous high contrast (bright) gypsum crystallites ($\sim$ 3-4 microns by 10 microns in length) are seen. The fresh fracture surface showing pyrite crystallites within coalified plant cell network in Fig. 4 is to be compared with Fig. 3. As the organic material is resistant to reaction with nitric acid, the encapsulated pyrite crystallites would not be dissolved in a standard pyritic sulfur determination. The arrow indicates a broken cell wall revealing a single pyrite crystallite within this particular cellular feature.

Other microstructural occurrences of pyrite crystallites and framboids are seen in Fig. 5a and b.

Figures 6a, b, and c are of great interest. Earlier work by investigators was performed without the high resolution and depth of field capabilities of a scanning electron microscope. "The aggregates of tiny spherical particles"[5,6,7] were obviously incompletely identified. As can be seen, the crystallites in this Iowa coal are most commonly octahedrons having crystallized within the coal. Occasionally the octahedra have their solid angles blunted (this seems to be more common in some Illinois coals which were also studied). Smithson[8] has indicated that the dominance of the octahedron may be favored by calcareous environment in grains collected from sedimentary rocks; however, no specific comments of this type have been made relative to pyrite habits found in coal deposits, although this is consistent with conditions required for the formation of pyrite (cubic $FeS_2$) and marcasite (orthorhombic $FeS_2$).[9,10,11,12] The character of the solution and the temperature of the solution seem to be the controlling factors.

A bacterial hypothesis for the origin of framboidal textures was held in favor in the early 1900's; however, in the 50's, A. R. Barringer, working at Imperial College, succeeded in producing similar structures experimentally from a gel. In the samples in Figures 6a-c, it is clear that these "pyrite

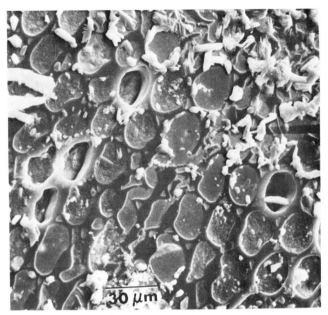

Fig. 3. Fusinite (Jude mine), secondary electron image showing cellular microstructure. Each cell is usually filled with an individual pyrite crystal. Scale bar is 30 microns.

Fig. 4. Fresh fracture surface showing fractured cell walls with pyrite crystallites exposed. Mich mine.

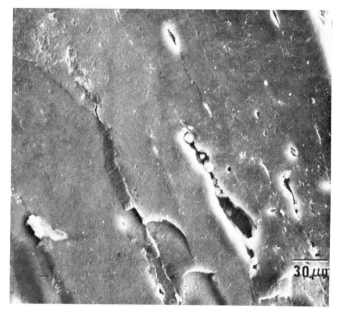

Fig. 5a. Pyrite generally occurs in massive aggregates, in radiating clusters and in reniform, globular, granular and stalactitic formations within coal. In a polished section, the central region of this cross section would appear reniform or globular, depending on the sectioning. ICO mine.

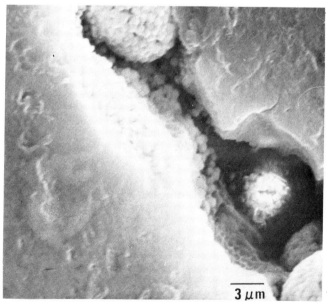

Fig. 5b. Higher magnification view of center of Fig. 5a.

Fig. 6a, b, and c. Increasing magnification series for an etched surface of a centimeter size pyrite nodule or concretion. The nodule is comprised of an assembly of various size framboids which in this case have been joined ("cemented") by an infilling of additional pyrite (established by wavelength dispersive X-ray analysis). Compare this with Fig. 2b where the framboids are just in contact without any additional pyrite filling in between them. As seen in Fig. 6, within each spherical assembly of crystallites (framboid) the pyrite crystallite size is remarkably uniform, and yet by comparing different framboids within this field of view the pyrite crystallite size varies from framboid to framboid. Independent nucleation and crystallization processes occurred for each framboid. Mich mine.

The specimen was electrolytically etched for a period of three minutes using an etchant solution of 24 ml Perchloric acid (70%), 300 ml Methanol, 180 ml Butyl Cellosolve, and 7 ml distilled water. Sufficient material was removed to see the primary microstructure which is representative of this massive inclusion or nodule. No conductive coating was required for viewing this surface using a scanning electron microscope. It is clear that numerous spherical forms are packed together to form the nodule. The pyrite material has condensed from a gel.

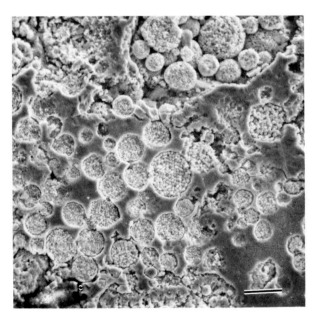

Fig. 6a. Scale bar is 20 microns.

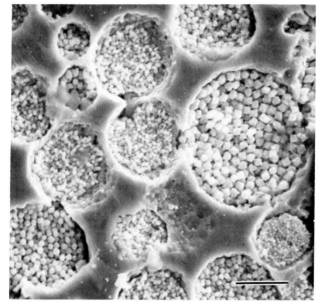

Fig. 6b.  Scale bar is 10 microns.

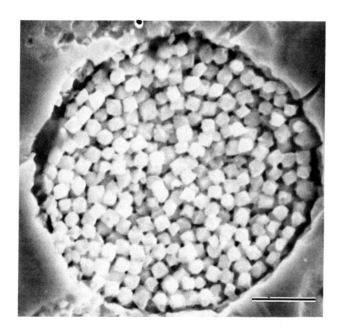

Fig. 6c.  Scale bar is 5 microns.

spheres" are the result of crystallization from a gel. Also, it is clear that there is no correlation between the size of the "sphere" and the size of the constituent crystallites comprising the "sphere". Solidification begins at a number of nuclei within specific regions of the gel. The crystallite size is remarkably uniform within the specific regions making up a framboid, and does vary between framboids. The remaining gel material solidified forming a cementing agent between the spherical framboidal forms.

## C. Microchemistry

Both energy dispersive X-ray and wavelength dispersive X-ray analyses were required to identify the material which fills in between the spherical framboidal forms which are viewed in cross section (Fig. 7). Detailed analyses of both relatively rough surfaces and polished surfaces revealed that two main materials were associated with surrounding the framboids: coalified material (positive carbon analyses) and in this case, pyrite material (positive iron and sulfur analyses by point, area scan, and relative ratio analytic techniques). Fig. 8a-c shows two dimensional X-ray scans for the etched surface near the edge of the nodule, representing iron, sulfur, and carbon respectively. These figures together with the backscattered electron image of Fig. 8d indicate that there is some intergrowth with the maceral material that originally surrounded the opening into which the pyrite was deposited. This is an important common association in that the pyrite seems to be predominantly found in inertinite, the maceral type in which the cellular features are well preserved. Numerous additional examples show both the individual crystallites as well as the framboids or assemblies of framboids to be associated with inertinite. Analyses performed in other areas of the nodule for both rough and polished surfaces indicated only pyrite material. Thus the framboids had filled a cavity and in this case additional pyrite material was available to fill in between framboids cementing the material to form a massive nodule.

## IV. SUMMARY

The pyrite in coal is seen to occur in several ways: as individual micron size crystallites within cracks and other openings of the coal (e.g., Fig. 1 and 2), as individual crystallites within coalified cellular features (e.g., Fig. 3 and 4), as spherical assemblies of the micron size crystallites forming framboids (e.g., Fig. 2), as assemblies

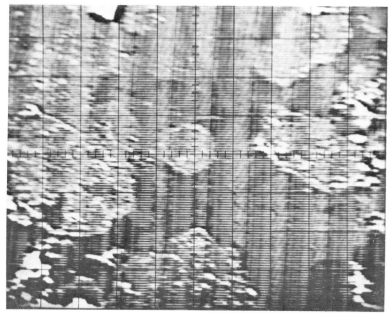

Fig. 7.    Sample current image of a polished surface of a
nodule.  This image was obtained by using an A.R.L. electron
microprobe X-ray analyzer.  The reticle scale is comprised of
1 cm x 1 cm blocks corresponding to a size of 2 microns/cm
for the sample field of view.

X-ray analyses were required to identify the material
which fills in between the spherical framboidal forms which
are seen in cross section in this micrograph.  For example,
in the center of the micrograph a 3 micron diameter spherical
feature is seen.  This corresponds to a section through a
framboid.  Notice the significantly lower microstructural
resolution in this micrograph compared with those of Fig. 6.
However, this sample current image is useful in conjunction
with the X-ray analyses to delineate features on a very smooth
surface.  In this case, the point analyses established that
all of the material in this view was pyrite ($FeS_2$), including
both the framboids and the material surrounding the framboids.
Mich mine.

As the edge of a nodule is approached, some interpene-
tration with the coal may be observed.  An example of this is
seen in Fig. 8a-d.

Analyses of this material in Fig. 7 shows only Fe and S.
No Zn or As was detected.  As both carbon and calcium X-ray
lines were also monitored, and no carbon or calcium was
detected, neither coal nor calcite ($CaCO_3$) was present in
this area.

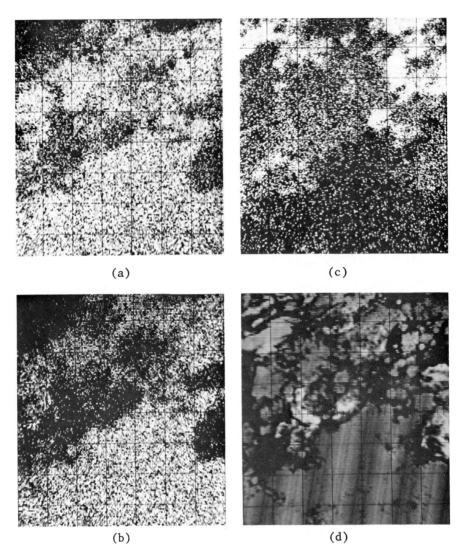

(a)　　　　　　　　　　　(c)

(b)　　　　　　　　　　　(d)

Fig. 8a-d. Identical field of view at the interface of a portion of a pyrite nodule (primarily the lower third of these micrographs; indicated by the Fe and the associated S displays) and the surrounding coal material (refer to the carbon display). Fig. 8a shows the two dimensional distribution of Fe; Fig. 8b of S; Fig. 8c of C; and Fig. 8d of the electron backscattering image which presents a display of mean atomic number variations within the field of view with the higher contrasts corresponding to higher mean atomic number materials. Mich mine. The reticle scale is 1 cm x 1 cm blocks for a size of 3 microns/cm for the sample field of view.

of framboids filling openings in the coal (e.g., Fig. 5), and as assemblies of framboids of the micron size crystallites surrounded by additional pyrite forming the large nodules and bands of pyrite observed in coal (e.g., Fig. 6 and 7). Thus, these crystallites are seen to be a primary constituent of the pyrite for a broad range of sizes. Insight into the distribution of the microscopic pyrite has led to the recognition of the presence of colloidal size pyrite as being a primary constituent of the various sized pyrite concretions found in coal samples of all of the operating Iowa mines. This has significant implications for removal of the pyrite from the coal as well as disposal of the pyrite that is removed. The association of pyrite with inertinite or as concentrations within cracks and other openings helps in planning removal schemes, but will make complete removal very difficult.

## V.  ACKNOWLEDGMENT

This work was supported by the Engineering Research Institute and the Energy and Mineral Resources Research Institute at Iowa State University.

## VI.  REFERENCES

1. Thiessen, R., Coal Age, 16, 668 (1919).
2. Thiessen, R., Trans. AIME, 63, 913 (1920).
3. Rust, G. W., J. Geol., 43, 398 (1935).
4. McCartney, J. T., O'Donnell, H. J., and Ergun, S., Bureau of Mines ROI 7231, pp. 1-18 (1969).
5. Bastin, E. S., Geol. Soc. Amer. Memoir 45, Waverly Press, Inc., Baltimore, Md., 1960.
6. Love, L. G., Quart. J. Geol. Soc., 113, 429 (1957).
7. Baker, G., Neues Jahrb. Min., Abh, 94, 564 (1960).
8. Smithson, F., Min. Mag., 31, 314 (1956).
9. Stokes, H. N., U.S.G.S. Bull., 186, (1901).
10. Allen, E. T., Crenshaw, J. L., and Johnson, J., Am. J. Sci., 33, 169 (1912).
11. Tarr, W. A., Amer. Min., 12, 417 (1927).
12. Smith, E. E., and Shumate, K. S., Fed. Water Pollution Control Admin. Report for Grant No. 14010 FPS, Water Pollution Control Series, Dept. of the Interior, 7 (1970).

USE OF MICELLAR SYSTEMS IN ANALYTICAL CHEMISTRY-
THEIR APPLICATION TO THE SPECTROPHOTOMETRIC DETERMINATION
OF SULFITE ION WITH ACTIVATED AROMATIC COMPOUNDS

Willie L. Hinze
*Wake Forest University*

*Sulfite ion undergoes essentially a quantitative equili-
brium reaction at room temperature in aqueous cationic micellar solutions with polynitroaromatic reagents to form the
corresponding anionic sigma complexes. The sulfite is determined by measurement of the absorption due to these sigma complexes. The reaction is useful for determining sulfur dioxide (as sulfite ion in solution) in the 0.3 to 9.0 μg/ml concentration range. Cationic surfactants are required in the
procedure because they solubilize the aromatic reagents and
completely shift the equilibrium reaction to the desired product side under the analytical conditions. Some possible
anion interferences can be eliminated by the use of an appropriate buffered solution system or prior neutralization of
the sample.*

I.  INTRODUCTION

Micellar systems have been the object of intensive investigations aimed at showing that they can serve as very
simple models for enzymatic, membrane-mediated, or related
processes.[1,2] These studies have shown that micellar systems
possess certain unique features and properties[1,3] such as:
ability to solubilize certain solutes; ability to serve as a
unique reaction media in which the rates, products, position
of equilibria, and in some instances, the stereochemistry can
be affected; ability to change microscopic properties of the
solvent; ability to alter the microenvironment about solubilized solutes; ability to affect or alter spectral parameters of solutes, etc.  In spite of these useful and well-
documented properties, only a few analytical applications of
micellar systems have been reported.

Many spectral analytical procedures (or potential procedures) often are limited for some of the following reasons: time consumption - due to slowness of reactions involved; solubility problems of the required analytical reagents; lack of chromophore stability; overlapping spectra of reagents and chromophore; and a lack of sensitivity among others. Up to now, analytical chemists have attempted to overcome some of these problems by resorting to different or mixed solvent systems. This results in a loss of sensitivity for the method involved due to dilution. An attractive potential alternate approach would be to employ micellar systems. This report demonstrates some of the useful applications and effects of micellar systems on a novel, new spectrophotometric method for the determination of sulfur dioxide (as sulfite ion) using activated aromatic nitro compounds.

Sulfite ion reacts with many activated (nitro) aromatic compounds to form anionic sigma complexes, commonly called Meisenheimer complexes.[4,12] This general equilibrium reaction scheme is shown in equation 1. Although this reaction

$$\text{(NO}_2)_n \quad + \quad SO_3^= \quad \rightleftharpoons \quad \text{(NO}_2)_n \qquad (1)$$

has been the subject of numerous kinetic, equilibrium, and structural investigations concerned with the mechanism of nucleophilic aromatic substitution,[12,13] no analytical applications have been reported. In this work, the activated aromatic reagents; 1,3,5-trinitrobenzene (135TNB), 2,4,6-trinitrobenzaldehyde (246TNBA), 1,3-dinitronaphthalene (13DNN), and 1,3,6,8-tetranitronaphthalene (1368TNN) have all been found to react rapidly and quantitatively at room temperature with sulfite ion yielding the corresponding absorbing anionic sigma complexes. These reactions have been evaluated as possible spectrophotometric methods for sulfite ion in which micellar systems were used to overcome potential problems in the methodology.


II. EXPERIMENTAL


A. Reagents

The activated aromatic reagents 135TNB (Eastman), 246-

TNBA (ICN Life Sciences), and 13DNN (Aldrich) were recrystallized several times before use following established procedures.[4,5,14] The 1368TNN was prepared as previously described.[14] Sodium sulfite was Baker Analyzed reagent grade.
Hexadecyltrimethylammonium bromide (CTAB) [Eastman, practical
grade] and sodium dodecyl sulfate (NaLS) [City Chemical
Corporation] were recrystallized[1] before use. All other
chemicals and reagents were the best available reagent grade
materials.

## B. Solutions

Stock sodium sulfite solutions (0.001 M with respect to
EDTA) in aqueous, mixed solvent, or micellar media were
usually in the $8-15 \times 10^{-4}$ M range and were prepared as
previously reported.[15] The activated aromatic reagent stock
solutions were approximately in the $10^{-3}$ to $10^{-2}$ M range and
were prepared by dissolving a weighed amount of the compound
in a small amount of DMSO and diluting to final volume using
the appropriate solvent system. These solutions appear to
be most stable in either pure DMSO or aqueous 0.10 M CTAB and
should be protected from light during storage. If these precautions are observed, the solutions are stable for at least
two weeks. The pH 6.86 buffer was 0.015 M with respect to
both $Na_2HPO_4$ and $KH_2PO_4$. The appropriate micellar solutions
were prepared via addition of a weighed amount of surfactant
to the appropriate aqueous solution.

## C. Instruments

Absorption measurements were made and spectra recorded
using either a Beckman DK-2A or a Cary 118C recording
spectrophotometer. The pH was measured with a Fisher Accumet
pH meter.

## D. Procedures

Reactions were conducted by adding a measured volume of
the standard sulfite solution (with Hamilton microsyringes)
to a solution of the appropriate aromatic reagent in the
particular solvent system. In the quantitative studies of
sulfite recovery, the effect of foreign ions, and the determination of Beer's Law plots, the activated aromatic compound was present in at least 2-9 fold excess over highest
concentration of sulfite ion used. The equilibrium reaction

is essentially complete and instantaneous for all activated aromatic reagents employed when the reaction media consists of either micellar CTAB (1.5-100.0 x $10^{-3}$ M) or greater than 50% (v/v) aqueous DMSO. The final concentration of aromatic was usually 1-9 x $10^{-4}$ M and that of sulfite ranged from 0.70-80 x $10^{-6}$ M. Either the aromatic or sulfite can be added last. Usually a small amount of the stock sulfite ion solution (5-500 μl) was added to appropriate solvent system and this constituted the synthetic sample to which a small amount (1.0 ml or less) of the desired activated aromatic was added. The final volume was 10.0 ml in most experiments. All readings were measured or recorded employing matched 1.0 cm cells versus a blank solution which contained all of the reagents except for sulfite ion.

III.  RESULTS AND DISCUSSION

A.  Study of the Polynitroaromatic-Sulfite Reaction

Although the equilibrium interaction of sulfite ion with 135TNB,[4,9,12] 246TNBA,[5,13] 13DNN,[17] and 1368TNN[17] have been reported with regard to mechanistic studies, the reaction apparently has never been used for the spectral determination of sulfite. These previously reported kinetic investigations were usually carried out under pseudo first-order conditions (i.e., [sulfite]>>[aromatic]) in either aqueous or mixed aqueous-DMSO media.[12,13] However, the analytical study of this system required that the concentration of the aromatic be greater than that of sulfite. It was found that aqueous solutions of the polynitroaromatic reagents at the desired analytical concentrations required (1-9 x $10^{-4}$ M) could not be prepared due to solubility problems. This potential problem was overcome merely by the additon of surfactants (CTAB, NaLS, Triton X-100) whose aqueous micellar solutions solubilized these aromatic reagents. Addition of appropriate amounts of dipolar aprotic cosolvents (such as DMSO, DMF, or DIOXAN) also eliminate those potential solubility problems.

It was found that the equilibrium reaction in an aqueous medium was incomplete (less than 8% formation of the sigma complex based on amount of sulfite ion present) when the concentration of both reactants were about equal or when the aromatic was present in up to 10 fold excess. In general, literature reports have shown that cationic surfactants increase, anionic surfactants decrease, and nonionic surfactants have no effect on the magnitude of the equilibrium constants of related systems.[1,18-21] In agreement with these reports, only cationic CTAB shifted the position of equilibrium so that essentially complete reaction of sulfite ion occurred with all the aromatics employed. As expected, anionic NaLS and nonionic Triton X-100 micellar solutions were of no use in the procedure since the equilibrium reaction did not proceed. Table I summarizes the solvent and media effects on the equilibrium reaction for the analytical procedure for sulfite ion using 135TNB. Similar results were observed for the other aromatic reagents employed.

TABLE 1
Solvent Effects on 135TNB Procedure[a]

| Solvent Composition | Effects on Reaction |
|---|---|
| Water (5% DMSO) | No Reaction[b] |
| Water (15% DMSO) | Incomplete Reaction[c] |
| 95% Methanol | Incomplete Reaction[c] |
| 95% Ethanol | Incomplete Reaction[c] |
| 50% DMSO/50% Water | Complete Reaction[d] |
| 90% DMSO/10% Water | Complete Reaction[d] |
| 5% NaLS | No Reaction[b] |
| 4% Triton X-100 | Incomplete Reaction[c] |
| 0.10 M CTAB (no buffer present) | Complete Reaction[d] |
| 0.10 M CTAB (pH 6.8 phosphate buffer) | Complete Reaction[d] |
| 0.05 M CTAB (no buffer present) | Complete Reaction[d] |
| 0.01 M CTAB (no buffer present) | Complete Reaction[d] |

[a] Concentration of 135TNN was $8 \times 10^{-4}$ M, sulfite ion concentration ranged from 3.0-70 $\times 10^{-6}$ M, and temperature was 25.0°C unless otherwise noted.

[b] Refers to essentially no equilibrium reaction of sulfite ion with the 135TNB. No Beer's Law data is possible.

[c] Refers to incomplete equilibrium reaction of sulfite ion with 135TNB. Beer's Law is not obeyed for these systems.

[d] Refers to essentially complete reaction of sulfite with 135TNB. Beer's Law is obeyed in these solvent systems.


Mole ratio and continuous variation studies for the equilibrium reactions of sulfite ion with the activated aromatic compounds indicate that the stoichiometry in every case is 1:1 as shown in equation (1) at low sulfite ion concentrations. Thus, the structures of the sigma complexes formed under the analytical experimental conditions are the same as those reported in the detailed kinetic and structural investigations.[4,5,9,13]

B. Absorption Spectra

Spectral data for the absorbing anionic sigma complexes formed upon the equilibrium interaction of sulfite with activated aromatic compounds are presented in Table II. The

TABLE II
Summary of Spectral Parameters for Anionic Sigma Complexes[a]

| Parent Aromatic | $\lambda_{max}$, nm[b] | $\varepsilon$, (X10$^{-4}$ M$^{-1}$ cm$^{-1}$)[b,c] |
|---|---|---|
| 246TNBA | 457.5 (458) | 2.12 (2.20) |
| | 558.0 | 1.02 |
| 135TNB | 461.5 (462) | 2.16 (2.30) |
| | 547.5 | 1.06 |
| 1368TNN | 487.0 (487) | 2.05 (2.10) |
| 13DNN | 537.0 (537) | 1.87 (1.95) |

[a] Concentration of parent aromatic was usually 1-9 x 10$^{-4}$ M, sulfite ion concentration ranged from 1-70 x 10$^{-6}$ M, and the reaction medium was 0.10 M aqueous CTAB at 25.0°C.

[b] Wavelength of maximum absorption for sigma complexes of indicated parent aromatic. Literature values in parentheses, ref. (6) for 246TNBA, ref. (4) for 135TNB, and ref. (17) for 13DNN and 1368TNN.

[c] Molar absorptivity values are the average of three separate experiments each of which employed at least seven different sulfite ion concentrations and were calculated from the concentration of sulfite in the samples.

position of maximum absorbance and molar absorptivities are essentially constant when the CTAB concentration is changed from 0.01 to 0.10 M. The spectra have been published elsewhere for 135TNB,[4,9] 246TNBA,[5] and 1368TNN[22] and their corresponding anionic sigma complexes due to reaction with sulfite ion. The absorption maxima for the anionic sigma complexes occur at longer wavelengths than those for the parent aromatic reagents and are usually far enough removed from the parent maxima so that the interference is quite small. Hence, the blank correction for absorption due to an excess of the parent aromatic is negligible or very small. However, in the case for 246TNBA, it was necessary for best results to correct for the absorbance due to the parent.

## C.   Concentration Range for Sulfite Determination

The molar absorptivity values for the determination of sulfite ion are listed in Table II. These values are based on the concentration of sulfite and confirm the 1:1 stoichiometry as shown in equation (1). The molar absorptivity values obtained are very similar for all of the activated aromatic compounds tested. Some typical Beer's law data for the determination of sulfite with 246TNBA is presented in Table III. In all cases, Beer's law is followed

TABLE III
Beer's Law Data for Determination of Sulfite Ion with 246TNBA[a]

| $[SO_3^=]$, $M(X10^6)$ | Absorbance[b] | Molar Absorptivity[c] |
|---|---|---|
| 8.48 | .193 | 22,800 |
| 12.72 | .286 | 22,500 |
| 21.20 | .478 | 22,500 |
| 31.80 | .795 | 25,000 |
| 53.0 | 1.16 | 22,000 |

[a]Conditions: [246TNBA] = 5 x $10^{-4}$M, [CTAB] = 0.05M, at 25.0°C.

[b]Wavelength, 457.5 nm, using 1-cm cells.

[c]Molar absorptivity values based on the molar concentration of sulfite ion.

and good linear calibration plots of absorbance due to sigma complexes formed versus concentration of sulfite ion are obtained.

The absorbance readings due to the sigma complexes formed from all parent aromatics were found to slowly decay with time. Hence, for best results, the absorbance must be measured within about 1 to 5 minutes or after some specific constant reaction time. This general procedure employing activated aromatic reagents should be useful for the determination of sulfite ion over the range of approximately 0.25 - 9.00 μg/ml as $SO_2$ (corresponding to approximately 1-70 x $10^{-6}$ M sulfite ion in solution). Recovery data of sulfite ion from weighted synthetic samples using this calibration procedure were reasonably good for all cases.

D.  Interferences

There are two likely categories of substances which might be expected to cause difficulty in this procedure. These are other nucleophiles (anions) which react similarly with the parent aromatic reagents to form corresponding absorbing sigma complexes and those substances which either oxidize or tie up the sulfite ion. The effect of various anions at the $1.5 \times 10^{-4}$ M level on the procedure for the determination of sulfite with 135TNB is shown in Table IV. The only anions

TABLE IV
Interference Study[a]

| Foreign Ion | Effect |
|---|---|
| Fluoride | None |
| Bromide | None |
| Chloride | None |
| Iodide | None |
| Acetate | None |
| Azide | None |
| Iodate | None |
| Mixture of Above ($1.5 \times 10^{-3}$ M) | None |
| Nitrate | None |
| Nitrite | None |
| Sulfate | None |
| Mixture of Above ($1.5 \times 10^{-3}$ M) | None |
| Borate | High[b] |
| Phosphate | High[b] |
| Carbonate | High[b] |
| Hydroxide | High[c] |

(Table IV footnotes)

[a] Conditions: 0.10 M CTAB, $1.6 \times 10^{-4}$ M 135TNB, $1.99 \times 10^{-5}$ M Sulfite Ion, $1.5 \times 10^{-4}$ M Foreign Ion, 25.0° C.

[b] No interference found if reaction is carried out in a buffered medium (pH 6.8 phosphate buffer).

[c] No interference noted if original sample is carefully neutralized with acid.

which initially showed any interference were borate, phosphate, and carbonate all of which generate some free hydroxide ion in solution which subsequently reacts. These interferences (as well as that for hydroxide ion) can be eliminated via either careful neutralization of original sample with acid or by carrying out the reaction in aqueous CTAB which is buffered to a pH of 6.85. The more reactive the parent aromatic, the lower the pH buffer must be to prevent these interferences. For example, the 246TNBA procedure requires a pH 4 buffer to prevent these difficulties. Since amines and thiols are known to react with activated aromatic reagents,[12,13] they would probably also interfere in certain cases. However, this could be solved in a similar fashion as already described for hydroxide ion.

The presence of cations of Na, K, Li, Ca, Ba, etc. do not interfere with the general procedure. However, excess acid decomposes the sigma complexes[13] causing very low results or no apparent reaction at all. This potential interference can be eliminated by careful neutralization of excess acid with base prior to the reaction forming the absorbing sigma complexes. Mercuric ion also interferes since it apparently preferentially reacts with sulfite.[15]

## IV. CONCLUSIONS AND COMPARISONS

The most commonly accepted spectral method for sulfur dioxide or sulfite ion appears to be the West-Gaeke procedure,[23] a variant of which is used as the EPA manual reference method.[24] Although the sensitivity of the West-Gaeke method is higher than that obtained in this procedure involving activated aromatic reagents, the sensitivity is probably high enough for most applications. In the West-Gaeke method, the pH is very critical, the color development is slow and temperature dependent, and the pararosaniline dye reagent must be carefully purified to obtain reproducible results.[25] In contrast to this, the procedure reported here using the activated polynitroaromatic reagents is relatively uncomplicated and may have advantages in terms of simplicity of method, instantaneous reaction, known stoichiometry, easy of preparation of reagent aromatic solutions, and adaptibility to automated schemes of analysis. The main disadvantage of the procedure is that the absorbing sigma complex decays with time, but the West-Gaeke procedure also suffers from a similar problem. Therefore, the determination of sulfite via use of polynitroaromatic reagents offers an attractive alternative procedure and should probably be applicable and adaptable to many types of sulfite ion analysis problems.

The use of micellar systems to successfully overcome potential problems in the analytical scheme has also been demonstrated. Since the properties of various surfactants are well documentated,[1] it should be possible, in many cases, to employ or at least attempt to use the appropriate micellar system to overcome potential problems in any analytical procedure. From the limited amount of work reported here, it appears that the use of micellar systems is more effective and less troublesome than is the changing of solvents in attempting to overcome problems associated with development of analytical procedures.

## V. REFERENCES

1. Fendler, J. H., and Fendler, E. J., "Catalysis in Micellar and Macromolecular Systems," Academic Press, New York, 1975.

2. For recent reviews see:
   (a) Cordes, E. H., and Gitler, C., Progr. Bioorg. Chem., 2, 1 (1973).
   (b) Fendler, E. J., and Fendler, J. H., Adv. Phys. Org. Chem., 8, 271 (1970).

3. Elworthy, P. H., Florence, A. T., and Macfarlane, C. B., "Solubilization by Surface Active Agents and its Application in Chemistry and the Biological Sciences," Chapman and Hall, London, 1968.

4. Norris, A. R., Can. J. Chem., 45, 175 (1967).

5. Marendic, N., and Norris, A. R., Can. J. Chem., 51, 3927 (1973).

6. Sasaki, M., Rev. of Phy. Chem. Japan, 45, 45 (1975).

7. Crampton, M. R., and El Ghariani, M., $\underline{J}$. $\underline{Chem}$. $\underline{Soc}$. ($\underline{B}$) 391 (1970).

8. Crampton, M. R., and Willison, M. J., $\underline{JCS}$ $\underline{Perkin}$ $\underline{II}$, 160 (1975).

9. Crampton, M. R., $\underline{J}$. $\underline{Chem}$. $\underline{Soc}$. ($\underline{B}$), 1341 (1967).

10. Bernasconi, C. F., and Bergstrom, R. G., $\underline{J}$. $\underline{Amer}$. $\underline{Chem}$. $\underline{Soc}$., $\underline{95}$, 3603 (1973).

11. Crampton, M. R., and El Ghariani, M., $\underline{J}$. $\underline{Chem}$. $\underline{Soc}$. ($\underline{B}$), 330 (1969).

12. Crampton, M. R., in "Advances in Physical Organic Chemistry" (V. Gold, editor), Vol. 7, p. 211, Academic Press, New York, 1969.

13. Strauss, M. J., $\underline{Chem}$. $\underline{Rev}$., $\underline{70}$, 675 (1970).

14. Hinze, W. L., Liu, L.-J. and Fendler, J. H., $\underline{JCS}$, $\underline{Perkin}$ $\underline{II}$, 1751 (1975).

15. Humphrey, R. E., Ward, H. M., and Hinze, W. L., $\underline{Anal}$. $\underline{Chem}$., $\underline{42}$, 698 (1970).

16. Butler, A. R., $\underline{JCS}$ $\underline{Perkin}$ $\underline{I}$, 1557 (1975).

17. Hinze, W. L., Fendler, J. H., and Fendler, E. J., unpublished results.

18. Fendler, J. H., Fendler, E. J., and Merritt, M. V., $\underline{J}$. $\underline{Org}$. $\underline{Chem}$., $\underline{36}$, 2172 (1971).

19. Casilio, L. M., Fendler, E. J., and Fendler, J. H., $\underline{J}$. $\underline{Chem}$. $\underline{Soc}$. ($\underline{B}$), 1378 (1971).

20. Bunton, C. A., and Robinson, L., $\underline{J}$. $\underline{Amer}$. $\underline{Chem}$. $\underline{Soc}$., $\underline{90}$, 5972 (1968).

21. Chang, S. A., Dissertation, Texas A&M University, August, 1974.

22. Fendler, J. H., Fendler, E. J., and Casilio, L. M., $\underline{J}$. $\underline{Org}$. $\underline{Chem}$., $\underline{36}$, 1749 (1971).

23. West, P. W., and Gaeke, G. A., $\underline{Anal}$. $\underline{Chem}$., $\underline{28}$, 1816 (1956).

24. Logsdon, O. J., and Carter, M. J., Environ. Sci. Tech., 9, 1172 (1975).

25. Scaringelli, F. P., Saltzman, B. E., and Frey, S. A., Anal. Chem., 39, 1709 (1967).

## VI.  ACKNOWLEGMENTS

Partial support of this project by the Research and Publication Fund of Wake Forest University is gratefully acknowledged.

# RECOVERY OF CRYSTALLINITY IN GROUND CALCITE*

R. B. Gammage
*Oak Ridge National Laboratory*

D. R. Glasson
*Plymouth Polytechnic*

## ABSTRACT

*Recovery processes by thermal treatment and recrystalli-
zation are examined in a calcite specimen severely disordered
by ball milling. As the annealing temperature is increased,
restructuring in the bulk lags behind the recovery of crystalline perfection in the surface regions. Surface reordering
is significant at temperatures as low as 150-175°C and is rapidly
completed at 400°C. Annealing at 600°C is required for removal
of all lattice strain. Before loss of surface can occur by sintering, the temperature needs to exceed 300°C. The corresponding temperature for a high-area precipitated calcite is 400°C.
Recovery of crystallinity is also promoted by light-etching
with aqueous acid when extensive whisker growth occurs. Aging
over a period of twelve years has led to loss of the ultra-
reactive characteristics.*

## I. INTRODUCTION

Previous researches (1-7) have shown that the ball milling
of calcium carbonate has a pronounced effect on the surface and
structural properties of the crystals. In detailed studies of
the effect of grinding on large crystals of Iceland spar, the
crystal structure became disordered with the appearance of non-
crystalline and strained conditions. The grinding operation
proceeded in three distinct and clearly defined stages with
respect to the degree of crystalline perfection of the calcite,
viz, (1) crystal fracture by cleavage, with the degree of crystallinity remaining unchanged; (2) plastic deformation of the
crystals (shown by electron microscopy), with some production

---

*Research sponsored by the Energy Research and Development
Administration under contract with Union Carbide Corporation
and the Science Research Council, United Kingdom.

of noncrystalline material (contributing to the absorption of X-rays, but not to the reflected intensity), and (3) progressive distortion of the crystal lattice (indicated by X-ray line-broadening and infrared absorption measurements). The transformation of calcite into aragonite, by ball milling under "thin film" conditions, can be regarded as a further stage in the disturbance of the crystal structure. The X-ray measurements indicated that the crystals of calcite had to be worked until the amount of lattice distortion reached a critical level before any phase transformation occurred.

Development of defective crystal structure in the Iceland spar enabled water molecules to penetrate increasingly into the interior of the grains. The crystals, in which the lattice was believed to be distorted, recrystallized slowly in the presence of water, and whiskers of nail-head spar calcite were formed. Within the deformed crystals, the structure was most disordered in the superficial regions which modified the surface properties. The structural degradation of the superficial regions increased the apparent solubility of the calcite, so that it increased stepwise, most considerably at the start of the second and third milling stages.

These results strongly supported the suggestion that dislocations were the defects primarily responsible for the structural degradation of the calcite crystals and that the particles of the samples milled for longer times, e.g., 1000 h, were strain-hardened. The development and storing of lattice imperfections - especially dislocations - within the strained crystals increase the free energy compared with that of the unstrained material. The process by which strain-hardening diminishes is called recovery and is enhanced by rise in temperature.

In the present research, further information on the nature and location of the defects has been obtained by studying the recovery of crystallinity of the milled material by heat treatment and light-etching with aqueous acid.

## II. MATERIALS AND EXPERIMENTS

The precipitated calcite was obtained, as a reagent grade material, from the John and E. Sturge Co., Birmingham, England. The optical grade crystals of natural Iceland spar were ground for 1000 h in a stainless steel ball mill, with the pot filled to 40% of its volume by a 1-kg charge. X-ray fluorescence analysis of the milled sample showed it to contain only 60 ppm of iron; thus, little wearing of the steel surfaces had occurred. X-ray diffraction showed that there had been no phase transition of calcite into aragonite. The large amount of calcite (1-kg charge) had produced sufficiently thick compacted layers on the internal surfaces of the mill to cushion impacts and prevent the attainment of the high pressures and shear necessary for conversion of calcite into aragonite.

Separate samples of the 1000-h milled Iceland spar were annealed for 2 h and 6 h at each of a series of fixed temperatures between 150 and 600°C. The samples were annealed in air at temperatures below 250°C, in dry nitrogen between 250°C and 450°C, and in carbon dioxide between 500° and 600°C (to prevent dissociation of $CaCO_3$ which becomes appreciable in air above 500°C). Techniques used to examine the samples were:

(a) Surface area determination by the adsorption of isotopically-labelled krypton at -196°C (8). Normally, the adsorption of krypton was measured after outgassing a sample for several hours in vacuo at 150°C followed by immediate cooling to -196°C (9). The cross-sectional area of the krypton atom was taken as 19.5 $A^2$, and the monolayer capacity was determined from the B.E.T. plot (10). The specific surface was reproducible within a standard deviation of 3%.

(b) X-ray diffraction by means of a Solus-Schall diffractometer with a copper target on the X-ray generator. The method of Adams and Rowe (11) was used to prepare specimens with a flat surface for X-ray analysis. The angular width of the $10\bar{1}4$ calcite line in the diffraction patterns was measured at half-peak intensity and corrected by the method of Jones (12) to give the intrinsic broadening from which the apparent internal lattice strain normal to the $10\bar{1}4$ planes could be calculated.

(c) Scanning and transmission electron microscopy.

(d) Meigen test (13) for comparing water solubility of the 2-h annealed calcitic samples in terms of the more soluble polymorph of calcium carbonate, aragonite; the defective and strained calcitic material represents a metastable condition which thermodynamically gives an increased solubility, rising towards that of aragonite.

(e)  Thermogravimetric analysis in an atmosphere of dry nitrogen at a constant rate of temperature rise, 3.1°C/min.

(f)  Light-etching with dilute nitric acid, for dissolution of controlled amounts of calcium carbonate and determination of any sharpening of X-ray lines as in (b). Samples were dried at 150°C and weighed before treatment with sufficient nitric acid to dissolve away 1, 10, or 100 molecular layers of $CaCO_3$ from each particle. The amount of acid required was estimated approximately from the equivalent spherical diameter of the particles as calculated from the specific surface (0.32 μm for the 1000-h milled sample), assuming the particles to be of uniform density 2.71 $g/cm^3$ for calcitic $CaCO_3$.

## III.  RESULTS AND DISCUSSION

### A.  Recovery of Crystallinity by Thermal Treatment

In Fig. 1, variations in specific surface, $S$, are shown for samples of the 1000-h ground Iceland spar and precipitated calcitic calcium carbonate annealed for 6 h at a series of temperatures. There is a decrease in $S$ for the ground Iceland spar of about 1 $m^2g^{-1}$ on raising the annealing temperature from 150 to 175°C. There is evidently some preliminary surface reordering at these surprisingly low temperatures. Then $S$ remains constant for stepwise increases in the annealing temperature to 300°C, before progressively decreasing at higher temperatures to give a much less active solid ($S$ = ca. 1 $m^2g^{-1}$) after annealing for 6 h at 600°C. The sharp fall in $S$ at temperatures above 300°C is ascribed, in the first instance, to sintering; the difference in free energy between the active and sintered solids provides the driving potential for the process. By comparison, the active precipitated calcite (of similar initial specific surface) begins to sinter at a temperature of 450°C (14). Thus, the effect of grinding the Iceland spar for 1000 h is to reduce the temperature for the onset of rapid sintering by about 100°C (from 450° to 350°C). The degraded crystal structure, indicated by X-ray line broadening, and line intensity reduction, was very probably the cause of this temperature reduction; the production of point defects and dislocations would raise the free energy of the milled calcite above that of the precipitated calcite. The

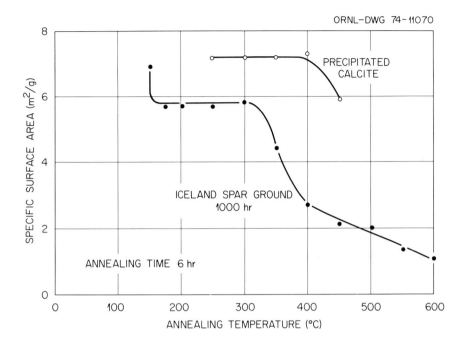

ORNL–DWG 74–11070

*Fig. 1. Specific surface as a function of annealing temperature, using nitrogen or krypton adsorbate for the precipitated and ground calcite, respectively. The mean value of S for Iceland spar ground 1000 h, after annealing and outgassing at 150°C, is 6.92 ± 0.19 m²/g for ten separate determinations.*

range of temperature (300°-350°C) over which sintering was first detected for the milled sample corresponds to 0.37-0.40 of the melting point of the calcite, 1560°K (15), i.e., not much above one-third of the m.p. (in °K) when sintering is expected to become possible by surface diffusion of ions (16,17) for a highly active surface condition.

The recovery of crystallinity in ground calcite by thermal treatment is analogous to that of lunar dust grains. Those grains damaged by solar flare and wind particles start to sinter at about 600°C. In comparison, fresh, undamaged dust grains sinter appreciably only at temperatures above 1100°C (18).

In support of the notion that surface reordering commences between 150°-175°C, the "surface solubility" also falls sharply (Fig. 2) in this temperature region. After annealing at 400°C,

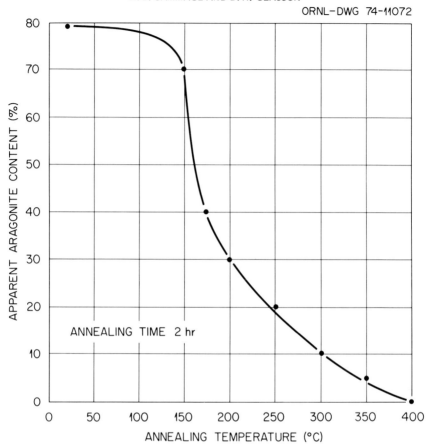

ORNL-DWG 74-11072

*Fig. 2. The solubility of surface material in 1000-h milled Iceland spar, measured by the Meigen color test, as a function of annealing temperature.*

the milled Iceland spar shows the normal Meigen test reaction for calcite, indicating a normalization of the surface condition. The lattice strain, a bulk rather than a surface condition, requires higher temperatures for reduction and elimination as shown in Fig. 3. At 400°C, for example, about one quarter of the apparent lattice strain still remains. At higher temperatures, 400-600°C, the specific surface and apparent lattice strain decrease less rapidly, with the material being rendered nearly strain-free at 600°C.

The overall recovery process is accompanied by increased crystalline outline in the microstructure, as illustrated in the electron micrographs of Figs. 4 and 5. At these magnifications, however, changes in the external appearance of the grains are not detectable until the annealing temperature exceeds 350°C. Micrographs of the 1000-h Iceland spar prior to annealing appear elsewhere (5).

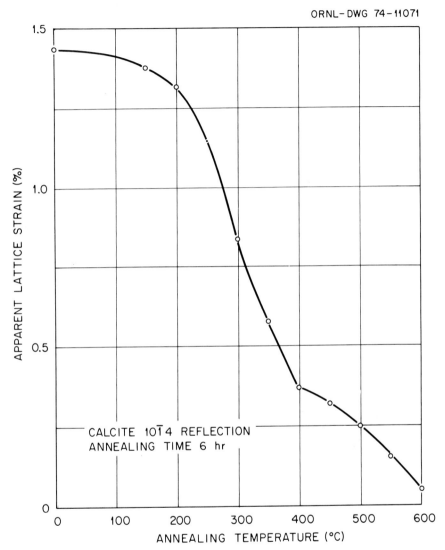

ORNL-DWG 74-11071

*Fig. 3. Apparent lattice strain of 1000-h milled Iceland spar from X-ray line broadening measurements as a function of annealing temperature.*

Earlier researches on vacuum thermogravimetric analysis of calcareous materials (19) have shown the need to use temperatures of at least 200°C in vacuo and 250°C in air for the isothermal outgassing of precipitated calcite samples. The difficulties of outgassing the precipitated and ground calcites used in this study are revealed in the thermogravimetric curves of Fig. 6.

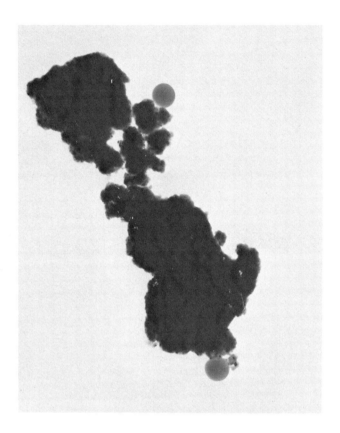

*Fig. 4. Electron micrograph of 1000-h milled Iceland spar annealed at 250°C for 10 h; latex beads 0.26 μm in diameter.*

The manner in which water escapes from these samples is also informative of the recovery processes which are occurring.

The ground and precipitated calcites contained 2 and > 3% by weight of water, respectively. Most of the water in the ground Iceland spar had been adsorbed and then buried along fault lines within the grains during the plastic flow which took place throughout all but the first few hours of the milling (2,5). The water in the precipitated calcite had been occluded during crystal growth in aqueous solution. As might be anticipated, escape of these buried waters was easier from the ground Iceland spar because of numerous fault lines linking the bulk to the surface. Because of the reluctance of the precipitated calcite to yield

*Fig. 5. Electron micrograph of 1000-h milled Iceland spar annealed at 600°C for 6 h, same magnification as in Fig. 4. Grain growth has been extensive.*

all of its occluded water, water was still being lost at about 550°C, the temperature at which carbonate decomposition caused a measurable weight loss under these experimental conditions. In contrast, the weight of the ground Iceland spar was invariant from 450° to 550°C.

Each curve of Fig. 6 contains three peaks. The low temperature peak is associated with loosely held water not buried in the interior of the grains. Between 120-200°C, water escapes quite rapidly, especially from the ground Iceland spar. At these temperatures, healing occurs rapidly in the surface region (Fig. 2). The third peak at 250-300°C coincides with the rapid elimination of strained material (Fig. 3). Venting of the water seems to be linked closely with the different healing processes taking place in different temperature ranges.

445

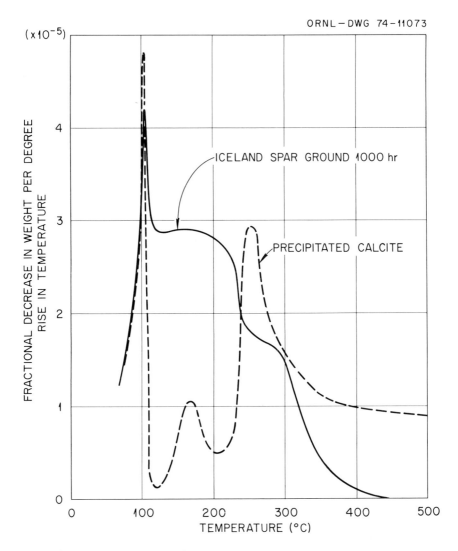

ORNL–DWG 74–11073

*Fig. 6. Thermogravimetric analysis in a stream of dry nitrogen at a constant rate of temperature rise of 3.1°/min.*

Access to the surface is more restricted in the precipitated calcite because of its greater degree of perfection. At 250°C, this leads to nearly explosive expulsion of the remaining occluded water.

The healing processes occur in the following stages and with considerable overlapping. Reordering in the surface occurs first and is very marked between 150°-200°C. Relaxation of bulk lattice strains occurs primarily between 200° and 400°C.

Sintering and loss of specific surface via the diffusion of surface ions is first noted at 350°C. Healing of the surface is completed at 400°C whereas 600°C is required to give a nearly
strain-free product. These final stages in the recovery of
crystallinity are promoted by crystal lattice diffusion of ions
and dislocations which become prominent above the Tammann temperature (20), which for calcite is about 780°K (half of the melting point in °K). These growth movements permit the grains to
assume the distinctly crystalline appearance shown in Fig. 5.

By analogy with the general behavior of cold-worked fabrics,
the ground Iceland spar undergoes polygonization below the
Tammann temperature, i.e., the single strained grains tend to
subdivide into several less strained subgrains having different
orientations. This involves the migration of dislocations to
form planar arrays at subgrain boundaries. Above the Tammann
temperature, new nuclei develop in annealing recrystallization
to produce unstrained crystals. The release of all stored
strain energy completes the recovery from the initial strain
hardened condition. With yet further prolonged heating, certain
grains grow at the expense of other grains, and there is a
general coarsening of the fabric of the material. Thus, in a
strain-hardened solid of high specific surface, such as the
ground Iceland spar, loss of specific surface by sintering and
recovery from lattice defects can be expected to occur side by
side. In several respects, they arise from the same causes,
primarily the combined effects of ionic diffusion and movement
of dislocations.

## B. Recovery of Crystallinity by Light-Etching

The recovery of crystallinity is promoted also by light-
etching for a few seconds with an aqueous mineral acid, such
as nitric acid. Rapid removal of calcium carbonate, in amounts
equivalent to 1 or 10 outer monolayers from each grain, reduced
the apparent lattice strain by 17 and 22%, respectively, as
measured from the X-ray line broadening. Electron micrographs
of the 1000-h ground Iceland spar treated with dilute nitric
acid, Figs. 7, 8, and 9, reveal that the light etching was
accompanied by nearly spontaneous and extensive recrystallization.
The products of recrystallization were whiskers (with the
morphology of nail-head spar crystals of calcite) of the order
0.1 μm in thickness, together with rhombs of calcite.

There are two prerequisites for this rapid and extensive
recrystallization. First, the calcite must be in a structurally-
degraded and lattice-distorted condition. Only in the Iceland
spar ground in excess of 500 hours was this condition met (2,5).

447

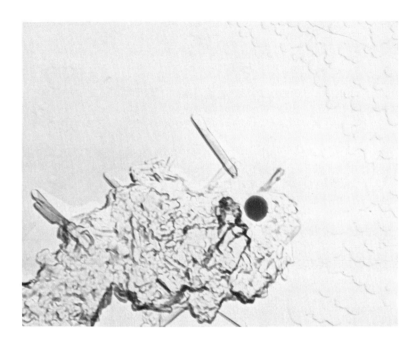

*Fig. 7. Grain of 1000-hr milled Iceland spar, after treatment with dilute nitric acid, showing predominantly whiskers of nailhead spar calcite. The latex beads are 0.264 μm in diameter.*

This type of recrystallization could not be induced in Iceland spar samples ground for lesser times or in any samples of precipitated calcium carbonate. Furthermore, after the 1000-h Iceland spar had been in storage for 12 years, it had lost the capacity for etching-induced recrystallization. Presumably, relaxation of the severely-strained condition had taken place during this time period. These findings are in accord with those of Keith and Gilman (21) who etched moderately deformed calcite crystals with various acids; only dislocation etch pits were produced. There was no whisker growth or any other type of recrystallization.

The second prerequisite is a _rapid_ attack of the surface. It appears that there must be an extensive and rapid perturbation of the surface condition to permit formation of the sites necessary for whiskers to grow from. In water, rather than dilute nitric acid, when the dissolution is much less drastic, whisker growth is rather slow. Exposure to water vapor at saturation pressure and at a temperature of 50°C for 50 days was necessary to produce the same amount of whisker growth as that promoted by a few seconds in dilute nitric acid.

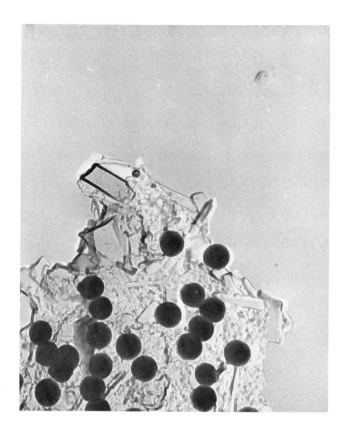

*Fig. 8. Grain of 1000-h milled Iceland spar, after dilute nitric acid treatment showing the mixed recrystalli-zation products of nail-head spar and calcite rhombs. Latex beads 0.26 μm in diameter.*

Amounts of acid sufficient to remove 1, 10, or 100 layers of $CaCO_3$ produce nearly the same amount of recrystallized material. It seems to be the rapid attack of the surface which triggers formation of the maximum number of growth sites rather than the total amount of $CaCO_3$ removed.

One can speculate that the whisker growth is taking place by a modified dislocation growth mechanism (22,23). Lattice strain is caused by pile-ups of dislocations against barriers, and evidently the surface acts as a particularly effective barrier due, perhaps, to the effects of space charge repulsion. Rapid dissolution of $CaCO_3$ in the surface would release the disloca-tion pile-ups enabling steps to form on the surface about which the whiskers can grow.

449

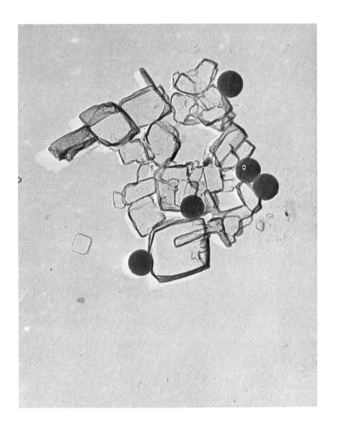

*Fig. 9.  Cluster of recrystallization products, mainly calcite rhombs, in 1000-h milled Iceland spar treated with nitric acid.  Latex beads 0.26 µm in diameter.*

IV.  SUMMARY

Iceland spar structurally degraded by 1000 hours of grinding exhibits recovery processes starting at temperatures as low as 150°-175°C.  Surface reordering starts first and is detected as a rapidly falling solubility (Meigen test) and a 15% decrease in apparent specific surface determined by krypton sorption.  Surface reordering is completed at 400°C.  Return of crystallinity to the bulk, by the measures of growth in particle size and reduction in lattice strain, requires higher temperatures. A temperature of 600°C is necessary to produce a strain-free product.

Perturbation of the surface by an etchant triggers sive recrystallization and whisker growth in 1000-h Iceland spar. Storage of the reactive material for 12 years results in loss of the ultrareactivity presumably through relaxation of internal strains.


V. REFERENCES

1. Gammage, R. B., and Glasson, D. R., Chem. Ind., London, 1466 (1963).

2. Gammage, R. B., Ph.D., Thesis, Exeter University, England (1964).

3. Lewis, D., and Northwood, D. O., Amer. Min. 53, 2089 (1968).

4. Gammage, R. B., and Gregg, S. J., J. Colloid Interface Sci., 38, 118 (1972).

5. Gammage, R. B., Holmes, H. F., Fuller, E. L., Jr., and Glasson, D. R., J. Colloid Interface Sci., 47, 350 (1974).

6. Gammage, R. B., and Glasson, D. R., J. Colloid Interface Sci., (1976) in press.

7. Lewis, D., and Northwood, D. O., Canadian Min. 10, 216 (1970).

8. Aylemore, D. W., and Jepson, W. B., J. Sci. Instr., 38 156 (1961).

9. Gammage, R. B., and Holmes, H. F., in press, "Recent Advances in Colloid and Surface Science", Academic Press.

10. Brunauer, S., Emmett, P. H., and Teller, E., J. Amer. Chem. Soc., 60, 309 (1938).

11. Adams, L. H., and Rowe, F. A., Amer. Min., 39, 215 (1954).

12. Jones, F. W., Proc. Roy. Soc., (a) 166, 16 (1938).

13. Meigen, H., Ber. Naturforsch. Ges., 15, 38 (1905).

14. Glasson, D. R., J. Appl. Chem., London, 11, 28 (1961).

15. Boeke, H. E., Neues Jahrb. Min., 91 (1), 1 (1912).

16. Hüttig, G. F., Kolloidzschr., 98, 6 203 (1942).

17. Gregg, S. J., J. Chem. Soc., 3940 (1953).

18. Maurette, M. and Price, P. B., Science, 187, 121 (1975).

19. Glasson, D. R., "Analysis of Calcareous Materials", Soc. Chem. Ind., Monogr., No. 18, 401 (1964) (London: The Society).

20. Tammann, G., Z. Anorg. Chem., 176, 46 (1928).

21. Keith, R. E., and Gilman, J. J., Acta Met., 8, 1 (1960).

22. Eshelby, J. D., Phys. Rev., 91, 755 (1953).

23. van Bueren, H. G., "Imperfections in Crystals", North Holland (1960).

# THE SURFACE STATE OF THERMOTROPIC LIQUID CRYSTALS

BUN-ICHI TAMAMUSHI
*NEZU CHEMICAL INSTITUTE, MUSASHI UNIVERSITY*

## I. INTRODUCTION

In my early studies on the two-dimentional equation of state for the liquid surface in 1934(1), it was noticed that some liquid crystalline substances have anomalies in the surface tension vs. temperature relationship, showing more or less a sharp maximum at the anisotropic-isotropic transition, and this anomaly was discussed from the viewpoint of the molecular interaction forces among oriented molecules at the surface. The same problem was recently treated again on the basis of some later obtained experimental data(2).

The discussions in these papers were made on the experimental data on p-azoxyanisole, p-azoxyphenetole, etc. obtained by Jeager(3), on the same substances obtained by Ferguson and Kennedy(4), on cholesteryl myristate by Churchill and Bailey(5) and on ammonium alkanoates and p-methoxybenzylidene-p'-butylaniline(MBBA) obtained by myself and collaborators(5), all of which exhibited such anomalies as above mentioned, independent of the nature of liquid crystals whether they are nematic, cholesteric or smectic. On the other hand, however, there have been known the experimental results of Naggiar(6) and Schwartz and Moseley(7) on similar nematic liquid crystals to those in Jeager or Ferguson and Kennedy, who could not find such an abrupt change in the surface tension vs. temperature relation at the transition as found by the former investigators.

The measuring method of the surface tension applied by the former investigators was either bubble -pressure, meniscus levelling -pressure, capillary -rise, or hanging plate method, all of which concern the glass surface, while the method used by the latter investigators was either drop-curveture, or wire-ring method. It was therefore considered that the anomalies found in the former cases have been due to the molecular orientation effect of the glass surface on the surface tension of liquid crystals(8). However, the more recent data of Langevin(9) obtained by measuring the light scattering from the free surface of MBBA showed a distinct discontinuity of the surface tension at the transition temperature. Therefore, the appearance of the maximum in the surface tension vs. temperature curve might have been due to a metastable or non-equilibrium state of the substance. At any rate, in order to decide whether there exists or not such an abrupt change or discontinuity of the surface tension at the transition temperature, it is desirable to make more careful measurements with the same well-purified substance by applying various methods, and such a trial is now being carried on in our laboratory.

In this presentation, however, I would like to mention the most common feature appearing in the surface tension vs. temperature relation which has been found in almost all the experimental data until now obtained.

## II. EXPERIMENTAL DATA AND DISCUSSION

The above mentioned common feature is clearly demonstrated if we apply to the experimental data the equation of Eötvös (10) which is expressed as follows:

$$\sigma V^{2/3} = K(T_c - T) \qquad [1]$$

where $\sigma$ means the surface tension, $V$ the molar volume, so that $\sigma V^{2/3}$ means the molar surface energy. $T_c$ is the critical temperature, $T$ the temperature (Kelvin), and $K$ a constant (Eötvös constant). For so-called "normal liquids" the constant $K$ has a value of nearly equal to 2, but the value can be greater or smaller than this value according to the nature of liquids.

When we apply this equation for liquid crystalline substances, we get such a figure as shown in Fig. 1 for p-azoxyanisole.

In this figure the experimental data of Jeager as well as Schwartz-Moseley are plotted for comparison, taking the molar surface energy in ordinate and the temperature in abscissa, transition point $T^*$ being set at zero. We can notice here the difference between Jeager's and Schwartz-Moseley's results: namely, there is a sharp rise of the molar surface energy at the transition in Jeager's curve, while only a slight rise in Schwartz-Moseley's curve. In Jeager's data, $K$-value is 8.37 for the anisotropic phase and 2.70 for the isotropic one, the difference between these two values being pretty large, whereas in Schwartz-Moseley's data the difference of the

454

corresponding values is not so remarkable, although the tendency that $K$-value for the anisotropic phase is greater than
that for the isotropic one is common for the both cases. A
similar result is illustrated in Fig. 2 for p-azoxyphenetole,
where $K$-value for the anisotropic phase in Jeager is greater
than that in Schwartz-Moseley, but the tendency that $K$-value

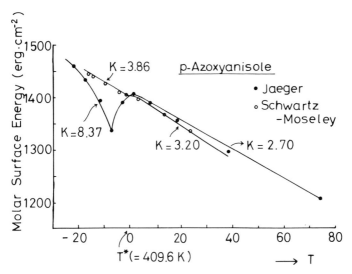

Fig. 1. The molar surface energy vs. temperature
relation for p-azoxyanisole.

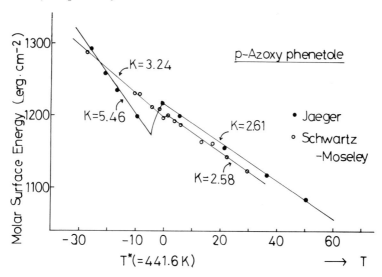

Fig. 2. The molar surface energy vs. temperature
relation for p-azoxyphenetole.

for the anisotropic phase is greater than that for the isotropic phase is also evident for both cases.

Fig. 3 shows the result for ammonium myristate measured by
the capillary rise method, in which the maximum is pretty
sharp at the transition and $K$-value for the anisotropic phase
is greater than that for the isotropic one.

Fig. 4. illustrates the result for p-methoxybenzylidene-
p'-butylaniline obtained by Wilhelmy's hanging-plate method,
where a special care was taken for attaining the equilibrium
and also for wetting the glass plate during the measurement.
This figure shows only a small maximum or plateau at the anisotropic-isotropic transition, but $K$-value for the anisotropic phase is also greater than that for the isotropic one,
although both values are smaller than those previously expected.

Now, applying the Gibbs-Helmholtz equation to the liquid
surface, for unit area, we have

$$\sigma = h - T\frac{d\sigma}{dT} \qquad [2]$$

where $\sigma$ is the surface free energy, $h$ the surface enthalpy,
and $-d\sigma/dT$ means the surface entropy. For molar surface area,
we obtain

$$\sigma V^{2/3} = H - T\frac{d(\sigma V^{2/3})}{dT} \qquad [3]$$

in which $\sigma V^{2/3}$ is the molar surface free energy, $H$ the molar
surface enthalpy, and $-d(\sigma V^{2/3})/dT$ the molar surface entropy.

These thermodynamic quantities for the surface are the
excess quantities at the surface of the liquid in comparison
with the corresponding quantities in the bulk of the underlying liquid.

Eötvös constant $K$ is the temperature coefficient of the
molar surface free energy, and therefore, this is equal to
the molar surface entropy $S$, as expressed by

$$K = -\frac{d(\sigma V^{2/3})}{dT} = S \qquad [4]$$

If we define the molar surface area after Defay-Prigogine-
Bellemans(11) as follows:

$$a = N\left(\frac{V}{N}\right)^{2/3} = N^{1/3}V^{2/3} \qquad [5]$$

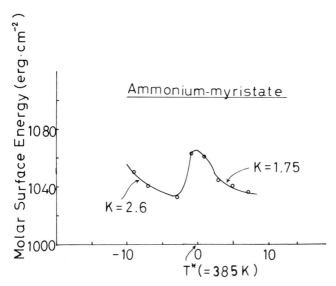

*Fig. 3. The molar surface energy vs. temperature relation for ammonium myristate.*

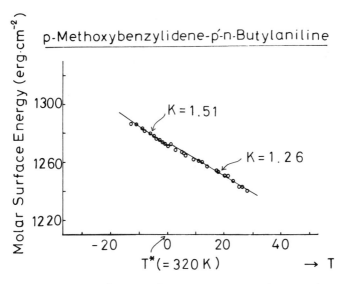

*Fig. 4. The molar surface energy vs. temperature relation for p-methoxybenzylidene-p'-butylaniline.*

then the molar surface entropy will be

$$S^m = N^{1/3} K \qquad\qquad [6]$$

where $N$ is Avogadro's number. For the $K$-value = 2.1 erg cm$^{-2}$ deg$^{-1}$mol$^{-1}$, $S^m$ = 4.2 cal cm$^{-2}$deg$^{-1}$mol$^{-1}$. Therefore, it is easy to calculate $S^m$-values from $K$-values by doubling the latter values and changing the energy unit from erg to calorie.

In table 1 the results for p-azoxyanisole from Jeager's as well as Schwartz-Moseley's data are shown in one table for comparison with each other. In Jeager's data, in the temperature range from 115 to 211°C, the surface tension varies from 40.1 to 31.4 erg cm$^{-2}$ and molar surface free energy varies from 1463 to 1209 erg cm$^{-2}$mol$^{-1}$, having a maximum at the transition, while the molar surface enthalpy which is practically constant for each anisotropic and isotropic phase, drops from 4714 to 2518 erg cm$^{-2}$mol$^{-1}$, and the molar surface entropy also drops from 8.37 to 2.70 erg cm$^{-2}$deg$^{-1}$mol$^{-1}$ at the transition. The variations of the corresponding thermodynamic quantities in Schwartz-Moseley's data are also evident, but the change of the molar surface enthalpy as well as the molar surface entropy are not so great as in the case of Jeager.

TABLE 1
*Surface Thermodynamic Properties of p-Azoxyanisole(M = 258.3, T\* = 136°C)*

| Investigator | Jeager | Schwartz-Moseley |
|---|---|---|
| Method | Bubble-pressure | Platinum wire-ring |
| Temperature range(°C) | 115 - 211 | 120 - 160 |
| Surface tension (erg cm$^{-2}$) | 40.1 - 31.4 | 39.1 - 35.4 |
| Molar surface free energy(erg cm$^{-2}$mol$^{-1}$) | 1463 - 1209 | 1440 - 1328 |
| Molar surface enthalpy (erg cm$^{-2}$mol$^{-1}$) | 4715(anisotropic) 2518(isotropic) | 2972(anisotropic) 2715(isotropic) |
| Molar surface entropy (erg cm$^{-2}$deg$^{-1}$mol$^{-1}$) | 8.37(anisotropic) 2.70(isotropic) | 3.86(anisotropic) 3.20(isotropic) |

Similar surface thermodynamic properties obtained for p-azoxyphenetole from Jeager's and Schwartz-Moseley's data are

listed in table 2, where we can notice that the molar surface enthalpy as well as the molar surface entropy are
greater in the anisotropic phase than in the isotropic phase,
although the numerical values are somewhat different for both
investigators.

TABLE 2
*Surface Thermodynamic Properties of p-Azoxyphenetole(M=286.3,
T\* = 169°C)*

| Investigator | Jeager | Schwartz-Moseley |
|---|---|---|
| Method | Bubble-pressure | Platinum wire-ring |
| Temperature range(°C) | 142 - 219 | 141 - 198 |
| Surface tension | | |
| (erg cm$^{-2}$) | 31.6 - 25.2 | 31.5 - 26.4 |
| Molar surface free | | |
| energy(erg cm$^{-2}$mol$^{-1}$) | 1293 - 1084 | 1288 - 1123 |
| Molar surface enthalpy | 3557(anisotropic) | 2630(anisotropic) |
| (erg cm$^{-2}$mol$^{-1}$) | 2352(isotropic) | 2338(isotropic) |
| Molar surface entropy | 5.45(anisotropic) | 3.24(anisotropic) |
| (erg cm$^{-2}$deg$^{-1}$mol$^{-1}$) | 2.61(isotropic) | 2.58(isotropic) |

In table 3 and table 4 the results for ammonium myristate
and p-methoxybenzylidene-p'-butylaniline are listed, respectively. In these tables we can also notice that remarkable
changes in the molar surface enthalpy as well as the molar
surface entropy take place at the anisotropic-isotropic transition.

TABLE 3
*Surface Thermodynamic Properties of Ammonium Myristate (M =
245.3, T\* = 112°C)*

| Investigator | Tamamushi et el. |
|---|---|
| Method | Capillary-rise |
| Temperature range(°C) | 102 - 118 |
| Surface tension(erg cm$^{-2}$) | 20.5 - 20.0 |
| Molar surface free energy | 1050 - 1038 |
| (erg cm$^{-2}$mol$^{-1}$) | |
| Molar surface enthalpy | 2024(anisotropic) |
| (erg cm$^{-2}$mol$^{-1}$) | 1725(isotropic) |
| Molar surface entropy | 2.60(anisotropic) |
| (erg cm$^{-2}$deg$^{-1}$mol$^{-1}$) | 1.75(isotropic) |

TABLE 4
*Surface Thermodynamic Properties of p-Methoxybenzylidene-p'-Butylaniline (M = 267.4, T\* = 47°C)*

---

| Investigator | Tamamushi et al. |
|---|---|
| Method | Hanging plate |
| Temperature range (°C) | 20 - 75 |
| Surface tension (erg cm$^{-2}$) | 32.3 - 29.9 |
| Molar surface free energy (erg cm$^{-2}$mol$^{-1}$) | 1298 - 1241 |
| Molar surface enthalpy (erg cm$^{-2}$mol$^{-1}$) | 1749 (anisotropic) 1678 (isotropic) |
| Molar surface entropy (erg cm$^{-2}$deg$^{-1}$mol$^{-1}$) | 1.51 (anisotropic) 1.26 (isotropic) |

---

In all these cases, therefore, it is to be remarked that $K$-values for the anisotropic mesophases are always greater than those for the corresponding isotropic liquid phases. It means that the excess entropy in the surface in comparison with that in the bulk liquid is greater for the anisotropic mesophase than that for the isotropic liquid phase. Moreover, the positive values of $K$ or $S$ indicate that molecules in the surface are more randomly oriented than in the underlying bulk phase, for both liquid crystals and isotropic liquids. And the fact, that $K$-values for the anisotropic mesophases are greater than those for the corresponding isotropic liquid phases, indicates that the degree of the molecular disorder in the surface phase in compariosn with that in the bulk phase is more pronounced for liquid crystals than for isotropic liquids.

In conclusion, we can say that there is a certain discontinuity in the surface tension or the molar surface energy vs. temperature relation for liquid crystals at the anisotropic-isotropic transition, as far as we find the fact, that $K$-value for the anisotropic mesophase is greater than that for the corresponding isotropic liquid, which fact is very common for the experimental data so far obtained by different investigators and by different measuring methods. The more or less anomalies found in the molar surface energy vs. temperature relationships are therefore considered to be connected with the decrease of the molar surface enthalpy or ontropy at the anisotropic-isotropic transition.

Acknowledgment

The author thanks to Mr. Y. Kodaira and other colleagues in our laboratory for their assistance in experimental and other technical works.

REFERENCES

1. Tamamushi, B., *Bull. Chem. Soc. Japan*, 9, 473 (1934).
2. Tamamushi, B., "Chemie, physikalische Chemie und Anwendungstechnik der grenzflächenaktiven Stoffe" Ber. VI. Int. Kongr. grenzflächenaktive Stoffe, Bd II, 1. 431. Zurich, 1972; Carl Hanser Verlag, München, 1973.
3. Jeager, F.M., *Z. anorg. allg. Chem.*, 101, 1 (1917).
4. Ferguson, A. and Kennedy, S.J., *Phil. Mag.*, 26, 41 (1938).
5. Churchill, D. and Bailey, L.W., "Liquid Crystals", Proc. 2nd. Int. Liq. Crys. Conf. 1968, Part II, 269, Gordon and Breach Science Publishers, London-New York-Paris, 1969.
6. Naggiar, V., *Ann. Physique*, 18, 5 (1943).
7. Schwartz, W.M. and Moseley, H.W., *J. Phys. Coll. Chem.*, 51, 826 (1947).
8. Gray, G.W., "Molecular Structure and Properties of Liquid Crystals", p.121, Academic Press, London, 1962.
9. Langevin, D. and Bouchiat, M.A., *J. Phys.*, 33, C1-77 (1972).
10. Eötvös, v.R., *Wied. Ann.*, 27, 448 (1886).
11. Defay, R., Prigogine, I. and Bellemans, A., translated by Everett, D.H., "Surface Tension and Adsorption", p.154, Longmans, London, 1966.

# TRANSPORT OF LIQUIDS THROUGH CELLULOSE MEMBRANES

J. Fred Hazel
University of Pennsylvania

Uniform sections of dialysis tubing containing a measured volume of the following liquids: water, methanol, ethanol, 2-propanol and acetone were suspended in air in closed 1000 ml graduated cylinders at room temperature, $20^{\circ}C$. Volume changes were measured with the aid of the scale markings on the outside of the cylinders. The scale was calibrated for each system. Readings were taken daily. Averages of the volume changes were calculated for seven-day periods extending over 82 days, or until the liquid was transported across the membrane to the cylinder. Calculated rates of transport, in mols transported per sq cm per unit time, were found to be proportional to factors that included the vapor pressure of the liquid, and suggest that the mechanism of transport is diffusive in character.

In some experiments, sections of the dialysis tubing containing the various liquids were suspended in air in a quiet room at room temperature. The time for the volume to decrease to one-half its value, $t_{\frac{1}{2}}$, was measured and compared with a calculated value based on an equation containing vapor pressure as a parameter.

## I. INTRODUCTION

Cellulose membranes in the form of dialysis tubing is available commercially. Typically, it is employed to purify colloidal systems. It also finds application in making colloidal systems. An example is the preparation of hydrous oxides by membrane hydrolysis. Another use is based on its efficacy in promoting the exchange of organic solvents for water. This initial report deals with the behavior of a liquid when placed on one side of a cellulose membrane.

Liquids may be forced through membranes by applying a pressure difference across the membrane. If the liquid is in contact with both sides of the membrane, addition of a non-diffusable solute to the liquid, on one side only, produces the same effect as an applied pressure. Mauro (1) has compared osmotic pressure measurements with permeability data in an analysis of the nature of the fluxes in solvent transfer. Ticknor (2), making use of the extensive data of Madras,

McIntosh and Mason (3), and Sohoegl (4) have evaluated the
factors affecting the contributions of the diffusional com-
ponent-individual molecules undergoing a chaotic drift- and
the viscous component- or mass movement of molecules, to the
permeability. The type of flow was correlated with the relative sizes of the permeating molecules and the pore size of
the membrane and on the extent of the bonding between the
transient molecules and the membrane material. A readable
summary on the nature of flow through membranes is available
(5). From the estimates that have been made (1) the pore
radius would have to be about equal to the molecular radius
for the flow to be diffusional. If the pore radius is very
much larger than the radius of the molecules, viscous flow is
indicated.

## II. EXPERIMENTAL

The liquids used in the experiments are shown in Table 1.
Distilled water and Baker Analyzed Reagents were employed.
Molecular volumes (V) and cross-sectional areas (A) of these
liquids were calculated from their respective molecular
weights and densities at $20^{o}C$. The number of molecules of
each compound required to cover one sq cm of membrane-liquid
interface as judged by the cross-sections area of the molecule
also was calculated. This number, called the specific surface
number (sp n), is multiplied by $10^{-14}$ in Table 1. Also tabulated are the vapor pressures (VP) at $20^{o}C$ in atmospheres, and
$\Delta G^{o}$ of vaporation in the standard state in kcal/mol. The latter permit a comparison with the vapor pressures at $25^{o}C$.
These liquids are non-ideal or associated as confirmed by the
Trouton Rule constants calculated from latent heats of vapori-
zation at the boiling points, and are listed in the column
headed $\Delta S$. Acetone is not associated as judged by this criterion.

TABLE 1
Molecular Parameters

| Liquid | V,Cu$\mathring{A}$ | A,Sq $\mathring{A}$ | Sp n | VP | $\Delta G^{o}$ | $\Delta S$ |
|---|---|---|---|---|---|---|
| $H_2O$ | 30 | 11.7 | 8.5 | 0.0230 | 2.055 | 26.0 |
| $CH_3OH$ | 68 | 20 | 5 | 0.1184 | 1.05 | 24.9 |
| $C_2H_5OH$ | 97 | 25.3 | 3.95 | 0.0618 | 1.47 | 26.7 |
| $i-C_3H_7OH$ | 128 | 30.5 | 3.3 | 0.0421 | | 26.8 |
| $(CH_3)_2CO$ | 123 | 30 | 3.3 | 0.2263 | 0.72 | 21.9 |

The dialysis tubing which served as the membrane was obtained from the Arthur H. Thomas Company and had the specifications: inflated diameter 1.125 inches; wall thickness 0.0010 inch; average pore diameter 4.8 millimicrons based on the rate of flow of water through the film. Eighteen inch lengths were cut from a roll of tubing and washed with water at room temperature for several hours followed by washing in a mixture of water and the liquid to be used in the subsequent experiment and again in water. One end of the tubing was closed with a double knot followed by inflation of the tubing and drying in a current of air.

## A. Closed Systems

One hundred twenty-five ml of liquid at 20°C was measured into the tubing while it was being held in a suspended position in a modified 1000 ml graduated cylinder. The cylinder was closed promptly and the positions, on the graduated scale, of the liquid meniscus and of the bottom of the membrane tube containing the liquid were recorded. The scale was calibrated for each liquid. Readings of the meniscus were made at 24 hour intervals and from these the changes in the volume and in the number of mols were calculated each day. Averages were calculated for successive seven-day periods which extended over 12 periods or 82 days in most cases.

## B. Open Systems

In other experiments, the tubing containing 125 ml of liquids at 20°C were suspended in air in a quiet room. The changes in volume and in the number of moles present were evaluated by weighing the systems at lengthening intervals but at either 15 minute or 20 minute intervals initially. This made it possible to estimate the times for the volumes to decrease to one-half their original values.

## III. RESULTS AND DISCUSSION

## A. Closed Systems

A material balance was determined, Table 2, from the volume changes measured as outlined above and as shown schematically in Fig. 1. The results indicate that the mass was conserved in each case.

The average rate of transport of the liquids across one sq cm of membrane-liquid interface in a seven-day period (1) and of the corresponding average values of moles of liquid present per sq cm of membrane surface (2) multiplied by one-thousand are given in Table 3.

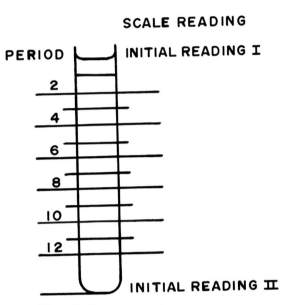

Fig. 1 Measurements

TABLE 2
Material Balance

| Liquid | Ave n/d[a] | Moles Left | Total Moles | Initial Moles |
|--------|-----------|------------|-------------|---------------|
| $H_2O$ | 5.20 | 1.61 | 6.81 | 6.93 |
| $CH_3OH$ | 2.906 | 0.172 | 3.078 | 3.08 |
| $C_2H_5OH$ | 1.771 | 0.361 | 2.132 | 2.14 |
| $i-C_3H_7OH$ | 1.278 | 0.306 | 1.584 | 1.63 |
| $(CH_3)_2CO$ | 1.613 | 0.027 | 1.64 | 1.70 |

a. The average of the number of moles diffusing through the membrane per day multiplied by the number of days over which the event occurred.

TABLE 3
Rate of Transport of Liquids Across Liquid-Cellulose Membrane Interfaces in Closed Systems.

| Period | $H_2O$ (1) | (2) | $CH_3OH$ (1) | (2) | $C_2H_5OH$ (1) | (2) | $i-C_3H_7OH$ (1) | (2) | $(CH_3)_2CO$ (1) | (2) |
|---|---|---|---|---|---|---|---|---|---|---|
| 1 | 2.82 | 39.6 | 1.69 | 17.6 | 0.85 | 12.2 | 0.54 | 9.3 | 1.52 | 9.7 |
| 2 | 2.67 | 39.7 | 1.83 | 17.7 | 0.91 | 12.2 | 0.62 | 9.3 | 2.10 | 9.7 |
| 3 | 2.66 | 39.7 | 1.90 | 17.7 | 0.89 | 12.3 | 0.53 | 9.1 | 2.03 | 9.8 |
| 4 | 2.48 | 39.7 | 1.64 | 17.7 | 0.83 | 12.2 | 0.61 | 9.0 | 2.56 | 9.8 |
| 5 | 2.17 | 39.8 | 1.80 | 17.6 | 0.85 | 12.3 | 0.54 | 9.5 | 3.54[a] | 9.3 |
| 6 | 2.08 | 39.8 | 1.50 | 17.7 | 0.85 | 12.2 | 0.55 | 9.4 | | |
| 7 | 1.81 | 39.9 | 1.86 | 17.7 | 0.98 | 12.3 | 0.76 | 9.4 | | |
| 8 | 2.06 | 40.3 | 2.21 | 17.4 | 1.05 | 12.4 | 0.81 | 9.4 | | |
| 9 | 1.79 | 40.1 | 1.85 | 17.0 | 0.73 | 12.5 | 0.57 | 9.4 | | |
| 10 | 1.94 | 39.8 | 1.90 | 16.9 | 0.80 | 12.3 | 0.52 | 9.3 | | |
| 11 | 2.20 | 40.8 | 3.11[a] | 16.7 | 1.275[a] | 12.3 | 0.94[a] | 9.3 | | |
| Mean | 2.24 | 39.9 | 1.82 | 17.5 | 0.87 | 12.3 | 0.60 | 9.30 | 2.06 | 9.7 |
| $S_{Dev.}$ | 0.09 | 0.36 | 0.19 | 0.37 | 0.09 | 0.09 | 0.10 | 0.15 | 0.42 | 0.20 |

a. These values were not included in the averages.

The mean values of mols of liquid passing across one sq cm of membrane-liquid interface per seven-day period $n_{tr}$/sq cm and the corresponding mean values of mols of liquid present per sq cm of membrane-liquid interface $n_p$/sq cm given at the bottom of Table 3 are repeated in Table 4. We note that $n_{tr}$ is directly proportional to $n_p$ when one sq cm of surface is considered $n_{tr} \alpha n_p$ or $n_{tr}$ = a constant x $n_p$. The constant varies with different liquids and is approximately equal to the vapor pressure of the liquid expressed in atmospheres. We use the constant $f_m$ with units $t^{-1}$ and write $n_{tr} = f_m n_p$ or $(n_{tr}/n_p) = f_m$. The numerical values of this constant, called the "membrane vapor pressure function", are compared with the vapor pressures of the liquids in Table 4. Formally, $f_m \alpha VP$ or $f_m = h^*VP$. The constant $h^*$ has units of $P^{-1}$ and $t^{-1}$. It is shown in Table 4 that $h^*$ equals unity when the vapor pressure of the liquid is expressed in atmospheres and time in days and the cellulose membrane has the same geometry. The data in Tables 3 and 4 support the view that liquids change phase during transport through cellulose membranes.

TABLE 4
"The Membrane Vapor Pressure Function, $f_m$" and the Vapor
Pressure of Liquids.

| | $n_{tr}$/sq cm$^a$ | $n_p$/sq cm$^a$ | $f_{m,}$ t$^{-1}$ | VP(20$^o$C) atm |
|---|---|---|---|---|
| $H_2O$ | 2.24 | 39.9 | 0.0561 | 0.0230 |
| $CH_3OH$ | 1.82 | 17.5 | 0.1040 | 0.1184 |
| $C_2H_5OH$ | 0.87 | 12.3 | 0.0707 | 0.0618 |
| $i-C_3H_7OH$ | 0.60 | 9.3 | 0.0645 | 0.0421 |
| $(CH_3)_2CO$ | 2.06 | 9.7 | 0.2124 | 0.2263 |

a. Multiplied by $10^3$.

From the number of molecules required to form one sq cm
of membrane - liquid interface given in Table 1 and the
transport rate given in Table 3, we may calculate the number
of times the interface must be formed per second to maintain a
steady state. This is called the replenishment rate. Dividing the latter by thirty gives the membrane vapor pressure
functions of the liquids as shown in Table 5.

TABLE 5
The membrane-liquid interface.
a = the specific surface number.
b = number of molecules transported per sq cm per second
c = replenishment rate
d = c/30

| Liquid | a | b | c | c/30 | $f_m$ |
|---|---|---|---|---|---|
| $H_2O$ | 8.5 x $10^{14}$ | 2.2 x $10^{15}$ | 2.58 | 0.086 | 0.0561 |
| $CH_3OH$ | 5 x $10^{14}$ | 1.8 x $10^{15}$ | 3.6 | 0.120 | 0.1040 |
| $C_2H_5OH$ | 3.95 x $10^{14}$ | 8.7 x $10^{14}$ | 2.2 | 0.073 | 0.0707 |
| $i-C_3H_7OH$ | 3.3 x $10^{14}$ | 6 x $10^{14}$ | 1.8 | 0.060 | 0.0645 |
| $(CH_3)_2CO$ | 3.3 x $10^{14}$ | 2 x $10^{15}$ | 6.1 | 0.203 | 0.2124 |

A preliminary study was made of the transport of the same liquids through the walls of 5/8 inch dialysis tubing having a rated thickness of 0.0008 inch. The tubing was washed and dried and then was looped to form a double column. One hundred ml volumes of the liquids were employed instead of 125 ml as with the larger tubing. Because of the geometry of the tubing, 100 ml of liquid formed 252 sq cm of membrane-liquid interface. In comparison with the larger tubing, 125 ml of liquid covered 175 sq cm of membrane surface. The effects of tube radius and wall thickness are shown in the following tables. The percentage of the liquid present that was transported per day was about 25% greater in the case of the thinner membrane for all of the liquids, Table 6.

TABLE 6
Effect of Membrane Thickness on Permeability by Liquids in a Closed System.

| | % tr/day[a] | | Ratio 5/8" | Thickness[b] | % |
| | 1 1/8" | 5/8" | 1 1/8" | Factor | Deviation |
|---|---|---|---|---|---|
| $H_2O$ | 1.67 | 2.19 | 1.31 | 1.25 | 4.8 |
| $CH_3OH$ | 3.31 | 4.36 | 1.32 | 1.25 | 5.6 |
| $C_2H_5OH$ | 2.11 | 2.89 | 1.37 | 1.25 | 9.6 |
| 2-Prop. | 2.03 | 2.52 | 1.24 | 1.25 | 0.8 |
| Acetone | 6.79 | 9.36 | 1.23 | 1.25 | 1.6 |

a. Average values for 12, 7-day periods. The values for acetone are for 5, 7-day and 3, 7-day periods.
b. Thickness of 1 1/8" dialysis tubing (0.001") divided by thickness of 5/8" dialysis tubing (0.0008").

When the comparison is based on transport per sq cm of surface, the results are reversed and the larger tube with more bulk liquid present per sq cm of liquid-membrane interface is favored as shown in Table 7.

TABLE 7

Effect of Tube Diameter on the Membrane Vapor Pressure Function.

| | $n_{tr}$/sq cm[a] | | $n_p$/sq cm[a] | | $f_m$ | |
|---|---|---|---|---|---|---|
| | 1.125 in | 0.625 in | 1.125 in | 0.625 in | 1.125 | 0.625 |
| $H_2O$ | 2.24 | 1.86 | 39.9 | 22.1 | 0.0561 | 0.0839 |
| $CH_3OH$ | 1.82 | 1.30 | 17.5 | 9.9 | 0.1040 | 0.1319 |
| $C_2H_5OH$ | 0.87 | 0.60 | 12.3 | 6.8 | 0.0707 | 0.0891 |
| $i-C_3H_7OH$ | 0.60 | 0.40 | 9.3 | 5.2 | 0.0645 | 0.0762 |
| $(CH_3)_2CO$ | 2.06 | 1.81[b] | 9.7 | 5.4 | 0.2124 | 0.3343[b] |

a. Multiply by 1000.
b. Based on three periods.

B. Open Systems

When the same liquids were placed inside 1 1/8 inch dialysis tubing, treated as described above, and suspended in air in a still room they were transported rapidly but at different rates through the walls of the tubing. The interface on the outside of the membrane in the case of open systems was membrane-air (outside) compared to the interface membrane-vapor (outside) for closed systems. Inside the tubes the interfaces at the transport sites were liquid-membrane (inside) and vapor-membrane (inside) for both closed systems and open systems. As shown below, our experiments indicate that transport occurs primarily at the liquid-membrane interface.

The rates of transport of the liquids are shown in Fig. 2. The half-lives or times for one-half a given number of mols of liquid to pass through the membrane are indicated by dashes on the curves.

We have developed an empirical equation designed to calculate the half-life, $t_{\frac{1}{2}}$ for the transport of liquids through the walls of dialysis tubing suspended in the air. Included in the relation are the vapor pressure of the liquid at $20°C$ expressed in atmospheres P, the molar volume of the liquid at $20°C$ expressed in liters V, a term t* which has units of seconds $mol^{-1}$ or of reciprocal rate, and of other constants. The complete equation has the general form of the equation of state for a perfect gas: PVt* = nRT. Comparisons of the observed half-life values of the liquids based on the data in

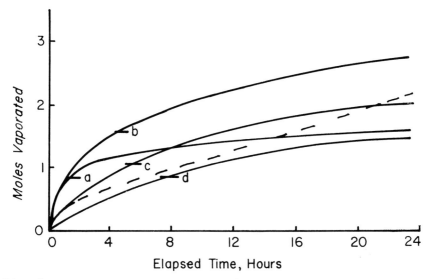

Fig. 2. Evaporation Rates and Half-lives.
Half-life (hrs): a≡acetone(1.5); b≡methanol(4.3);
c≡ethyl alcohol(5.5); d≡2-propanol(7.6); dashes≡water(45)

Fig. 2 and the calculated values for the same liquids based on the above equation are given in Table 8.

TABLE 8
Half-Life Values of Liquids in Open Systems

| Liquid | $t^*$ Hrs/Mol | Total Mols | $t_{\frac{1}{2}}$, calc[a] | $t_{\frac{1}{2}}$ obs. |
|--------|------|------|------|------|
| $H_2O$ | 16.1 | 6.93 | 58.5 | 45 |
| $CH_3OH$ | 1.39 | 3.08 | 2.14 | 4.3 |
| $C_2H_5OH$ | 1.85 | 2.14 | 2.0 | 5.5 |
| $i-C_3H_7OH$ | 2.07 | 1.63 | 1.7 | 7.6 |
| $(CH_3)_2CO$ | 0.40 | 1.70 | 0.34 | 1.5 |

a. Using $PVt^* = nRT$      $t^* = 3600$ sec $mol^{-1}$
  $n = 1$, $t = 293$, $r = 0.082\ell$ atm $mol^{-1}$ degree$^{-1}$

## C. Vapor Contact Systems

In tests designed to give information on the path taken by the liquid through the membrane, or more specifically whether the vapor in equilibrium with the liquid was transported, 8-inch pyrex test-tubes of the same diameter as the 1 1/8 inch dialysis tubing were encased in tubing that had been knotted at one end. The membranes were processed as usual except that they were not dried. Excess water was minimized by pressing. Since the tubing was damp and pliable it could be teased over the glass test-tubes. Further drying was not attempted. It is estimated that about one ml of water was present in each of the membranes.

Sixty ml of each of the liquids was placed in an encased glass test-tube. The dialysis tubing was knotted at the upper end and the system suspended in air in a closed cylinder. A 6-cm length of tubing extend above the glass tube and was exposed to the vapor of the liquid. Daily readings were taken of the liquid meniscus. The results are recorded in Table 9 where a comparison is made with the systems in which the liquids were also in contact with the membrane as described above, Tables 2-8.

The times chosen for the comparison of the volume changes were those shown in column two under the heading Equilibrium Time (1) or the time after which there were no changes in volume. At these stated times the transport from the vapor phases were about 10% of the total transport shown in the last column. The absence of a volume change in the case of water might be attributed to the presence of some water on the outside of the membrane initially due to procedural requirements.

TABLE 9

Comparison of Transport Sites in Closed Systems

| Liquid | Equilibrium Time, Days | Duration of Trial Days | $\Delta V$ Vapor Site | $\Delta V$ Liquid Site |
|---|---|---|---|---|
| $H_2O$ | - | 10 | - | - |
| $CH_3OH$ | 24 | 33 | 8.5 | 66.0 |
| $C_2H_5OH$ | 9 | 19 | ?.8 | 22.6 |
| $i-C_3H_7OH$ | 9 | 19 | 1.0 | 20.4 |
| $(CH_3)_2CO$ | 47[a] | 240 | 15.5 | 123.8 |

a. V-values in the bottom line are for 47 days.

(1) Acetone required 210 days to reach equilibrium.

D. Summary

The present studies involved placing one liquid at a time on one side of a cellulose membrane and measuring the volume changes over extended periods under a variety of conditions. The rates of transport of different liquids bore a correspondence to their vapor pressures suggesting a random molecular drift through the membrane. Whether the molecules were present in the membrane as a gas or as a solute with the membrane acting as a solvent is a moot question. According to theory the kinetic units in gases and in solutions have similar behaviors. It is the function of the membrane to promote this state of the system during transport. Bonding between the transient molecules and the membrane affected the transport less than the vapor pressures of the liquids. It is believed that a more favorable condition for observing mole-cule-membrane interactions is when there are two different molecular species competing for the membrane as when two different liquids are present on opposite sides of the membrane.

The relationship developed to fit the rate data for the transport of liquids through cellulose membranes in closed systems $n_{tr} = f_m n_{\ell_1}$ where $n_{tr}$ is the time differential in units of mols sq cm$^{-1}$ of liquid transported and $f_m$ is the specific transport rate constant in units of $t^{-1}$, bear a striking resemblance to the rate law for a first order chemical process. The data for open systems in which half-life values were calculated, also, seem to fit a first-order process.

Figures 1 and 2 were drawn by Miss Laurie Mitchell as was the sketch of the molecular models shown in a slide at the meeting in San Juan 1976.

IV REFERENCES
1. A. Mauro, Science 126, 252 (1957).
2. L. B. Ticknor, J. Phys. Chem. 62, 1483 (1958).
3. S. Madras, R. L. McIntosh and S. G. Mason, Can. J. Res. B27, 764 (1949).
4. R. Schoegl, Protoplasma 63, 288 (1967).
5. N. Lakshimnarayaniaha in "Transport Phenomena in Membranes, Academic Press, New York, 1969, p. 319.

# MICROEMULSIONS AND MICELLAR SOLUTIONS

M. Rosoff and A. Giniger

College of Pharmaceutical Sciences, Columbia University, N.Y.C.

## I. INTRODUCTION

Current interpretations of microemulsions either emphasize their emulsion like properties and attempt to build
molecular models of the interface (1-5) or consider them as
solubilized micellar solutions which can be systematically
studied by phase diagrams (6-8). Two recent reviews summarize
these respective schools of thought (9), (10). Such divergence of complementary viewpoints is not uncommon and their
differences are not so great except over the question of the
kinetic versus thermodynamic stability of these disperse
systems. A proposal (11) that the term "micellar" be reserved
for a stable system of swollen micelles and "microemulsion" be
used for those transparent dispersions which are not thermodynamically stable, exemplifies some of the confusion that
exists regarding this point. Recent investigations show that
under certain conditions spontaneous dispersion to equilibrium emulsion systems is theoretically possible at positive
interfacial tensions (12), (13), (14). The concept of a
thermodynamically stable two-phase disperse system is of
fundamental importance in colloid science and should be taken
into account in categorizing microemulsions.

The association of microemulsions with regions bordering
on phase separation correlates with the possibility of the
formation of thermodynamically stable disperse systems close
to the critical region (15), (16). Nucleation in the metastable regions can take an increasingly long time as the
critical point is approached and such mixtures may also take
on the aspect of thermodynamically stable systems. Long-lived
metastable microemulsions gave been produced through the
influence of temperature (17), the addition of cosurfactant to
a ternary system (18), and the addition of liquid crystalline
phase to $L_1$ or $L_2$ phases (8). There has, however, been
insufficient experimental work to confirm the existence of
stable emulsions. Nevertheless, thermodynamically stable or
metastable disperse systems that behave like microemulsions
significantly blur the current classifications. Phase studies
assume equilibrium to have been reached and due to the

slowness of kinetic processes may miss possible equilibrium phases. Since only stable states are described, little information is given of the unstable or metastable phases through which a system may transform during its approach to equilibrium. Surface effects of small systems are also neglected in conventional phase diagrams and local micro-heterogeneities within the same phase may not be accounted for.

The purpose of this investigation was to examine the "interfacial" and "phase" approaches, particularly with regard to the influence of non-equilibrium states as the source of their apparent incompatability.

## II. EXPERIMENTAL

Toluene (Fisher) distilled, fraction collected 110-111° C (b.p. 110.6°C)

Benzene (Fisher) distilled, fraction collected at 79-81°C

Water passed through an ion exchange resin and then distilled through a Corning all glass still

Potassium Oleate (K & K Laboratories)

n-Hexanol (Union Carbide) b.p. 115-158°C

Sodium Octanoate prepared in the laboratory from octanoic acid and sodium ethylate. The sodium salt was precipitated, washed, filtered and dried at 78°C over phosphorous pentoxide

Octanoic Acid (Fisher) distilled under vacuum at 4.0 mm Hg and the fraction collected at 109-111°C. Refractive index at 24.5°C was 1.4260.

Sodium Dodecyl Sulfate (Eastman Kodak)

Pentanol (Eastman Kodak) b.p. 136-138°C

Sodium Hexanoate (Pfaltz and Bauer). The salt was purified by precipitation from water with acetone two times. The salt was dried at 78°C under vacuum after washing and filtering.

p-Xylene (Eastman Kodak) distilled, fraction collected at 138-139°C. Refractive index at 21°C, 1.4954

n-Octylamine (ICN) distilled under vacuum at 2.3 mm Hg and the fraction collected at 36-38°C. Refractive index at 21°C was 1.4273.

The light scattering measurements were carried out on a Brice-Phoenix model 2000 DM attached to a Westronic recorder model LS 11B. The instrument was calibrated using a National Bureau of Standards sample of polystyrene. Depolarization ratios were checked for benzene and toluene. In some experiments a temperature controlled light scattering cell was used immersed in a bath of dimethyl sulfoxide of refractive index 1.474. Solutions were rendered dust free by centrifugation in a Sorvall centrifuge, model SS-1. For the temperature controlled experiments the whole unit could be

refrigerated permitting centrifugation of samples at constant
temperature between $-5^{\circ}$C and $30^{\circ}$C.


III.  RESULTS

A.  Na Dodecyl Sulfate - Water - Benzene - Pentanol

In Fig. 1 the tetrahedron has been opened up to give the

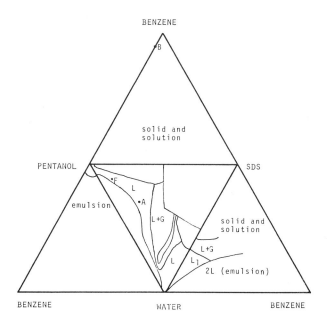

Fig. 1.  Ternary phase diagrams for the system Na Dodecyl
Sulfate - Water - Benzene - Pentanol at $26^{\circ}$C.

four ternary phase diagrams for this system.  The SDS corner
was not studied as it contains mainly undissolved soap in the
form of crystalline gel and probably liquid crystalline mixtures.  Fig. 2 gives the partial phase diagram taken from
Fig. 1 so that the systems can be clearly identified.  This
phase diagram was obtained at $26^{\circ}$C by preparing solutions
containing equal amounts of water and pentanol and varying
percentages of SDS.  These solutions were then titrated with
water or with pentanol.  Gerbacia and Rosano (19) investigated
the role of the cosurfactant in lowering the dynamic interfacial tension.  They proposed that on titrating with pentanol
it is possible for the interfacial tension of such systems to

477

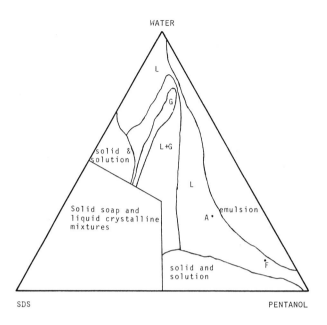

Fiç. 2. Partial phase diagram from Fig. 1.

drop to zero by diffusion of the amphipathic component across the interface. To show this they mixed two solutions: one containing SDS, water and part of the pentanol equivalent to the amount found in the interface at equilibrium (point A, Fig. 1), the other consisting of benzene and pentanol at concentrations found in the continuous phase at equilibrium (point B, Fig. 1). It was claimed that microemulsions were reported for pairs of solutions where the oil was hexadecane We have found that for the benzene system, the two solutions when mixed gave a clear solution which did not separate. In the system containtng hexadecane as the oil, the solutions on vigorous shaking gave no single homogeneous phase but became clear on mild sonication. Both equilibrium solutions when heated became turbid and separated, but behaved reversibly as they cleared again on cooling. Since under certain conditions a transition did not take place, metastable states are involved. It is possible that as A and B are mixed to give the final microemulsion (80% benzene, above point F in Figs. 1 and 2) liquid crystalline phase regions are traversed which present a kinetic barrier to the formation of $L_2$ micelles that is not overcome by ordinary mixing.

## B. K Oleate - Water - Benzene - Hexanol

Fig. 3 shows the changes of shape and size of the three

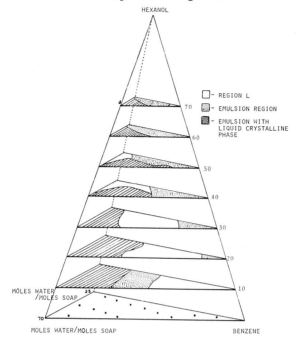

Fig. 3. Quaternary phase diagram for K Oleate - Water - Benzene - Hexanol

regions of the phase diagram with the addition of alcohol. This system was studied earlier by Bowcott and Schulman (20) who determined the number of moles of alcohol needed to titrate an emulsion to clarity. The molar ratio of water to soap was 26.25 - 70, since it was claimed that above a ratio of 75 it was impossible to obtain clear systems. When translated to weight percentages, the area encompassed by the above ratios is indeed a small part of the total phase diagram as shown in Fig. 4B. Furthermore, in the original study, the molar ratio of benzene to soap was larger than 25 and below 100. This also greatly limits the area on the phase diagram that was actually studied. Figure 5 shows titration curves for this system. In curve (a) the linear relation for the molar ratio of water to soap 26.25 - 35 has been reproduced from Bowcott and Schulman's results. One can question the validity of the extrapolation to zero benzene content of this line since, whereas the region studied covers a benzene composition rage of 20%, the region of the phase diagram included in the extrapolation (molar ratio 0-20) amounts to a

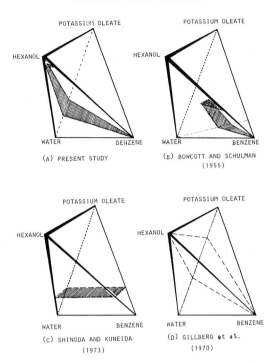

POTASSIUM OLEATE

HEXANOL

WATER       BENZENE

(A) PRESENT STUDY

POTASSIUM OLEATE

HEXANOL

WATER       BENZENE

(B) BOWCOTT AND SCHULMAN
(1955)

POTASSIUM OLEATE

HEXANOL

WATER       BENZENE

(C) SHINODA AND KUNEIDA
(1973)

POTASSIUM OLEATE

HEXANOL

WATER       BENZENE

(D) GILLBERG et al.
(1970)

Fig. 4.  Quaternary phase diagrams for the system
K Oleate – Water – Benzene – Hexanol on a weight percent basis.

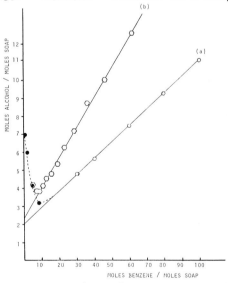

Fig. 5.  Microemulsions for the system K Olate – Water –
Benzene – Hexanol

480

range of 55% benzene content. The solid circles show results
that were obtained in the present study for the same water to
soap molar ratios but for the low benzene content range.
Curve (b) shows a titration curve for a water to soap molar
ratio of 58.5. The coressponding area in the phase diagram is
given in Fig. 4A. It is clearly seen that in both cases a
minimum exists below which the amount of alcohol needed for
clarity increases with a decrease in benzene content of the
system. The minimum which occurs at approximately 30-35 wt %
of benzene may indicate a region where $L_1$ and $L_2$ phases are
at equilibrium; to the right of the minimum, $L_2$ structure is
dominant, whereas to the left, $L_1$ is dominant. It can be
noted from curve (b) of Fig. 5 that clear systems could not be
obtained for molar ratios of benzene to soap greater than 7
in good agreement with Fig. 3, where it can be seen that as
the relative water concentration increases, areas in the
phase diagram are reached that do not form microemulsions.

Shinoda and Kuneida (21) in looking for the optimal
conditions to form microemulsions plot the same system on a
two dimensional phase diagram. One coordinate represents the
weight fraction of the oil and water portion, which totals
75% and the other coordinate is the weight fraction of soap
and alcohol equal to 25%. The area covered by the dark
rectangle in Fig. 4C is a lateral cut through the volume of
the tetrahedron and represents only a limited region of the
phase diagram.

## C. n-Octylamine - p-Xylene - Water

Fig. 6 shows the phase diagram at 26$^{\circ}$C and the same
system (dashed lines) at 20$^{\circ}$C as reported by La Force and
Sarthz (22). It is likely that a small liquid crystalline
area still exists along the water-octylamine tie line. Such
liquid crystals have been reported by Collison and Lawrence
(23) to exist up to a temperature of 35$^{\circ}$C.

The phase diagram consists of a continuous isotropic
solution area covering the whole p-xylene - octylamine   axis
and reaching into the water and p-xylene corners. Two series
of solutions with molar ratios of p-xylene to octylamine 1.47
and 0.2 were selected for light scattering measurements.
The systems are represented by the points marked 6 and 7
respectively in Fig. 6. The intensity and depolarization of
light scattering was measured at a wavelength of 546 mμ at
angles of 35$^{\circ}$ to 135$^{\circ}$. For series 7, Figs. 7-10 give $R_\theta$, $V_u$,
$H_u$ and $V_v$ vs. sin $\theta/2$, where $R_\theta$ is the intensity ratio of
scattering of unpolarized light, and $V_u$ and $H_u$ are the vertical and horizontal components of the scattering for an

unpolarized incident beam, and $V_v$ is the vertical component
of the scattered light for a vertically polarized incident
beam.  As the concentration of water increases there is an
increase in the intensity of scattering, but no change in the
shape of the cattering curves can be observed.  The results

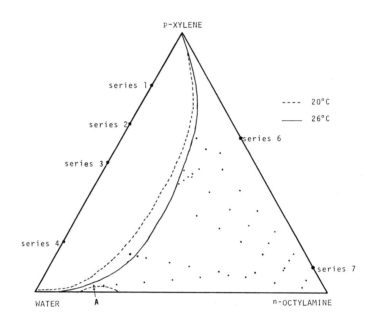

Fig. 6.  Phase diagram for system n–Octylamine – p-Xylene
– Water at 25°C.  Dashed lines are for the same system at 20°C.

for series 6 were similar.  For the last sample of this series
the horizontal component of the scattering for vertically
polarized light was measured as a function of temperature
(Fig. 11).  The temperature was increased from 21 to 26.5°C at
a rate of 0.25°C/min., and as is seen from Fig. 6 this results
in the almost total elimination of the liquid crystalline
phase from the vicinity of the composition studied (point A,
Fig. 6).  At 21°C, $H_v$ shows a large increase in intensity
which decreases as the liquid crystalline phase disappears.
Dissymmetry values estimated from the angular scattering
curves were approximately 6 and indicate large particles of
spherical shape approximately 2500 Å–3000 Å in diameter.
Values of $\varrho_h$ for series 7 calculated from $\varrho_u$ and $\varrho_v$, were much
smaller than unity, again indicating large particles.  A com-
parison of the calculated theoretical values of $H_v$ with experimental values using an equation derived for anisotropic

rods (24) did not agree, thus ruling out the possibility of a solution containing only rodlike particles.

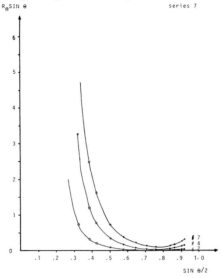

Fig. 7. Intensity of angular scattered light for a constant molar ratio of p-xylene : octylamine of 0.2. Solutions #2, #4, and #7 represent increasing molar percentages of water, 20.1, 40.2, and 70.2 respectively.

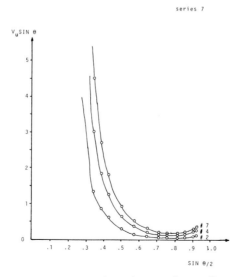

Fig. 8. Angular scattering intensity of vertically polarized light for unpolarized incident light.

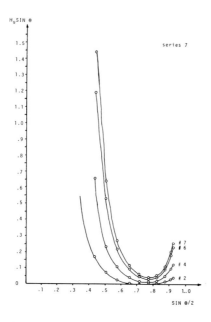

Fig. 9. Angular scattering intensity of horizontally polarized light for unpolarized incident light.

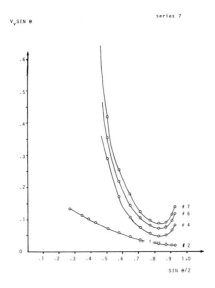

Fig. 10. Angular scattering intensity of the vertical component of scattered light for a vertically polarized incident beam.

Plots of $1/H_V$ and $1/V_V$ vs. $\sin^2 \theta/2$ were nonlinear when extrapolated to $\theta^V = 0$ and gave $\rho_V$ values which tended toward zero, confirming that the systems may be mixtures of more than one particle shape and size. Large spherical isotropic clusters account for the large dissymmetries, while the anisotropic rods would contribute mainly to the observed $\rho_V$ of the order of 0.01 to 0.05.

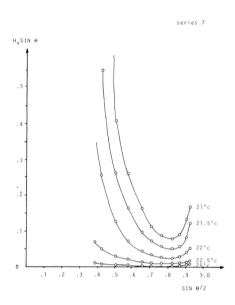

series 7

Fig. 11. Angular scattering intensity of the horizontal component for vertically polarized incident light as a function of temperature. Composition designated by point A in Fig. 6. corresponds to 78 mole % water.

Fig. 12 shows the relative intensity of scattering at $90^O$ as a function of mole % of n-octylamine for series 1-4 (Fig. 6) each having a constant mole ratio of xylene to water. As expected the curves merge in the octylamine rich area, but it is somewhat surprising to find that the curves close to the phase transition boundary differ very little from each other, and the difference between those from the xylene rich area and water rich solutions is small. It appears that the micelles tend to aggregate on approaching phase separation and this aggregate attains a similar size at a constant distance from the phase boundary curve (22). The exception is within the channel region in the vicinity of the liquid crystalline phase

485

(series 7) where the clusters close to the liquid crystalline phase transition show anisotropic ordering.

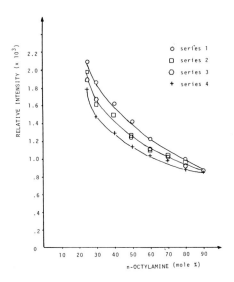

Fig. 12. Relative scattering intensity at 90° as a function of mole % of octylamine for constant ratios of xylene to water. Series 1 : 4.0, Series 2 : 1.9, Series 3 : 1.0, Series 4 : 0.4.

## D. Sodium Hexanoate - Hexanoic acid - Water

A phase diagram for this system at 25°C is presented in Fig. 13. The nature of the L region is very likely a homogeneous equilibrium system of $L_1$ and $L_2$ since no phase transitions have been detected and nuclear magnetic relaxation data give no indication of a change in relaxation rate in the concentration regions of similar systems where a change from normal to reversed micelles has been proposed (25). The intensity of light scattering in different parts of this region was studied for three series of samples shown in the figure. Soap to water ratios were kept constant at 1:3, 3:7 and 2:3, while the acid concentration was varied between 0 to 90%. These compositions represent points entirely in the L region, or interruptions by phase transitions either into two clear liquids, or a clear liquid and a clear liquid crystalline phase. It can be seen from Fig. 14 that in the

L region there is a maximum in the intensity of the scattering at 90° at about 40% wt % hexanoic acid. Such a maximum may be indicative of a transition from normal to reversed micelle structure. Dissymmetry values through this region range from 1.02 - 1.23 and $\rho_h$ values between 0.85 and 1.00 confirming that the particles are relatively small. Together with non-zero $\rho_v$ values these data are consistent with cylindrical particles having a long axis of approximately 600 Å-800 Å.

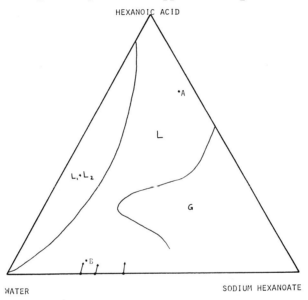

Ekwall, Mandell and Fontell (1969)

Fig. 13. Phase diagram at 25°C for Water - Hexanoic acid -Sodium hexanoate.

In Fig. (15) are shown $\rho_v$ values for the three titration paths. It is apparent that the ratio increases as the water to soap ratio decreases i.e. as the liquid crystalline region is more closely approached. Points A and B refer to compositions of two clear solutions which were mixed at different ratios. Solution was instantaneous and no birefringence was detected. In another case two solutions, one an isotropic liquid and the other a liquid crystalline phase were mixed and gave either homogeneous or emulsion systems with no evidence of metastability. No induction period before the onset of a turbidity increase was observed as was also the case for an L→ emulsion transition. The increase in scattering in the former case was lower, however. Presumably, the occurrence of metastability depends on locations in particular areas near phase boundaries.

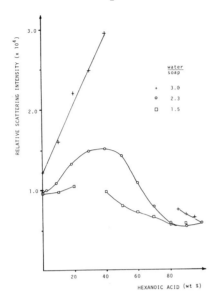

Fig. 14.   Relative scattering intensity at 90° as a function of hexanoic acid concentration for various molar ratios of water to soap.

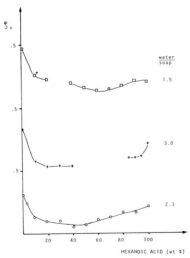

Fig. 15.  $\rho_v$ values as a function of hexanoic acid concentration for various molar ratios of water to soap.

## IV.  DISCUSSION AND CONCLUSIONS

The need to understand phase relationships before drawing conclusions from a study of a particular oil - water - amphiphile system has been confirmed in this study.  Non-linear relations for the amount of alcohol vs. oil required to produce microemulsions are obtained when a more extensive portion of the phase diagram is examined.  Since the intercept is normally taken to be the number of moles of alcohol at the interface per mole of surfactant, such a result implies that the formation of microemulsions cannot solely be attributed to penetration of cosurfactant into the interface.  Similarly phase diagrams even of only equilibrium phases can indicate the changes and sources of metastability which occur as the system moves towards equilibrium along a dilution path. Whether spontaneous microemulsification will take place depends ultimately on the intermediate stable or unstable states through which a system may transform.

On the other hand phase diagrams do not account for microscopic heterogeneities within an isotropic phase as was found when a liquid crystalline region was in the vicinity of a phase boundary.  Since an $L_2 \rightarrow L_1$ transition occurs close to this region it may be that the mechanism of reversal involves a liquid crystalline intermediate region or a transition to a lamellar micelle.  The formation of micellar clusters when a phase boundary is approached was also demonstrated by light scattering.  Such clusters differ in size and structure depending on their location within the phase diagram.

On approaching a phase boundary line through a change in temperature,  the intensity of scattered light increases many fold and then decreases abruptly as droplets of the new phase form and the system ultimately layers into two homogeneous solutions.  Within a stable critical emulsion the light scattering maximum is less intense and broader and there is continuous transition between coexisting normal and inverted emulsions (26).  This situation corresponds to the light scattering found in L regions at compositions where inversion is supposed to take place.  Evidence for stable emulsions has been advanced for a neat phase in a small amount of micellar solution (the so-called lamellar C phase) (27).  Wagner (28), however, points out that experimental verification of stable emulsions is difficult particularly in the presence of micelles.  Light scattering may be a useful tool in determining new boundaries of stable and metastable disperse systems within conventional phase diagrams which neglect surface effects.

Current theories of micelle formation provide for a nucleation step and a conformational change (29).  The

metastability of supercooled micellar solutions has been
ascribed to the former process (30), while solubilizates
cause metastability in the latter (31). Liquid crystalline
phases which require the penetration of water to transform
are another source of metastability within regions corresponding to microemulsions. Katchalsky (32) has emphasized that
the existence of metastable steady states is "more prevalent
than realized by those indoctrinated with thermodynamic
dogma". Whether a microemulsion is a "frozen in" metastable
equilibrium or a true equilibrium is a dubios distinction.
Seen from this viewpoint micellar and emulsion views of microemulsions can be reconciled.

## V. REFERENCES

1. Hoar, T.P., and Schulman, J.H., Nature, 152, 102 (1943)
2. Stockenius, W., Schulman, J.H., and Prince, L.,
   Kolloid - Z. 169, 170 (1960)
3. Prince, L.M., J. Colloid Interface Sci., 23, 165 (1967)
4. Prince, L.M., J. Colloid Interface Sci., 52, 182 (1975)
5. Shah, D.D., Tamjeedi, A., Falco, J.W., and
   Walker, Jr., R.D., Aiche J., 18, 116 (1972)
6. Gillberg, G., Lehtinen, H., and Friberg, S., J. Colloid
   Interface Sci., 33, 40 (1970)
7. Shinoda, K., and Kuneida, H., J.Colloid Interface Sci.,
   42, 381 (1973)
8. Ahmad, S.I., Shinoda, K. and Friberg, S., J. Colloid
   Interface Sci., 47, 32 (1974)
9. Shinoda, K., and Friberg, S., Advances Colloid Interface
   Sci., 4, 281 (1975)
10. Rosoff, M., in "Progress in Surface and Membrane Science"
    (J.F. Danielli, M.D. Rosenberg, and D.A. Cadenhead, Eds.)
    Vol. 11. Academic Press, N.Y., 1976
11. Gerbacia, W.E., Rosano, H.L., and Zajac,M., J. Amer.
    Oil Chem. Soc., 53, 101 (1976)
12. Rusanov, A.I., Kuni,F.M., Shchukin, E,D,, and
    Rebinder, P.A., Kolloid Zh. 30, 744 (1968)
13. Ruckenstein, E. and Chi, J.C., JCS Faraday II, 71, 1690
    (1975)
14. Reiss, H., J. Colloid Interface Sci., 53,61 (1975)
15. Frenkel, J., "Kinetic Theory of Liquids", Oxford Press,
    London, 1947
16. Abrosenkova, V.F., and Vlodavets, I.N., Kolloid Zh. 37,
    962 (1975)
17. Elbing, E. and Parts, A.G., J. Colloid Interface Sci.,
    37, 635 (1971)

18. Larsson, K., Chem. Phys. Lipids, 9, 181 (1972)
19. Gerbacia, W. and Rosano, H.L., J. Colloid Interface Sci., 44, 242 (1973)
20. Bowcott, J.E. and Schulman, J.H., Z. Elektrochemie, 59, 283 (1955)
21. Shinoda, K., and Kuneida, H., J. Colloid Interface Sci., 42, 381 (1973)
22. LaForce, G. and Sarthz, B., J. Colloid Interface Sci., 37, 254 (1971)
23. Collison, R. and Lawrence, C.S., Trans Faraday Soc., 55, 662, (1958)
24. Horn, P., Benoit, H., and Oster, G., J. Chimie Physique, 48, 530 (1951)
25. Gustavsson, H. and Lindman, B., Proc. Int'l Conf. on Colloid and Surface Sci., Budapest, 1975. V. 1, p. 625
26. Shchukin, E.D., Fedoseeva, N.P., Kochanova, L.A., and Rebinder, P.A., Doklady Akadem. Nauk. USSR, 189, 123 (1969)
27. Persson, N.O., Fontell, K. Lindman, B., and Tiddy, G.J.T., J. Colloid and Interface Sci., 53, 461 (1975)
28. Wagner, C., Colloid and Polymer Sci., 254, 400 (1976)
29. Eicke, H.F., Christen, H. and Hofmann, R., Proc. Int'l Colloid and Surface Sci., Budapest, 1975. Vol 1, p. 489
30. Mazer, N.A., Benedek, G.B. and Carey, M.C., J. Phys. Chem., 80, 1075 (1976)
31. Hoffmann, H., NATO Adv. Study Inst. Ser. Ser. C., 18, 181 (1975)
32. Katchalsky, A. and Spangler, R., Quart. Rev. Biophysics, 1, 127 (1968)

# THE SCATTERING OF LIGHT BY HOLLOW SPHERES

Dr. Yi-Chang Fu
The Procter and Gamble Company

The theory of light scattering from a dilute suspension of monodisperse hollow spheres has been derived using the Rayleigh-Debye approximation. Scattering intensity and particle scattering functions were expanded in series whose terms are functions of the polarizability tensor, the particle sizes, and the scattering angles. Particle scattering functions were evaluated numerically for monodisperse distearoyl lecithin vesicle systems. Dissymmetry was also computed as a function of vesicle size at various polarizability ratios, $\alpha_{\perp}/\alpha_{\parallel}$, where $\alpha_{\perp}$ and $\alpha_{\parallel}$ are the polarizability of the scattering elements in the bilayer membrane perpendicular and parallel to the molecular chain axis, respectively.

The particle scattering function and scattered light intensity have been determined on fractionated distearoyl lecithin vesicles. The scattered light intensity drops sharply as the temperature is increased past the transition temperature. All these experimental data are in excellent agreement with the theoretical predictions of the anisotropic hollow sphere model.

Experimental measurements of the vesicle sizes of monodisperse distearoyl lecithin vesicles by electron microscopy (EM) compared favorably with the values derived from light scattering data by application of this theory.

# DIFFUSION AND SORPTION OF SIMPLE IONS IN CELLUSOSE ACETATE - SEMIPERMEABILITY

M. Bender, B. Khazai and T. E. Dougherty

Fairleigh Dickinson University

The nature of the interaction between cellulose acetate membrane (CA) and the simple salts NaCl, NaI,KCl and KI, as interpreted from sorption and diffusion studies (B. Khazai, M. S. Thesis, May 1974; B. Khazai, B. S. Thesis May 1972; T. E. Dougherty, B.S. Thesis, May 1968) is considered. A similar approach had been utilized on the systems organic anionic electrolyte in cellulose (1) and simple salts in cellulose (2).

The dielectric constant of CA is estimated to be about 13 (3), quite lower than that for cellulose. Thus its swelling in water is appreciably less (14% vs. 92%) and the restrictions to salts are greater. CA partition coefficient and diffusion constant studies are reported by Anderson, Hoffman and Peters (4) and by Heyde, Peters and Anderson ("Factors Influencing Reverse Osmosis Rejection of Inorganic Solutes from Aqueous Solution", ms. of the Scientific Research Staff, Ford Motor Co., Dearborn, Mich.), While they allow for the possibility of the presence of a small concentration of ionizable carboxyl groups which would contribute to the CA weak ion-exchange character and its electrical charge, they conclude that dielectric exclusion of ions is the essential factor in the rejection of salts. Ullman develops this mathematically ("The Rejection of Salt by Microporous Membranes" and "Electrostatic Models of Salt Solubility in Membranes", Feb. 7, 1974 mss. of the Scientific Research Staff, Ford Motor Co., Dearborn, Mich.) utilizing the Debye-Huckel shielded potential concept in interionic attraction theory. He indicates that under conditions of lower concentration there should be a greater range of ionic atmosphere, the salt rejection by the membrane being greater. Spiegler's (5) observations of a streaming potential for CA membrane as a result of reverse osmosis or hyperfiltration are evidence of its electrokinetic nature.

Sorption for the CA and simple salts was found not to be reversible. Residual solute remains which can only be removed by ion exchange, and $H^+$, $K^+$ and $Na^+$ membranes were prepared accordingly. This was the case for cellulose and simple salts (2). But the system (1) utilizing aqueous organic anionic electrolyte in cellulose was reversible with respect to the organic solute except at relatively low concentrations. However, this solute, the disodium salt of 16, 17-dimethoxy-violanthrone leuco (technically known as Vat Jade Green, C. I. No. 59825) had to be maintained in sodium hydroxide and sodium hydrosulphite. Here, throughout the whole range of sorption runs, electrolyte concentration was high, the ionic strength being 0.424 at zero (organic) sorbate concentration and 0.436 at the highest concentration considered.

In CA the sorption of K was almost twice that of Na while in cellulose it was only 10% greater. This increased preference for K with lower dielectric constant is of interest in that biological cells are high in K content vs. Na. Also, it was observed that H was preferentially sorbed over K in the CA. The line-up of H, K, and Na in CA thus (also seen in cellulose as well (2) although less exaggerated) follows the "lyotropic series" (Heyde, Peters and Anderson - vide ante, and Heyde and Anderson "Ion Sorption by Membranes from Binary Salt Solutions" ms. of the Scientific Research Staff of Ford Motor Company).

The total sorption of K in CA (10 moles $m^{-3}$) is seen to be half that in cellulose. Meanwhile the concentration of water in wet CA (6.8 x $10^3$ moles $m^{-3}$) is about 700 times as much as the K and in the neighborhood of 1300 that for K in the cellulose. The preferential solubility of water vs. the salt whether in CA or cellulose, is very striking and can be seen to be an important factor in the membrane semi-permeability properties. The greater semi-permeability for the CA vs. the cellulose is associated with the lower solubility in this medium, understandable in terms of the respective dielectric constants of salt, water and membranes.

Likewise, the semi-permeability, including the difference between CA and cellulose according to "dielectric exclusion" is appreciated in terms of the differences in the diffusion of the salts and water. For instance, the self diffusion coefficient of water is 2 x $10^{-9} m^2 s^{-1}$. In CA the coefficient of water is 2 x $10^{-10}$ (6) and it is expected that the value in cellulose is somewhere in between. Meanwhile NaCl in water is close to that of water in water. In cellulose the coefficient for KCl (at higher concentrations and based on solution concentration difference) is 5.92 x $10^{-11} m^2 s^{-1}$ and NaCl is 8.80 x $10^{-13}$ in CA under similar experimental circumstances. It can thus be seen that while water may be impeded as much as 10 fold by the CA, salt is impeded about thirty-five fold in the cellulose and 2000 fold in the CA.

Examination of the amount of simple salt absorbed in CA and in cellulose in terms of the number of cellobiose units indicates 1 cation to 120-200 units. While one might lean toward the idea of "accidental" COOH groups being sites for sorption, consideration of all the experimental information tends to preclude the idea of specific solute sites. The results obtained with the organic anionic electrolyte (1) did not seem to point to specific sites and Anderson, Hoffman, and Peters (4) and Heyde, Peters, and Anderson (vide ante) refer to dielectric exclusion rather than sites.

In this light, and with consideration of the other experimental observations, suitable explanation offers itself via an electrokinetic approach. For instance, both K and Na sorption were found to decrease with decreasing pH indicating $H^+$ and $OH^-$ as potential determining ions. Heyde and Anderson's work (vide ante) on ion sorption from binary salt solutions likewise implies that potential determining ions are affecting the uptake of simple salt.

The origin of the electrokinetic potential is in the electronegative nature of the oxygen in the cellulose(acetate), the hydration and the potential determining ions. Double layer thickness is greatest and the potential highest at low membrane electrolyte content and increased dielectric constant. Higher concentration tends toward swamping.

At increased membrane concentrations, the cations are held less strongly in the double layer and desorption can take place. But at less concentration the coulombic attraction is sufficient to counteract the random translational energy of the cation, interfering with its migration to contiguous solution despite the concentration difference in the membrane vs. solution. Further evidence for such increased interaction at lower concentration comes from our recent electrical conductivity measurements on cellulose membrane in NaCl solutions (R. Klima, B. S. Thesis, May 1976 and J. Marshall, B. S. Thesis, May 1976). Conductivity in the membrane was found to parallel membrane NaCl absorption over the range of solution concentrations employed in the present cellulose acetate absorption experiments and cellulose (2) as well, the same kind of Freundlich-Langmuir curve being obtained.

Therefore the irreversibility of desorption vs. absorption is understood, given species at low concentration being only removable by cation exchange where the increased ionic strength which is applied in terms of the exchanging ion enables redistribution of the cations by lowering their double layer interaction.

It is expected thus that diffusion would be lowest when interaction in the double layer is greatest, i.e., at the lower salt concentrations, which is what was found. This is in keeping with reverse osmosis salt retention generally being more efficient for lower concentration solutions.

Finally there is support for electrokinetic theory in terms of the relative appearance of the measured diffusion concentration gradients in organic anion in cellulose, vs. simple salts in cellulose, vs. simple salts in CA. In all cases the diffusion (differential coefficient) was highly concentration dependent. With the organic anionic electrolyte in cellulose, swamping is expected due to the high ionic strength and the double layer should be minimal. The concentration gradient is bow-shaped and explanation is offered in terms of less interaction of organic solute with the membrane the higher its concentration. The simple salt in cellulose gradient however is much more extreme such that dc/dx (concentration over distance) takes an acute drop at very low concentration indicating a sharp barrier where the double layer would be expected to be greatest. This barrier at very low concentration is also evident for NaCl in CA. But other barriers are found at higher concentrations in the different runs of simple salt in CA where different overall concentration drops are considered. The location of these extra barriers moved towards lower concentration as the overall concentration drop was increased, the gradient curve shape approaching that for simple salt in cellulose.

Thus, the salt retention properties of cellulose acetate in reverse osmosis are understandable in terms of an electrokinetic barrier system superimposed on its dielectric restriction properties, the semi-permeability decreasing with increasing solute concentrations. The much greater effectiveness of cellulose acetate vs. cellulose is evident in terms of the reduced absorption and diffusion of salt in the more compact less swollen molecular chain network of lower average dielectric constant in which electrostatic barriers can exist on a "multi" basis.

REFERENCES

(1)    Bender, M. and Foster, W. H., Jr., Trans Faraday Soc. 61, 159 (1965); 64, 2549 (1968).
(2)    Bender, M., Moon, J. K., Stine, J., Fried, A., Klein, R., and Bonjoulkian, R., Trans Faraday Soc. 71, 491, (1975).
(3)    Barker, R. E., Jr. and Thomas, C. R., J. Appl. Phys. 35, 3203 (1964).
(4)    Anderson, J. E., Hoffman, S. J. and Peters, C. R., J. Phys. Chem. 76, 4006 (1972).

(5)     Spiegler, K. S., Preprint of Paper Presented at 155th
        ACS Nat. Meeting, San Francisco, Calif., Div. of Water,
        Air and Waste Chem., 1968 (Spring), 8 No. 1,79.
(6)     Lonsdale, H. K., Merten, U., and Riley, R. L., J. Appl.
        Polymer Sci.  9,1341 (1965).

# ELLIPSOMETRIC INVESTIGATIONS ON THE ADSORPTION OF POLY(VINYL ALCOHOL) ON SILICA OBTAINED BY THERMAL OXIDATION OF SILICON

G. J. Fleer
Laboratory for Physical and
   Colloid Chemistry
Agricultural University
Wageningen

L. E. Smith
Institute for Materials Research
National Bureau of Standards

The adsorption of poly(vinyl alcohol) from aqueous solution on the surface of a silica layer of about 200 nm thickness supported on silicon has been studied by ellipsometry. Ellipsometric measurements of thin films adsorbed on bulk silica are generally rather insensitive and measurement uncertainties are correspondingly large. In this system the adsorption process is entirely governed by the properties of the silica film while the optical properties of the underlying silicon provide considerably enhanced ellipsometric sensitivity. In addition, as the silica is obtained by high temperature (1000°C) oxidation of silicon, its surface is composed solely of siloxane groups. The surface can then be hydrated to conveniently vary the surface silanol content. The adsorbance and layer thickness of the adsorbed poly(vinyl alcohol) were measured as a function of acetate content, pH and degree of surface hydration.

For a polymer with a weight average molecular weight of $2.7 \times 10^5$ and 12% acetate groups, the adsorbance decreases from 3.7 mg/m$^2$ on a hydrophobic surface (only siloxane bonds) to about 0.7 mg/m$^2$ on a hydrophilic surface (containing silanol groups), while the layer thickness remains approximately constant. For a polymer with the same molecular weight but containing only 2% acetate groups, the adsorbance decreases similarly while the thickness more than doubles. With increasing pH, the adsorbance decreases and the thickness increases for the polymer with 12% acetate groups.

The results are interpreted in terms of the conformation of the adsorbed layer which is determined by the segment-surface interaction. It is suggested that on a hydrophobic surface the attachment takes place mainly through acetate and methylene groups, while on a hydrophilic surface hydrogen bonding may occur between the surface silanols and the carboxyl of the acetate groups.

# SURFACE PROPERTIES AND CLAYS IN THE EARTH SCIENCES

Gary R. Olhoeft
*U. S. Geological Survey*

## ABSTRACT

Clays and similar materials are extremely important in many areas of the earth sciences. In the search for minerals and energy resources in places as diverse as geothermal areas and permafrost regions, they dominate certain types of exploration. In permafrost, clays are often the dominant constituents of the material, and the interactions at the surfaces of the clays dominate and determine the physical properties. In geothermal areas, the alteration products lining the pore walls and cracks of the host rock behave analogously, sometimes drastically altering the thermodynamic behavior of the system. In sulfide mineralization, uranium/thorium deposition, and others areas, similar clay/rock systems arise. The electrical properties of many earth materials are particularly sensitive to the behavior of the clays. Because many geophysical exploration techniques rely upon measuring electrical properties, pertinent examples and the major mechanisms will be briefly reviewed.

# SURFACE PROPERTIES OF SILOXANE BLOCK COPOLYMERS

James J. O'Malley and John VanDusen
*Xerox Corporation*

*ABSTRACT*

Siloxane polymers are well known surface active agents
in solution and in the solid state. In this study we
examined the effect of block copolymerization on the surface
properties of siloxane containing polymer films as determined
from contact angle and ESCA measurements. Two block copolymers systems were studies, one containing polystyrene (A)
and polydimethylsiloxane (B) and another containing polyhexamethylene sebacate (A) and polydimethylsiloxane (B).
The two A polymers differ markedly in that polystyrene is
amorphous whereas polyhexamethylene sebacate is a highly
crystalline polyester. In particular, AB, ABA, BAB and
multi-block type siloxane copolymers were synthesized and
the effects of copolymer composition, block length, block
copolymer structure and morphology on surface properties
were examined.

# ADHESION OF THERMOPLASTIC ELASTOMERS

A. N. Gent and G. R. Hamed
*The University of Akron*

## ABSTRACT

*Measurements have been made of the strength of adhesion of thermoplastic elastomers (Shell Kraton 1101, 1102, 1107 and 3202, consisting of block copolymers of styrene with either butadiene or isoprene center blocks) applied as hot melts or from solution to Mylar or brass sheets. In comparison with corresponding random copolymers, these triblock elastomers were found to adhere tenaciously, with energies required for detachment ranging up to 5 kj/m$^2$. However, these bond strengths depended strongly upon the mode of detachment, reflecting energy dissipated in bending the substrate sheets, or in deforming the elastomer layer, in addition to the intrinsic work of detachment. In consequence, they also depended upon the thickness of the substrate (when it was bent during detachment) and upon the thickness of the elastomer layer (when it was deformed extensively) and upon the speed of detachment and the test temperature, and upon the conditions of application of the elastomer to the substrate. These experimental results are discussed in terms of the work of detachment of an ideal elastic-plastic material adhering to a rigid substrate with varying degrees of intrinsic adhesion.*

A 6
B 7
C 8
D 9
E 0
F 1
G 2
H 3
I 4
J 5